ILLUSTRATED DICTIONARY OF
BUILDING MATERIALS AND TECHNIQUES

AN INVALUABLE SOURCEBOOK OF THE TOOLS, TERMS, MATERIALS, AND TECHNIQUES USED BY BUILDING PROFESSIONALS

PAUL BIANCHINA

JOHN WILEY & SONS, INC.
New York · Chichester · Brisbane · Toronto · Singapore

In recognition of the importance of preserving what has been written, it is a policy of John Wiley & Sons, Inc., to have books of enduring value printed on acid-free paper, and we exert our best efforts to that end.

Copyright © 1986 by TAB Books, 1993 by Paul Bianchina
First John Wiley & Sons, Inc. edition 1993

This publication is designed to provide accurate and authoritative information in regard to the subject matter covered. It is sold with the understanding that the publisher is not engaged in rendering legal, accounting, or other professional service. If legal advice or other expert assistance is required, the services of a competent professional person should be sought. *From a Declaration of Principles jointly adopted by a Committee of the American Bar Association and a Committee of Publishers.*

Library of Congress Cataloging-in-Publication Data

Bianchina, Paul.
　　Illustrated dictionary of building materials and techniques: an invaluable sourcebook of the tools, terms, materials, and techniques used by building professionals / Paul Bianchina.
　　　　p.　　cm.
　　Includes bibliographical references and index.
　　ISBN 0-471-57656-5 (alk. paper)—ISBN 0-471-57657-3 (pbk. : alk. paper)
　　　1. Building—Dictionaries.　　2. Building materials—Dictionaries.
　　I. Title.
　TH9.B466　1993
　690'.03—dc20　　　　　　　　　　　　　　　　　　　92-15437

Printed in the United States of America

10　9　8　7　6　5　4　3　2　1

To Rose,
Whose constant support and encouragement made this book possible.

ACKNOWLEDGMENTS

I wish to extend my appreciation to the following individuals, manufacturers, and organizations for their tremendous help. I am especially grateful to Rose Bianchina for the literally hundreds of hours she spent editing, organizing, cross-referencing and typing the huge volume of material that went into the creation of this book. Also, many thanks to Deborah Schneider of John Farquharson, Ltd., New York.

American Institute of Timber Construction, Englewood, CO
American National Standards Institute, New York, NY
American Plywood Association, Tacoma, WA
American Wood Preservers Institute, Vienna, VA
California Department of Consumer Affairs, Sacramento, CA
Channellock, Inc., Meadville, PA
Georgia-Pacific Corporation, Atlanta, GA
Glidden Coatings & Resins, Cleveland, OH
International Association of Plumbing & Mechanical Officials, Los Angeles, CA
International Conference of Building Officials, Whittier, CA
ITT Corporation, Phillips Drill Division, Michigan City, IN
K C Metal Products, Inc., San Jose, CA

Lawrence Berkeley Laboratory, University of California, Berkeley, CA
Louisiana-Pacific Building Products, Barberton, OH
Manina, Steve, Contractor, Manina Construction, Sacramento, CA
Maze Nails, W.H. Maze Co., Peru, IL
National Center for Appropriate Technology, Butte, MT
National Fire Protection Association, Quincy, MA
National Forest Products Association, Washington, DC
Oregon Department of Energy, Salem, OR
Pacific Power and Light, Bend, OR
Renewable Energy Information Service, Silver Spring, MD
Rice, Vern, Conservation Coordinator, Midstate Electric Cooperative, Inc., La Pine, OR
Stanley Tools, Division of the Stanley Works, New Britain, CT
Universal Forest Products, Inc., Grand Rapids, MI
Western Wood Products Association, Portland, OR
Wykes, R. Thomas, Energy Extension Agent, Oregon State University Energy Extension, Bend, OR

HOW TO USE THIS DICTIONARY

This dictionary contains approximately 4,500 definitions and cross-references, arranged alphabetically using the same format as most standard dictionaries. Terms containing numbers, such as *1/8th bend,* will be found as though the number were spelled out, in this case as **eighth bend.**

Some words are listed both by abbreviation and definition. For example, the abbreviation *ac* will be found listed under **ac** and the full term, **alternating current.** A cross-reference under the abbreviation will direct you to the actual term. In this way, you can easily locate the term you are seeking, whether you know the entire word or just its common abbreviation.

Another feature of this dictionary is the grouping of like terms under a single heading, such as **window.** Under the main heading, you will find an alphabetical list of 24 subheadings—AWNING WINDOW, BAY WINDOW, BOW WINDOW, etc.—with a definition for each. This grouping allows for easy comparison of the different types of windows, and provides a means for learning about some window types you may not have been familiar with.

In some cases, a term you are seeking might be contained within one of the subheadings, for example *gliding window,* which is mentioned under SLIDING WINDOW in the **window** section. In this case, two cross-references have been used. Under G you will find **gliding window** See *window: sliding window,* and under the **window** section in W you will find GLIDING WINDOW See *window: sliding window.*

In an effort to provide an even clearer understanding of what a particular term means, almost 850 of the definitions have been illustrated in drawings. A listing at the end of an illustrated term, such as "See Fig. S-12," directs you to the illustration that contains that particular term. In most cases, the drawing will contain illustrations of several interrelated items shown in common usage.

CONTENTS

Introduction   xi

Dictionary of Building Materials and Techniques   1

Appendix A: Common Abbreviations   181

Appendix B: Conversions, Tables, and Weights   191

Appendix C: Building and Framing Information   203

Appendix D: Lumber and Plywood Information   209

Appendix E: Hardware Information   221

Appendix F: Electrical Information   231

Sources   237

INTRODUCTION

A tremendous variety and volume of books and magazines have sprung up in recent years, offering advice and instruction on everything from changing a faucet washer to building an entire house. But who explains the explanations?

The goal behind the creation of the *Illustrated Dictionary of Building Materials and Techniques* was to compile in one usable, easily understood volume virtually every building term you are likely to encounter, whether in other books, in magazines, on instruction sheets, or at your local lumberyard. Approximately 4,500 terms have been included, from the simplest to the most complex, from archaic and slang words to the very latest in technical information. Building, plumbing, electrical wiring, solar heating, hardware, architecture, heating, air conditioning and many other facets of the building industry are covered in depth.

If you are a do-it-yourselfer, you will find that basic terminology has been given careful and complete attention, allowing you to build a solid understanding of all of the many aspects that go into your home. The definitions have been designed to be as straightforward and easily understood as possible, often enhanced by common examples. In addition, approximately 850 of the terms are depicted in drawings, showing their usage in common, on-the-job applications.

If you are a professional, you will find the book equally useful because it offers an opportunity to clarify your existing understanding of everyday terms while expanding your knowledge into areas of the building industry that may be outside your normal expertise or exposure. In addition to the definitions and illustrations, you will find a carefully compiled appendix of useful charts, tables, and other information—from common abbreviations to lumber and plywood grades—all at your fingertips for easy reference.

This book is intended for everyone—homeowner or professional, newcomer or old pro. From the thousands of terms and illustrations to the handy and practical appendixes, this dictionary will quickly become a well-thumbed addition to your personal library.

A

AAA Abbreviation for *American Arbitration Association*, a nonprofit group that assists with the resolution of disputes.

above grade That portion of a foundation, wall, or other structure which is on or above the level of the surrounding ground.

abrasion resistant Refers to materials that have been hardened or coated to resist damage from scraping, rubbing, or other abrasive friction.

abrasive Any of a variety of materials used for grinding, sanding, or otherwise removing waste stock from an object.

ABS cement A specially formulated liquid solvent cement used to join ABS plastic pipe and fittings.

absolute zero The temperature at which all molecules stop moving and at which absolutely no heat is present; estimated to be -459.7° F, -273.16° C.

absorber plate Metal plate inside a solar collector, either painted black or coated with a selective surface, that absorbs the sun's shortwave radiation. See Fig. G-4.

absorptance See *collector efficiency*.

absorptive film A film applied to the interior side of a window to absorb certain wavelengths of sunlight, preventing their entry into the building.

absorptivity The ability of a material to absorb light or sound.

ABS pipe See *plastic pipe*.

abutment 1. The supporting pier from which an arch begins its curve, and which takes the arch's thrust. 2. A reinforcing structure, as at the end of a bridge or wall, designed to provide lateral support.

ac Abbreviation for *alternating current*.

ac cable See *armored cable*.

accelerator An ingredient added to paint, cement, or other materials to speed up the drying time.

accessible Any fixture, connection, appliance, or piece of equipment that can be reached by the removal of a door, panel, or other access cover.

accordion door See *door*.

ac/h Abbreviation for *air changes per hour*.

achromatic Having no color; colorless.

acid-resistant brick See *brick*.

A-coil See *central air conditioning*.

acoustical materials Materials used to absorb sound waves, thus reducing the amount of noise transmitted between rooms or buildings.

acoustical plaster See *plaster*.

acoustic ceiling A type of decorative, sound-absorbing ceiling finish, usually a combination of polystyrene beads and joint cement, that is sprayed wet over a sheetrock base.

acoustic glass A special type of laminated glass designed for soundproofing.

acoustics The transmission and effect of sound on the ear.

acoustic tile A decorative, porous tile made from a variety of materials in several standard sizes. Tiles are stapled or glued to a ceiling or wall in order to absorb sound waves and reduce reflected noise.

acre A measurement of land area equal to 43,560 square feet.

acre-foot A quantity, as of water, equal to 1 acre of area covered to a depth of 1 foot.

acrylic latex caulk See *caulking compound*.

acrylic resin See *paint resin*.

active door See *door*.

active solar system A system that collects and distributes solar-generated heat and/or cooling to a building by mechanical means, including collection and storage panels, ducts, fans, and pumps.

active solar water heater A solar heating system for domestic water. Solar collectors trap sunlight to heat the water, which is then pumped into a storage tank for distribution to the building. A conventional backup heater is often tied into the system to provide hot water during prolonged periods of bad weather. See Fig. C-6.

active water preheater A system of preheating cold water before it enters the water heater. Cold water en route to the heater flows into a preheater tank or heat exchanger, which is either a separate unit or is located within the building's solar heating storage tank. There the water is preheated before it enters the water heater, thus reducing the load on the heater itself.

actual size 1. The actual dimensions of certain building materials. 2. For lumber, the finished size after milling, as opposed to the rough size. 3. For masonry, the size of the masonry unit without allowances for mortar joints.

acute angle An angle of less than 90 degrees. See Fig. P-8.

ADA Abbreviation for *airtight drywall approach*.

adapter See *transition*.

addendum A document or written agreement added to a contract to amend or clarify the original contract.

adhesive Any material that will wet and join two surfaces, also commonly known as glue or cement. Some of the more widely used adhesives in the construction industry follow.

ALIPHATIC RESIN ADHESIVE A synthetic glue that combines the tack and durability of hide glue with the fast set and ease of use found in white glue. It is a strong, convenient adhesive for wood, paper, cloth, and other porous materials.

ANIMAL AND FISH GLUES The earliest of the woodworking glues, not extensively in use today. Available as flakes or powder, they must be mixed with water and used fairly quickly. They have a poor resistance to moisture and humidity.

CONSTRUCTION ADHESIVE A heavy-duty, generally waterproof mastic for such uses as gluing down subflooring and attaching furring strips and drywall.

CONTACT CEMENT A thick, liquid neoprene adhesive, handy for bonding large areas that can't be clamped. It is applied to both surfaces and allowed to dry. When the surfaces contact each other, the cement will bond immediately and permanently. Often used with veneers and plastic laminates, it is transparent, nonstaining, and resistant to water, heat, oil, grease, and most chemicals. It is available in lacquer or water base.

CYANOACRYLATE ADHESIVE Commonly known by the brand name Super Glue or other such names, types of this recently developed adhesive will provide a fast, solid, water- and chemical-resistant bond on most nonporous materials, such as metal and glass. Some types will also work on porous materials.

EPOXY RESIN GLUE A very strong glue used to form a reliable, water-resistant bond on a wide variety of porous and nonporous materials, including wood, metal, plastic, and concrete. It is packaged in two containers—one each of epoxy resin and catalyst, which must be mixed before use. It is available in both liquid and putty forms.

FURNACE REPAIR CEMENT See *refractory cement.*

HIDE GLUE A widely used glue in the woodworking industry. Originally it had to be heated before use, but now it is available as a ready-to-use liquid. It provides a strong, lasting wood-to-wood bond, and is used extensively in furniture and cabinetmaking. It is for interior use only.

HOT-MELT ADHESIVE A glue made of a mixture of polymers and applied hot for most wood, paper, glass, plastic, and metal surfaces. It is commonly sold in sticks for use in a special electric gun that melts and applies the glue, or in sheets which, when applied under a piece of veneer, can be activated by pressing over the veneer with a hot household iron.

PANEL ADHESIVE Similar to construction adhesive, but not quite as strong. It is used primarily for wood paneling, various types of tile, foam, and some forms of metal.

PLASTIC RESIN ADHESIVE An all-purpose, highly water-resistant glue for wood, made with urea-formaldehyde powder or urea-resin powder. It is good for boats, sporting equipment, and general woodworking.

POLYVINYL RESIN ADHESIVE PVA, the familiar household glue sold ready-to-use in plastic squeeze bottles. This type of glue sets quickly, dries clear, and forms a fairly strong joint for interior use on wood, paper, plastic, and other porous materials.

PVA See *polyvinyl resin adhesive.*

REFRACTORY CEMENT A material used to line or patch the inside of fireplaces, furnaces, and other high-heat areas.

RESORCINOL RESIN GLUE A glue that provides a very strong, solvent-resistant, waterproof bond on wood, cork, concrete, some fibers and plastics, and other materials. Liquid resin and powdered catalyst are provided together in one package and are mixed before use. The disadvantages of using this glue are its high cost and a dark glue line.

RUBBER CEMENT A liquid, rubber-based, pressure-sensitive adhesive for nonstructural bonds on wood, paper, rubber, and other materials. The bond might be permanent or temporary, depending on the type and application of the cement.

SPRAY ADHESIVE A type of liquid, rubber-based pressure-sensitive adhesive in a pressurized spray can. Depending on the thickness of application, it forms a temporary or permanent bond on paper, cardboard, fabric, and other porous materials.

VEGETABLE GLUE Among the most common of this group of adhesives are wheat starch, used as a wallpaper glue; and casein glue, a milk curd derivative used with wood in some types of paint.

WHITE GLUE See *polyvinyl resin adhesive.*

adhesive caulk See *caulking compounds.*

adjustable elbow A sheet-metal elbow for use with duct pipes and having a series of overlapped, interlocking plates that allow the elbow to be infinitely adjusted to any angle from 0 to 90 degrees.

adjustable gluing clamp See *hand screw.*

adjustable wrench A type of wrench having a movable jaw that is adjusted by turning a worm screw fitting. Designed to accommodate nuts and bolts of various head sizes.

admixture Materials added to the ingredients of a basic mixture to serve a specific purpose, such as adding coloring to concrete, or adding water-repellent agents to mortar.

adobe brick See *brick.*

advertisement for bids A written and published notification intended to solicit bids for a construction project.

adz A tool used in the hand-shaping and hand-forming of timbers from raw logs, and consisting of a long handle and wide, curved blade.

aerated concrete A lightweight form of concrete, created by mixing cement, sand, and lightweight aggregates such as hard coal, pumice, vermiculite, and others. Although not as strong as regular concrete, it offers more thermal and acoustical insulation.

aerator 1. A small device that attaches to the end of a faucet arm, mixing the outcoming water with air for a smoother flow. 2. Any device that supplies a charge of air or other gas.

aerosol A liquid that is dispersed in a gas, as with paint in a spray can.

A-frame A type of construction primarily for residential use, having steeply sloped framing, usually of heavy timbers, which is a combination of wall and roof. The name is derived from the structure's sectional resemblance to a letter A. See Fig. A-1.

aggregate 1. Strong, coarse materials used in the manufacture of concrete. Fine aggregate is usually sand; coarse aggregate is gravel, crushed rock, furnace slag, and other materials, usually ranging in size from 3/8 inch to 2 inches

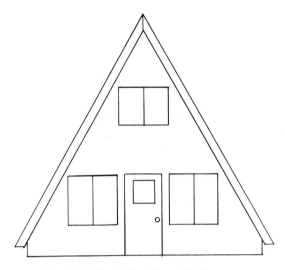

Fig. A-1. Typical residential A-frame

in diameter. 2. Similar materials used in plaster, stucco, roofing, and other applications.

aggregate, exposed Polished rock, gravel, or other decorative materials partially embedded in concrete as it dries, forming a decorative textured surface.

AIA Abbreviation for *American Institute of Architects.*

aiguille A slender drill for boring into stone or masonry.

air barrier Plastic sheeting, building paper, or similar materials used to wrap the outside of a building's exterior walls so as to cut down on the passage of air into and out of the building; also called an air infiltration barrier or an infiltration barrier.

air brick A standard-sized hollow brick or hollow metal box, used in brick construction to provide ventilation.

air chamber A capped length of coiled or vertical pipe near a plumbing fixture, partially filled with air. The trapped air cushions the abrupt stop in water flow that occurs when the fixture is shut off, thus helping to prevent noise and possible damage. Also called an air cushion. See Fig. A-2.

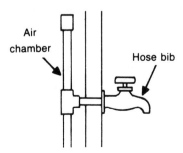

Fig. A-2. Typical air chamber

air changes per hour The number of times that the entire volume of air within a building changes during one hour; a measure of air infiltration or ventilation.

air compressor A gas- or electric-powered machine used to compress air, thereby reducing its volume while increasing its pressure. The resulting pressurized air is stored in a tank and used to power various types of tools and other accessories.

air conditioning 1. The process of controlling the temperature, humidity, and quality of air in a building, and directing its movement. 2. In common usage, any appliance or machine used to lower the temperature of the air within a building.

air-cooled slag Molten blast furnace slag that has been slowly cooled and crushed for various commercial uses, including aggregate and insulation.

air cushion See *air chamber.*

air dried See *stack dried.*

air duct A pipe or other passageway used to convey and direct air.

air entrained Concrete to which chemicals have been added so as to create a mass of tiny bubbles throughout the mixture. The air improves the concrete's workability and imparts various other favorable qualities.

air filter Any of a variety of devices and materials used to trap and remove harmful particles from the air; commonly used on heating and cooling systems, and on various types of machinery and tools.

air gap 1. A specially designed pipe fitting inserted into the dishwasher drain line at a height greater than the dishwasher itself. It vents the dishwasher and prevents the siphoning of waste water back into it. 2. The vertical distance between a faucet spout and the maximum water level of the sink or other fixture it serves. See Fig. A-3.

air gun Any of a variety of air-operated tools used for driving nails or staples.

air infiltration barrier See *air barrier.*

airless Of or relating to a type of paint sprayer that utilizes a high-pressure pump to atomize and propel the paint without the use of compressed air.

air lock 1. Air that is unintentionally trapped in a plumbing system, thus preventing the flow of water or waste. 2. See *vestibule.*

air pocket A cavity of dead air between two surfaces that is used for insulation.

air shutter An opening with a movable cover, located in front of each burner on a gas furnace. The cover is opened or closed as necessary to regulate the amount of air being provided to the burner.

air space 1. A space created in a wall or between various structural members to provide or assist circulation. 2. A dead air space created between two panes of glass or other materials to provide thermal insulation.

air system A solar heating system in which air is fan-forced through the collectors to absorb heat, and then circulated

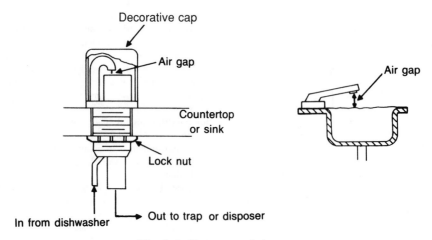

Fig. A-3. Two types of air gaps

within the building or routed through a heat storage area before returning to the collectors.

airtight   An enclosure sealed in such a manner as to prohibit the entrance or escape of air.

airtight drywall approach   A method of house construction in which gaskets or caulking are used between certain framing members and also between the framing and the drywall at specific locations. A continuous seal is created between the drywall and the outside of the building, allowing the drywall to act as a barrier against air infiltration.

airtight wood stove   A wood-burning appliance that, when the door is closed and latched, is airtight. Combustion air is introduced and regulated through one or more air inlets.

air-to-air heat exchanger   A mechanical device that draws fresh, outside air into a building while it exhausts stale inside air to the outside. A special heat-transfer medium is located between the two air flows, allowing the heat from the outgoing inside air to be transferred to the colder, incoming outside air. The device helps to maintain a flow of fresh air into a building while minimizing heat loss.

air tube   That part of an oil furnace in which air and atomized oil are combined before entering the furnace combustion chamber.

air uncoupled   Combustion appliances, such as wood stoves or furnaces, that utilize a supply of outside air for combustion rather than drawing in and wasting heated inside air.

air valve   A device used on a steam heating system to allow trapped air to be vented from the steam.

airway   An area between the roof boards and the roof insulation that allows air movement.

AITC   Abbreviation for *American Institute of Timber Construction*.

alburnum   The living sapwood of deciduous trees, located between the bark and the heartwood.

alcohol   In construction, a term commonly referring to a combination of ethyl and wood alcohol that is used to thin liquid shellac or as a heat source for melting shellac sticks.

alcohol base stain   See *stain*.

alcohol resistant   A finish that will not suffer any damage when contacted by alcohol.

alcove   A recess that is cut or built into the side of a larger room.

A-level   An ancient tool, still occasionally in use today, used to determine a level surface. It consists of an A-shaped frame of wood or other material, with an attached string and weight;

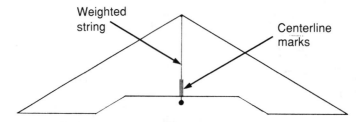

Fig. A-4. Simple A-level

when the feet of the frame are level, the string will hang across the centerline mark. See Fig. A-4.

aliphatic resin adhesive   See *adhesive.*

alkali   A water-soluble salt found in some water and soils that is capable of neutralizing acids.

alkyd resin   See *paint resins.*

alligatoring   A coarse, checkered pattern in a top coat of paint caused by poor preparation of the old paint surface to receive the new one. The new paint slips over the underlying layer, allowing the old paint to show through the resulting fissures.

alloy   A metal formed when two or more metals are fused, melted, or otherwise joined.

alteration   Any changes made to an existing building—structurally, mechanically, or otherwise—that do not increase the structure's cubic footage.

alternating current   Voltage that flows first in one direction, then reverses to flow in the opposite direction, changing several times each second. Alternating the current simplifies its transmission over long distances. Abbreviated ac.

altitude   In solar design, the angle from the horizon to the sun at any given time.

aluminum oxide paper   See *sandpaper.*

aluminum shingle   An aluminum sheet that has been formed into a shingle and textured to resemble wood. Available in various colors and patterns.

aluminum siding   Aluminum that is manufactured in various shapes and sizes for use as an exterior siding. It is usually factory painted with a baked-on enamel, and is suitable for new work or for re-siding existing buildings.

ambient lighting   A relatively uniform, low level of light provided throughout an entire area, intended more to decorate and highlight the area than to illuminate it.

ambient sound   The combination of general background noise in a building or other space. The overall noise is audible to the listener, but individual sounds cannot readily be identified.

ambient temperature   The temperature of the surrounding outside air.

American Institute of Architects   A national association of architects that establishes and regulates the principles and practices of its members. Abbreviated AIA.

American Institute of Timber Construction   The national technical trade association of the structural glued laminating (glulam) industry. Abbreviated AITC.

American National Standards Institute   The coordinating organization for the American system of uniform standards for materials and practices. It coordinates voluntary development of standards and represents the United States in the development of international standards. Abbreviated ANSI.

American Plywood Association   An association that represents the majority of American manufacturers of plywood and construction panels. The association determines minimum product standards, supervises quality control and testing, establishes installation procedures and applications for its products, and engages in ongoing research and promotion of new panel products. Abbreviated APA.

American Wire Gauge   A standardized method of describing wire diameter, using numbers from 18 to 0000 (or 4/0). The larger the gauge number, the smaller the wire.

American Wood Preservers' Bureau   An association that sets the standards for pressure-treated wood and oversees quality control in mills and treatment plants. Abbreviated AWPB.

ammeter   An electrical testing device used to measure current flow in amperes.

amp   Abbreviation for *ampere.*

ampacity   A combination of the words *ampere* and *capacity,* used to express in amperes the current carrying capacity of an electrical conductor.

ampere   The constant current that, if maintained in two straight parallel conductors of infinite length and of negligible circular cross section, and placed 1 meter apart in a vacuum, would produce between the conductors a force of 2×10^{-7} newton per meter. One ampere flows through a 1-ohm resistance when a potential of 1 volt is applied.

amplitude   1. The maximum departure from the average value of the value of an alternating current or alternating radio wave during one cycle. 2. The extent of a vibration to each side of the average position. 3. Maximum height or magnitude.

analogous harmony   Pertaining to colors that are next to each other on the color wheel. They are related to each other through one other color they have in common.

anchor block   A wooden block embedded in a masonry wall to serve as a point of attachment for various fixtures or framing.

anchor bolt   A steel bolt, usually L-shaped, that is set in concrete while the foundation is being poured and is used to anchor the wooden sill plate to the top of the foundation wall or to anchor the sole plate to a slab floor. Also called a J-bolt. See Fig. P-2.

anchor, hollow wall   Any of a variety of devices used to secure objects to walls, ceilings, or other areas that have no solid backing behind the finished wall surface, such as areas finished with plasterboard or plaster. Some of the most common types follow.

COLLAPSIBLE ANCHOR Most commonly referred to as a Molly bolt, this anchor consists of a formed, slotted, sheet-metal sleeve that is threaded at the bottom and a machine screw that is passed through the sleeve to engage the threads. After it is inserted into the wall, the screw is turned, causing the sleeve to bend and draw up against the back side of the wall. See Fig. A-5.

DRIVE ANCHOR A type of collapsible anchor having a plastic point over the end of the screw that allows the anchor to be driven into the wall with a hammer and eliminates the need to predrill a hole in the wall.

HOLLOW DOOR ANCHOR A short, collapsible anchor specifically made for the thin face veneers of a hollow core door.

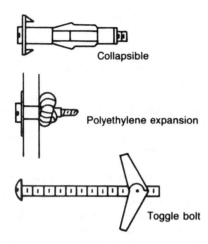

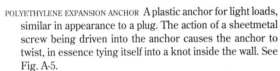

Fig. A-5. Common hollow wall anchors

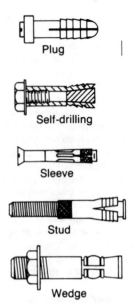

Fig. A-6. Common masonry anchors (Courtesy Red Head, Phillips Drill Division, Michigan City, IN)

POLYETHYLENE EXPANSION ANCHOR A plastic anchor for light loads, similar in appearance to a plug. The action of a sheetmetal screw being driven into the anchor causes the anchor to twist, in essence tying itself into a knot inside the wall. See Fig. A-5.

TOGGLE BOLT An anchor having a machine screw and two springloaded, collapsible metal wings attached to a nut. The wings are collapsed; the toggle is passed through the item being hung and then through a hole in the wall, and the wings snap open again once they are inside the wall. Tightening the screw locks the anchor against the back of the wall. The toggle is lost if the screw is removed. See Fig. A-5.

anchor, masonry A device used to secure an object to a masonry surface, such as concrete, brick, or stone. Some of the more common types follow.

EXPANSION ANCHOR Any of a variety of threaded metal anchors that are inserted into a predrilled hole. When a bolt is inserted and tightened, the anchor expands and secures itself by gripping the sides of the hole. Some types of nonthreaded plugs are also referred to as expansion anchors.

LAG SHIELD A metal cylinder split lengthwise and threaded inside. After the shield is inserted into a predrilled hole, a lag bolt of corresponding size is placed through the object to be fastened and into the shield, which is expanded by the bolt to grip the hole.

LEAD CAULKING ANCHOR A lighter-duty expansion anchor for use with machine screws and small bolts. It is set into a predrilled a hole with a special caulking tool that is usually provided when a box of anchors is purchased.

PLUGS Available in a variety of materials, including lead, fiber, plastic, and nylon, and used for light-duty fastening. A hole the size of the plug is drilled into the masonry, the plug is inserted, then a screw is tightened into the anchor to expand it inside the hole. See Fig. A-6.

SELF-DRILLING ANCHOR A type of female-threaded expansion anchor having case-hardened steel teeth on the bottom. The anchor is inserted into a special holder on a hammer drill and is used in place of a drill bit to create its own exact hole size. It is then used like a conventional expansion anchor. See Fig. A-6.

SLEEVE ANCHOR A general-purpose anchor, commonly available as a flat- or round-head screw, or as a male-threaded stud. The entire thread area is enclosed with a metal sleeve that is solid on top and split on the bottom. As the screw or nut is tightened, the split end of the sleeve expands in the hole. See Fig. A-6.

STUD ANCHOR A type of expansion anchor having male threads instead of female. See Fig. A-6.

WEDGE ANCHOR A male-threaded stud with a tapered bottom enclosed by two steel wedges. The stud is inserted into a predrilled hole, then a nut is placed over the threads. When the nut is tightened, the tapered area rises, expanding the wedges to grip the hold. See Fig. A-6.

andiron A thick, heavy metal bar supported by feet, used to hold burning logs up off the hearth. Sometimes called a firedog. Alternate spelling: endiron.

angle bead See *corner bead*.

angle bond Bricks or sheet-metal ties used to join and interlock the corners of a brick or concrete block wall.

angle brace 1. A steel support used to reinforce a right-angle joint and commonly screwed to the inside corner of the adjoining pieces. 2. A wooden support serving the same purpose as 1, although technically an incorrect usage of the term.

angle bracket A right-angled metal brace used to join and reinforce the corner where two boards meet.

angle brick See *brick*.

angle grain See *grain*.

angle iron An iron or steel bar L-shaped in section and used as supports or braces in steel construction.

angle of incidence The angle at which direct sunlight strikes a particular surface; a prime consideration in calculating the amount of energy absorbed by a solar collector.

angle stop A plumbing shutoff valve having a 90-degree bend between the inlet and outlet sides; commonly used as a shut-off under sinks and toilets.

angle trowel A trowel in which the blade has been bent to an angle of about 90 degrees; used for finishing plaster or joint cement in inside corners.

angle valve A water control valve, similar in design and operation to a globe valve but having the inlet at right angles to the outlet.

angular measurement A method of measuring an angle by relating it to the degrees of a circle. See Fig. C-3.

anhydrous A compound that has no water in its makeup.

anhydrous lime See *lime*.

animal glue See *adhesive*.

annealed wire A strong, pliable wire used for binding rebar, forms, etc.

annealing The process of heating a metal to a given temperature, holding it temporarily at that temperature, and then cooling it slowly so as to reduce the hardness of the metal and make it easier to machine.

annual ring The layer of wood that shows the growth of the tree for the year. Annual rings indicate the age of a tree and also affect the appearance of finished lumber, depending on how the board is cut relative to the rings.

annular threaded nail See *nails, shank types: ring shank*.

anodizing An electrochemical bath for aluminum that improves its hardness, corrosion resistance, and overall appearance. The anodized surface can also hold dye or enamel for coloring, a common process used in making window frames and other aluminum items used in building.

ANSI Abbreviation for *American National Standards Institute*.

anticorrosion anode A metal rod holding a positive charge and inserted into the tank of a water heater. It acts to draw minerals and other corrosive elements out of the water, thus helping to prevent corrosion of the tank. See Fig. W-2.

anticorrosion paint A paint for metal that has been blended with special additives to inhibit rust and corrosion.

antifouling paint A special coating, designed for ship bottoms, that releases poison at a controlled rate to prevent the attachment and growth of marine organisms.

antiquing A technique combining a variety of materials and application methods to produce an appearance of age or wear on new wood.

antisiphon valve A special valve containing a one-way rubber seal, and used to prevent contaminated water from siphoning back into the potable water system; commonly used on sprinkler systems.

APA Abbreviation for *American Plywood Association*.

aperture 1. An opening. 2. That portion of a south-facing glass or collector surface which contributes to a building's solar heat gain.

apex stone The triangular or wedge-shaped stone at the top of a masonry gable wall.

appliance 1. Any device utilizing electricity or combustible fuel to produce heat, power, refrigeration, or air conditioning. 2. An electrically powered device used to perform a household function. Appliances are commonly rated as follows.

FIXED APPLIANCE An appliance that is fastened in a given location, such as an oven.

PORTABLE APPLIANCE An appliance that is easily moved from place to place, such as a toaster or vacuum cleaner.

STATIONARY APPLIANCE An appliance that can be moved, but normally is not, in everyday use, such as a refrigerator.

appliance circuit See *branch circuit*.

appliance garage A kitchen storage feature consisting of a small enclosure with a tilt-up or folding door and located between the countertop and the bottom of the upper cabinets, usually in a corner. It is used to store and conceal portable appliances.

appliance plug A specially designed electrical cord and plug for use with small, portable appliances; identified by the two round receptacles in the female end of the plug.

approved A term used to show that an item or installation procedure has been deemed acceptable by the authority having jurisdiction over it, such as a building department, government code, or testing laboratory.

apron 1. A molding applied horizontally to the wall, directly below the windowsill. It is used to hide the rough edge of the sheetrock or plaster below the window framing. See Figs. M-4 and W-3. 2. A solid platform or drive, usually concrete, alongside a driveway, dock, or other similar area to provide a smooth, stable working or loading area.

aquastat A specialized type of thermostat used to regulate water temperature.

aqueous Pertaining to or containing water.

arbitration A method of hearing and resolving disputes, in which two parties present their claims to a neutral third party, who issues a decision.

arbor 1. A decorative wooden latticework used to support vines to create a shaded area; commonly used as a patio or path cover. 2. A revolving shaft to which a cutting tool is attached.

arc 1. A visible discharge of electricity across a gap between two conductors in a circuit. 2. A curve or portion of a circle. See Fig. C-3.

arcade A series of covered arches supported by columns, either attached to a wall or free-standing.

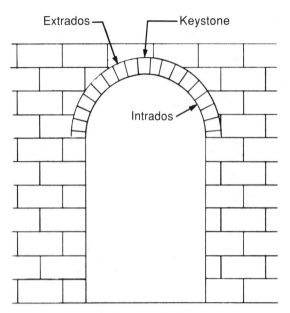

Fig. A-7. Arch

arch A compressive structural member, usually curved, that spans an opening and is capable of supporting a load over that opening. See Fig. A-7.

arch brick See *brick*.

architect A person qualified through specific education and experience and licensed by a state architectural board to design and construct buildings.

architectural drawings The detailed design and construction plans for a structure.

architectural sheet metal All visible, exterior sheet metal, including such things as sheet-metal roofs and decks, gutters and downspouts, flashings, trim, etc.

architectural symbols Standardized symbols used on building plans and drawings to depict various specifications and features. See Fig. A-8.

architrave 1. In the style of classical architecture, a horizontal beam resting on top of columns, forming the lowest part of an entablature. 2. The combination of moldings surrounding a door or window.

arch tone Triangular or wedge-shaped stones used in forming an arch or other circular shape.

archway The passage area underneath an arch.

arc welding A method of joining metal in which electricity is used to create an arc between the metal and a welding rod. The arc generates intense heat, which melts the rod and the edges of the metal, fusing the joint together.

areaway An open, below-grade space next to a building that provides light, ventilation, or access to a basement. See Fig. A-9.

armored cable Insulated electrical conductors enclosed in a flexible metal casing and used for interior wiring; also commonly called BX or ac cable.

arris In architecture, the ridge formed by the meeting of two surfaces, especially two moldings.

artificial stone A decorative veneer material manufactured to resemble various types of natural stone.

asbestos The fibrous form of several different minerals and hydrous silicates of magnesium. It is fireproof, a poor conductor of heat, and resistant to most common chemicals. It is widely used as an insulator in pipes, stoves, roofing materials, and much more.

asbestos cement board See *building board*.

asbestos cement shingle A type of roofing or siding shingle manufactured from asbestos fiber and portland cement. These shingles are durable and fireproof, but brittle; also called mineral fiber shingles.

asbestos tape A noncombustible asbestos-impregnated paper, usually sold in rolls and used for sealing joints in ductwork. Some types are applied with a special paste; others are prepasted and need only to be wet with water prior to application.

as-built A drawing depicting a structure, plumbing system, electrical system, or other building component as it was actually built and installed.

ash See *hardwood*.

ash dump A covered opening provided in the floor of a fireplace that allows accumulated ashes to drop down into the ash pit for later retrieval. See Fig. F-4.

ashlar Blocks of cut stone that are laid up in regular or irregular courses for a variety of structural and decorative uses.

ashlar brick See *brick*.

ash pit An accessible area under a fireplace or furnace in which ashes are collected and stored for later removal. See Fig. F-4.

ASHRAE Abbreviation for the *American Society of Heating, Refrigerating and Air Conditioning Engineers, Inc.*

askarel A group of nonflammable synthetic hydrocarbon compositions from which electrical insulation materials are produced.

asphalt In general, a residue from evaporated petroleum that is insoluble in water and melts when heated. It has a great number of applications, primarily waterproofing, and is used in roofing materials, building paper, floor tile, and much more.

asphalt shingle Any of a variety of sizes and shapes of roofing shingles made from a thick felt or fiberglass mat that is saturated and coated with asphalt, then covered with mineral granules in a variety of colors; often categorized by weight per 100 square feet of roof-deck coverage.

asphalt tile A relatively thick, resilient, durable floor tile suitable for above- or below-grade use.

assembly drawing A type of drawing in which all of an object's parts are shown separately and in proper relation to each other; used to illustrate the order in which an object is

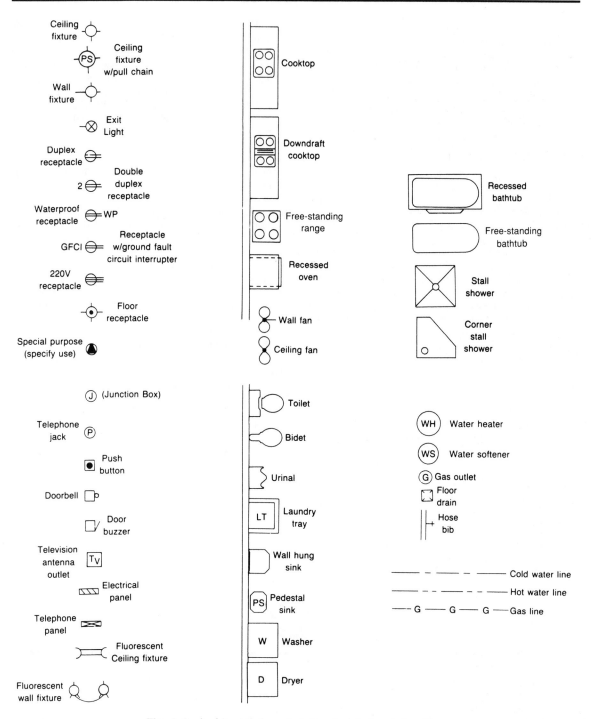

Fig. A-8. Architectural symbols for electrical and plumbing

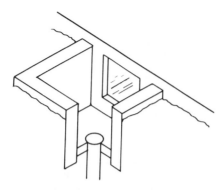

Fig. A-9. Areaway around basement window

to be assembled and the relationship of its parts to each other.

ASTM   Abbreviation for *American Society for Testing Materials*.

astragal   A T-shaped wooden or metal molding used between double doors. It is attached to the lock edge of the inactive door and acts as a stop against which the active door closes.

asymmetrical   A figure or object which is not symmetrical. See *symmetrical*.

atomize   To reduce a liquid into a vaporlike state so as to allow it to be sprayed.

atrium   A courtyard, hallway, or lobby, often several floors in height and having windows or galleries that open onto it at each floor.

atrium door   See *door*.

attic   The enclosed area between the ceiling and the underside of the roof. See Fig. C-15.

attic fan   See *attic ventilation*.

attic vent   1. Screened openings in the sidewalls of a building,

and located near the underside of the roof; used to provide an exit for air in the building's attic. 2. Covered openings cut into the roof to serve such a purpose.

attic ventilation   The process of removing stale air from an attic. Cooler air enters the attic from outside through frieze or soffit vents, causing the hotter attic air to rise naturally and exit the building through gable or ridge vents. This movement is sometimes supplemented through the use of thermostatically controlled power attic fans, which are cut into the roof or placed over the gable vents.

auger bit   A long, screw-tipped, wood boring bit.

autoclave   An airtight vessel or chamber in which various chemical curing and testing procedures are conducted under pressure.

autumnal equinox   The day on which the sun has traveled half the distance south on its yearly north-south path, and so crosses the equator; occurs on or about September 21st, traditionally the start of autumn. See Fig. S-23.

auxiliary heating system   See *backup heating system*.

auxiliary view   A supplemental drawing on a set of plans, shown in true relation to one part of another, larger drawing. It is used to provide greater or additional detail of some aspect of the larger, principal view. See Fig. O-4.

avoirdupois weight   The common system of weights used in the United States, Great Britain, and many other countries in which 16 ounces = 1 pound and 16 drams = 1 ounce.

AWG   Abbreviation for *American Wire Gauge*.

awl   A small, pointed hand tool used for making or boring small holes; also called a scratch awl.

awning window   See *window*.

AWPB   Abbreviation for *American Wood Preservers' Bureau*.

axis   A real or imaginary centerline passing through an object, about which the object could rotate.

azimuth   In solar design, the angle between true south and a point on the horizon directly beneath the sun at any given time.

B

backband   A narrow molding applied around the outside edge of interior window and door casings to widen and decorate the casing. See Fig. M-4.

backblocking   Scraps of wood, plasterboard, or other backing material applied with adhesive to the back of unsupported plasterboard seams.

backbone   See *stringer*.

backdraft   Air that flows back down a flue or chimney.

backdraft damper   A metal flap mounted inside a flue or chimney that is designed to close in the presence of a backdraft to prevent air from flowing back into the building.

backfill 1. The act of replacing earth or rock into an excavation. 2. The dirt, rock, or other material that is placed back into an excavation.

back filling 1. In masonry, rough brick or other masonry units, placed behind the finish facing bricks, as in a fireplace. Also called a backup or backing tier. 2. See *brick nogging.*

backflow Contaminated water or waste that flows back into a potable water system because of a cross-connection, siphoning, or other plumbing problem.

backflow preventer See *backflow valve.*

backflow valve A one-way valve or similar device used in a piping or solar system to prevent the reverse flow of liquid.

backing Wooden strips or blocks placed behind thin or hollow materials either for strength or as a provision for the later attachment of other objects.

backing tier See *back filling.*

backlash The amount of play between loosely fitting parts in a machine, particularly the distance or play between the teeth of a gear.

backsaw See *handsaw.*

backset The horizontal distance from the strike edge of a door to the center of the cylinder or keyhole. See Fig. D-3.

backup See *back filling.*

backup heating system A conventionally fueled furnace, wood stove, water heater, or other heat source used to provide backup space or water heat for a solar heating system during prolonged periods without full sunlight; also called an auxiliary heating system. See Fig. C-6.

back vent A plumbing vent from a fixture which connects with another vent at a higher level; also called a loop vent or revent. See Fig. D-14.

backwired terminal A type of terminal located on the back of an electrical device such as a switch or receptacle. It consists of a hole with a contact inside, into which the stripped end of the wire is inserted to make the connection. See Fig. G-6.

baffle 1. A partition inside a fireplace or appliance that directs the flow of air, smoke, or flame. Also called a baffle plate. 2. Any partition, plate, or other device that directs the flow of air, water, or other materials.

bag hook Any of a variety of T-shaped tools having a wood handle and a curved, pointed metal shaft, used for lifting sacks and for other workshop uses.

Bakelite The trade name for a thermosetting plastic made from phenol-formaldehyde resin and used in the manufacture of telephones, electrical boxes and components, and many other items.

balancing valve Small valves that are installed in the pipe which supply the individual room heaters in a steam or hot water heating system. They control the flow of steam or water entering each unit, allowing the entire system to be balanced for best operation.

balk A heavy, squared beam or timber, often used as a collar beam or brace.

ballast An electrical device used with fluorescent, mercury vapor, and other types of gaseous-discharge lights to stabilize the amount of current entering the tube.

ball bearings A series of small, hardened steel balls placed in a circular frame called a bearing race and used as a bearing for revolving shafts, such as those in a motor or power tool; considered superior to sleeve bearings in most applications because of their longer wear life.

ball catch A small cylinder placed in the edge of a cabinet door and containing a spring-loaded steel ball. The ball engages a recessed metal plate on the cabinet, keeping the door closed.

ball cock The valve and float assembly inside a toilet tank that connects to the water supply line. It starts and stops the flow of water into the tank after flushing and controls the tank's water level. See Fig. F-9.

balloon bag A device used in combination with a garden hose to flush a clogged sewer line. The bag is attached to the hose and inserted into the line, where it fills with water to seal the pipe, thus allowing the hose pressure to be directed against the clog. See Fig. B-1.

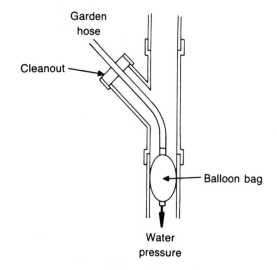

Fig. B-1. Balloon bag

balloon framing An older method of framing in which the wall studs extend full length from the foundation to the ceiling line and the joists for each floor are hung from the studs. See Fig. B-2.

ball-peen hammer See *hammer: peen hammer.*

balsa See *hardwood.*

baluster A vertical member, usually decoratively turned or cut, that supports a stair rail. See Fig. S-18.

balustrade A railing made up of balusters attached to a top rail, either with or without a bottom rail, and used along stairs and landings, porches, decks, and balconies; also called a banister. See Fig. S-18.

band Any of a variety of low, flat moldings, typically used to

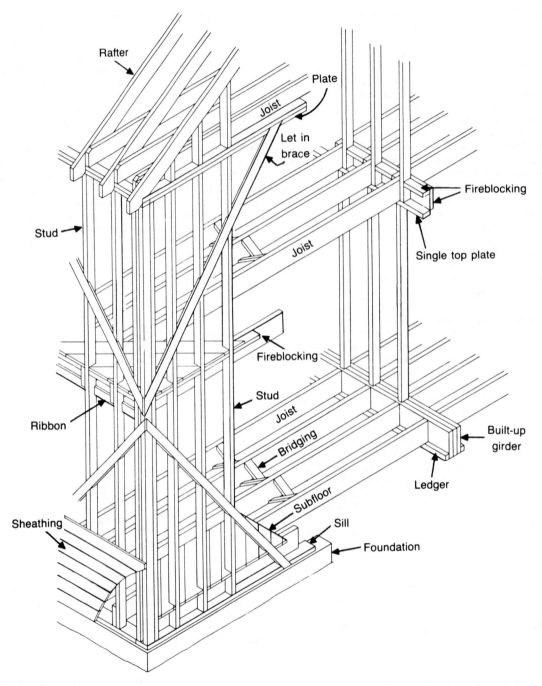

Fig. B-2. The main components used in balloon framing (Courtesy the National Forest Products Association, Washington, DC)

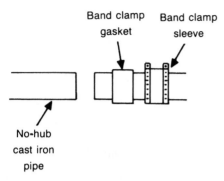

Fig. B-3. Band clamp for a no-hub pipe

conceal seams in other materials, as decorative dividing strips, or as a base underneath other moldings to add depth.

band board A flat, decorative board used to separate siding from one floor to another, or between the top floor and the gable end. Also called a belt board. See Fig. F-5.

band clamp A round metal sleeve with two screw-operated clamps and a removable neoprene gasket; used to connect and seal no-hub cast iron pipe. See Fig. B-3.

band joist See *rim joist.*

band saw A stationary power tool having an endless, toothed steel belt that revolves vertically over two pulleys. It is used primarily for the freehand cutting of curves and other irregular shapes in wood or metal, or for resawing lumber into smaller thicknesses.

banister See *balustrade.*

banjo A device that holds a roll of joint tape and a quantity of joint cement, applying both at the same time for taping plasterboard joints.

barbed nail See *nails, shank types.*

bar clamp A woodworking clamp consisting of two jaws and a flat steel bar available in a variety of lengths. One jaw is fixed at the end of the bar; the other contains a screw-activated block and is movable along the bar's length to adjust for the size of the work being clamped.

bare conductor An electrical wire that does not have a wrapping of insulation; used as a ground conductor.

barge board A decorative board used in gable roof framing when no roof overhang is present. It is used to cover the exposed face of the last rafter. Also called a verge board.

barge rafter In gable roof framing, the outside rafter that follows the last common rafter. It forms the support for the section of roof that overhangs the gable end wall. Also called a fly rafter. See Figs. C-13 and R-3.

bar hanger An adjustable metal bar having a threaded post attached; used for mounting electrical boxes. The bar is secured between two ceiling joists or two studs, then the box is attached to the post with a locknut.

bark See *wood.*

bark pocket See *lumber defect.*

baroque An ornate and heavily decorated style of architecture and decoration. Originated in Italy and flourished from the mid-sixteenth to mid-seventeenth centuries.

barrel bolt A latch mechanism consisting of a sliding steel bolt that engages a U-shaped catch; used on doors, gates, and similar applications.

barrel roof A rounded roof structure resembling a barrel cut in half lengthwise.

base 1. The bottom layer, coat, level, or other starting point for another operation. 2. See *baseboard.* 3. The bottom portion of a column, below the shaft. See Fig. C-8.

base bid A contractor's basic bid for a construction project, not including options or changes.

base block See *plinth block.*

baseboard A board or molding—available in a wide variety of sizes and styles—that is installed against the wall at the point where the wall and floor meet; also called a base, mopboard, or skirting. See Figs. B-4, D-6, and M-4.

baseboard heater A long, narrow space heater attached horizontally to the wall of a room at floor level; it can operate off electricity, hot water, or steam.

base cabinet A lower cabinet in a kitchen or other area upon which a countertop rests.

base coat See *prime coat.*

base course The first and lowest row of masonry in a wall, serving as a footing on which the remaining courses rest.

basement The lowest level of a house, commonly below grade and with minimum interior refinements. See Fig. C-15.

base molding A molding used to trim the area between the top of the baseboard and the wall. See Figs. B-4 and M-4.

base plate See *plate: sole plate.*

base sheet Asphalt-impregnated felt that is used as the first layer in a built-up roof.

base shoe A small, curved molding attached to the baseboard where it meets the floor; commonly used in conjunction with hardwood flooring. See Figs. B-4 and M-4.

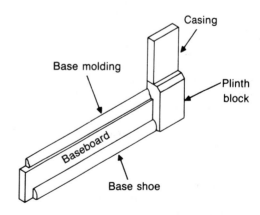

Fig. B-4. Baseboard and related parts

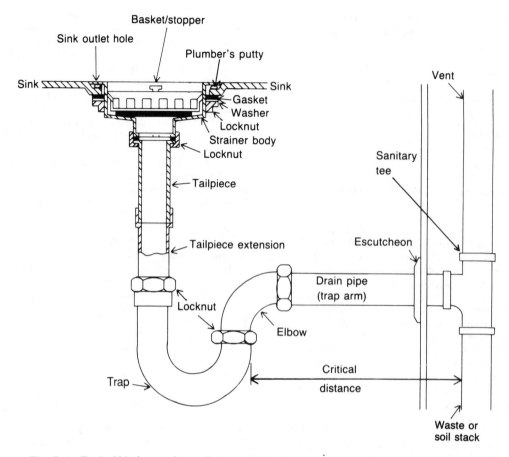

Fig. B-5. Typical kitchen sink installation, with basket strainer, tailpiece, trap, and trap arm

basin wrench   A long-handled tool with a serrated, pivoting arm at one end; designed for tightening or loosening the hard-to-reach locknuts on the underside of sink-mounted faucets.

basket strainer   A plumbing fitting that seats into the outlet hole at the bottom of a kitchen or service sink. It consists of the strainer body, a removable basket that acts as both a stopper and a trap for food particles, and various locknuts, washers, and rubber gaskets that seal it to the sink. Also called a sink strainer. See Fig. B-5.

basswood   See *hardwood*.

bat   A piece of brick that is whole on one end and broken off on the other end.

batch   The amount of concrete, mortar, glue, or similar material that is mixed up and used at one time.

bathroom sealant   See *sealant: silicone sealant*.

batt   1. An abbreviation for *batten*. 2. A strip of insulation. See *insulation, thermal*.

batten   1. A narrow piece of wood used to cover the joints where plywood or lumber edges meet; commonly applied vertically and used as a decoration on exterior walls. 2. Vertical boards used in a batten door. See Fig. B-6.

batten and board   A style of siding in which narrow, uniformly spaced battens are first installed vertically on the wall, then wider boards are attached to the battens to cover the intervening spaces. See Fig. S-11.

batten door   A style of door comprised of vertical boards called battens, secured to horizontal and/or diagonal wooden cleats, located on one or both sides of the door. See Fig. B-6.

batter   1. A masonry wall that slopes up and back, the thickest part being at the bottom; used primarily in retaining walls. 2. Any wall or material that slopes up and back.

batter boards   A temporary framework, usually consisting of two horizontal boards set on three vertical posts to form a 90-degree corner; used to establish and mark the corners of an area being excavated for a foundation. See Fig. S-19.

batter pile   A pile driven at an angle to brace a structure against lateral force.

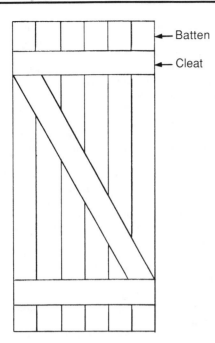

Fig. B-6. Batten door

battery A device that uses the interaction of certain chemicals to produce and store dc electricity.

bay 1. A principal part or division of a structure, usually marked by beams or columns. 2. A wing or extension of a building. 3. An alcove, especially one containing a window.

bayonet saw See *handsaw.*

bay window See *window.*

bead 1. A small rounded or grooved molding used to decorate the joints in a series of boards. 2. A continuous ribbon of caulk, adhesive, or other material.

bead board See *insulation, thermal: molded polystyrene.*

Beadwall The brand name for a window insulating device in which a vacuum pump is used to blow tiny polystyrene beads into the airspace between the two panes of glass in a window. When the sun is out, the motor reverses itself automatically and sucks the beads back out into a storage tank.

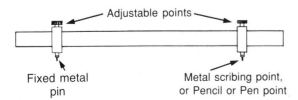

Fig. B-7. Beam compass

beam A main structural member, installed horizontally and supported by posts, walls, or columns, that acts to support other loads installed across it. See Figs. O-1 and P-10.

beam compass A horizontal bar with two sliding steel points that can be locked anywhere along the bar's length; used for scribing circles too large for an ordinary compass. See Fig. B-7.

beam fill Masonry or concrete used to fill the spaces between the floor joists at the basement level of a building, thus acting as a fire stop.

beam pocket A formed recess in the top of a masonry wall that is designed to receive and support the end of a beam.

beam radiation See *solar radiation.*

beam seat See *timber connector.*

bearing 1. Any structural member or combination of members, such as a wall, that supports a load in addition to its own weight. 2. That part or area of a machine upon which a shaft, pin, or other part moves or rotates.

bearing block A block or plate placed under a structural member to help distribute that member's load over a greater area; also called a bearing plate.

bearing partition Any inside partition that supports a load, such as a ceiling or a second floor, in addition to its own weight.

bearing plate 1. A metal plate used on the exterior of a building to distribute the load imposed by a tie rod. See Fig. T-5. 2. See *bearing block.*

bearing race See *ball bearings.*

bearing stress The compressive stress exerted upon a structural member by another member at the point of contact, such as that of a joist laid on top of a girder.

bearing wall A wall that supports a load, such as a second floor or a roof, in addition to its own weight.

bedding Any filling material, such as putty, cement, or mortar, that acts as a solid base for a subsequent part or installation.

bedding strip See *sill sealer.*

bed joint The horizontal layer of mortar upon which bricks or other masonry units rest. See Fig. B-10.

bed molding A molding installed to cover the inside corner where two surfaces intersect, as between an overhanging cornice and the sidewall.

bedplate A metal plate used as a foundation or anchor for a structural member or piece of equipment.

beech See *hardwood.*

bel A unit that expresses the ratio of two amounts of power; used as a measure of sound intensity. One bel equals 10 decibels.

bell and spigot See *hub.*

bell bit Any of a variety of long, slender drill bits, typically 18 inches or longer, originally used to drill long holes for running small-diameter bell wire.

bell reducer A bell-shaped plumbing fitting used to join two threaded steel pipes of different diameters; also referred to as a reducing coupling. See Fig. T-4.

bell wire   Small-diameter electrical wire used for doorbells, intercoms, and other low voltage wiring applications. 2. Strong, slender, nonelectrical wire, used in older-style mechanical bell systems.

below grade   That portion of a foundation, wall, or other structure or area which is below ground level; also called subgrade. See Fig. C-15.

belt board   See *band board*.

belt course   A row or line of stone, brick, wood, or other material carried horizontally along a wall to serve as a transition between two siding materials of different types or thicknesses.

belt sander   A motorized tool, either portable or stationary, having an endless sandpaper belt driven over two or more rollers.

bench dog   See *dog*.

bench hook   A flat board with two cleats, one on top along one end, and one on the bottom along the other end. It is hooked over the top of a workbench to hold another board. See Fig. B-8.

bench mark   A mark set in a permanent object, such as a stone or concrete block, and used as a reference point for surveying.

bench plane   See *plane*.

bench stop   A metal device recessed into a workbench top to act as a stop for boards being worked on the bench top.

benderboard   Thin, flexible strips of redwood lumber, usually 1/4 to 3/8 inch thick and 4 or 6 inches wide, commonly used as forms for concrete flatwork and for a variety of landscaping applications.

bent   A transverse framework in a building designed to carry both horizontal and vertical loads.

benzopyrene or benzpyrene   A tarlike, organic material that is a by-product of incomplete combustion.

berm   1. Earth, sometimes combined with rock, that is banked against the walls of a house. It acts as an insulator and helps to moderate temperature swings within the building. 2. A mound or small hill of dirt used to deflect wind currents or for landscaping effect.

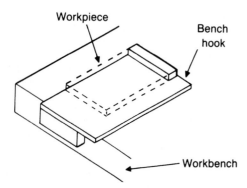

Workpiece

Bench hook

Workbench

Fig. B-8. Bench hook

bevel   1. Any angle or inclination of a surface other than 90 degrees. See Fig. D-3. 2. The act of cutting such an angle.

bevel block   See *V-block*.

bevel board   A board used as a guide in the framing of a roof or stairway. Also called a pitch board.

bevel siding   Wedge-shaped boards, usually 1/2 to 3/4 inch thick at the butt end and up to 12 inches wide, applied horizontally in overlapping rows as an exterior wall covering. Some types are rabbeted on the bottom edge.

bezel   The sloping or beveled edge on the blade of a cutting tool.

bid   An offer or proposal, usually in writing, to supply materials, labor, or equipment at a specified price.

bid bond   A bond furnished by an individual or company bidding on a construction project, intended to ensure that the bidder will agree to perform the work at the bid price if that bid is accepted.

bidet   A bathroom fixture somewhat resembling a toilet, but also equipped with hot water and a fixed spray head, used for washing the perineal parts of the body. See Fig. D-14.

bidirectional   1. Responding or acting in two different directions. 2. Having the capacity or ability to act in two directions.

bifold doors   See *door*.

billet   A small piece of metal or wood that has been roughly shaped to size, ready for final finishing.

bill of materials   See *material list*.

bimetal   Consisting of two metals; particularly alloys, which are formed from the combination of two metals.

bind   1. To stick or otherwise not move freely, as a door in its frame. 2. To fasten or tie together, as with a cord or wire. 3. To become stiff or hard, or to adhere together.

binder   An adhesive, particularly one mixed with wood chips in the making of hardboard, particle board, and similar panels.

binding stone   A stone that passes through one or both sides of a stone wall to form a bond with the intersecting walls.

birch   See *hardwood*.

bird mouth   The notch cut in the lower end of a rafter to fit the top plate of the wall. It consists of a horizontal seat cut and a vertical plumb cut. See Fig. R-5.

bird's-eye   The natural occurrence of small areas on a board's surface where the fibers are contorted in elliptical shapes, somewhat resembling the eye of a bird. Mostly found in sugar maple, and used for decoration.

bisect   To divide into two parts or angles of equal size. See Fig. P-8.

bit   Any of a variety of tools having a sharpened edge for cutting or boring and a shank for inserting into a drill press, brace, or similar tool.

bit roll   A flat, compartmented holder made of canvas or leather, designed to hold brace bits and then be rolled up for storage or transport.

bit stop   A plastic or metal clamp attached to a drill bit to control the depth of the hole being bored.

bituminous   Anything containing bitumen, a natural mixture of solid and semisolid hydrocarbons such as asphalt and naphtha.

black pipe Threaded steel pipe that is painted black rather than being galvanized; used for gas lines and hydronic heating systems, where its color differentiates it from the pipes carrying potable water.

black tools A generic term for tools that are left unpolished after manufacture.

blacktop Asphalt or other bituminous compounds used as a surfacing material, as for roads and driveways.

blacktop fix See *sealant.*

blade 1. The cutting edge of a tool. 2. The longer of the two arms of a framing square, at right angles to the tongue and commonly 24 inches long and 2 inches wide.

blanket insulation See *insulation, thermal.*

blank flue In masonry, an unused flue constructed on the opposite side of a working flue to save materials and balance the weight.

bleaching The use of various bleaching agents to lighten the color of wood or to restore discolored or stained wood.

bleeder tile A pipe placed in a footing that angles down from a drain tile outside the building to a drainpipe inside the building to allow surface water collected from around the foundation to drain off.

bleeding 1. Exuding of stains, paints, or other discolorations through succeeding top coats of finish material. 2. Removing excess air from a pipe. 3. Oozing of water to the surface or through the forms of freshly poured concrete.

blemish Any mark or scar that detracts from the appearance of a board while not significantly affecting its strength.

blind 1. Any portion of a joint or assembled object that cannot be seen after assembly. 2. Any of a variety of operable window coverings designed to block light or to provide privacy.

blind dovetail A woodworking joint in which the ends of the boards are mitered and the dovetails are held back from the front edge of the miter. The assembled joint shows a mitered line, but is much stronger.

blind header Decorative bricks or stones used to give the appearance of a structural header.

blind-nail To drive a nail in such a way that the head does not show on the board's finished surface; generally used with tongue-and-groove lumber by nailing down at an angle through the tongue.

blind stop A rectangular molding attached to the side and head jambs of a window to serve as a stop for storm windows and screens. See Fig. W-3.

blind story A story having no windows or other exterior openings that allow a view to the outside.

blistering Forming bubbles in a layer of paint or varnish; usually caused by moisture trapped between the paint layer and the underlying surface.

block 1. A type of masonry unit, usually hollow concrete, that is larger and heavier than a common brick. 2. A pulley. 3. A piece of cut-and-dressed stone. 4. An obstruction in a pipe, duct, or other opening. 5. Any relatively small or short piece of lumber.

block and tackle A device consisting of one or more pulleys, called blocks, through which rope or cable, called tackle, is passed; used to lift or otherwise move a load.

block plane A short, narrow plane commonly used to cut end grain and to work on small stock, such as moldings.

bloom 1. A varnished surface that appears cloudy. 2. Efflorescence that appears on a masonry wall.

blue board A commonly used name for Styrofoam brand insulation, made by Dow Chemical Company. See *insulation, thermal: extruded polystyrene.*

blued nail See *nails, finishes.*

blueprint 1. A method of reproducing original drawings. The drawing, done on tracing paper, is placed on light-sensitive paper and passed through a blueprinting machine. The exposed paper, after washing in water, turns dark blue, while the areas protected from light by the drawing wash out white. 2. A generic term often used to indicate any working drawings, regardless of color.

blue stain See *lumber defect.*

bluestone A blue-gray sandstone used primarily for exterior building decoration.

board See *lumber.*

board and batten 1. A style of siding in which wide boards are first installed vertically, followed by narrow battens, also applied vertically, that cover the joints between the wider boards. 2. Any object, such as a door, that is built or decorated in a board-and-batten style. See Fig. S-11.

board and batt siding See *board and batten.*

board and board A style of siding in which boards of the same width are installed vertically, with the outer boards covering the spaces between the inner boards. See Fig. S-11.

board foot A standard unit of measurement for lumber, equal to a board 1 foot square and 1 inch thick. See Fig. B-9.

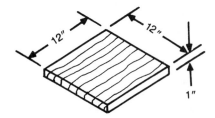

Fig. B-9. The basic dimensions of a board foot

board-foot measure The formula for converting the dimensions of any given piece of wood into board feet: T x W x L/ 12, where T = the thickness in inches, W = the width in inches, and L = the length in feet.

board in The process of applying horizontal or diagonal sheathing boards to the outside of wall studs.

boaster A type of chisel used to remove tooling marks or to otherwise smooth a cut piece of stone.

BOCA Abbreviation for *Building Officials and Code Administrators International, Inc.*, authors of BOCA National Building Code.

bodied linseed oil   See *linseed oil.*

body   See *viscosity.*

body coat   See *undercoat.*

boiled linseed oil   See *linseed oil.*

boiler   A central furnace using hot water or steam as a heat source.

boiling point   The temperature at which a liquid is converted into a vapor. The boiling point of water at sea level is 212° F, 100° C.

bolster   1. A short timber or steel beam installed horizontally on top of a column to support a beam or girder and decrease its span. 2. A steel chisel used for butting bricks or stone. 3. The enlarged portion of a chisel, below the handle.

bond   1. The adhesion achieved between two pieces that have been glued together. 2. The installation of bricks in various combinations of alternate courses to achieve a solid wall.

bond beam   A type of built-up beam that locks together an upper and lower wall section for rigidity.

bond breaker   A material used to prevent new concrete from sticking to the forms.

bond coat gypsum plaster   See *plaster.*

bond course   A row of masonry units that overlap and interlock two withes.

bonding   Using a mechanical bond such as a screw or wire to connect metal components of an electrical system, thus forming a continuous path for the ground.

book match   Two pieces of veneer that are matched along one common edge, similar to two pages of a book that are joined at the spine.

boot   A type of sheet-metal duct fitting available in various sizes and shapes and set into the floor, wall, or ceiling to make the transition between the duct pipe and the register.

border   A narrow strip of wall covering that matches or contrasts with a room's wallpaper and is used along the ceiling and around doors and windows to accent the room.

boring   Drilling or cutting a circular hole in wood or other materials.

borrowed light   Natural light that is brought from one room into another through the use of windows or other openings in the adjoining walls.

boss   1. A projecting ornament, especially at the meeting of the ribs of an arch. 2. The enlarged part of a shaft, usually designed to couple with a wheel.

Boston ridge   In roofing, a type of finished hip or ridge. Shingles are laid lengthwise in two rows, one on either side of the ridge, with the edges overlapping in alternating courses. Also referred to as a boston hip.

bow   See *lumber defect: warp.*

bow window   See *window.*

box   1. To surround partially or enclose completely, usually described as "box in." 2. To intermix the paint from two or more cans on the job site prior to application. Done to ensure consistent color between all the cans.

box bay window   See *window.*

box beam   1. A strong, lightweight beam constructed of one or more vertical plywood panels that are glued under pressure

to each side of a lumber frame. 2. A nonsupporting, decorative hollow beam constructed from three or four sections of lumber assembled into a long box.

box column   A supporting column, typically for a porch or other visible application, consisting of a load-bearing wood or steel post that is boxed in with decorative wood.

box cornice   See *closed cornice.*

box gutter   A type of rain gutter system that is built into a building's cornice and concealed from view.

box nail   See *nails, common types.*

box stairs   Stairs that are built between two enclosing walls. See *stairs: closed stairs.*

brace   1. An inclined or angled piece of wood or metal applied to walls, floors, corners, or other areas to stiffen the structure. Often used temporarily until other supporting work is completed. 2. A hand-operated, cranklike handle used for holding and turning a drill bit.

RATCHETING A type of brace having a revolving head in place of the standard crank handle, primarily used for work in areas with restricted clearance.

brace bits   Drill bits and other attachments designed specifically for use in a brace.

braced framing   An old, complex, rigid-frame construction, common in early New England. It used heavy, solid posts in the corners and at several intermediate points, connected by large horizontal members and a system of diagonal braces. Sometimes called eastern framing.

bracket   A projecting support for a shelf or other structure.

brad   A small finishing nail, up to 1-1/4 inches in length. Brads are specified by their length and gauge.

brad awl   A small hand tool used to create holes for the insertion of brads.

brad driver   A hand tool used to push brads into wood without the use of a hammer.

brake   Any of a variety of portable or stationary tools, hand or power operated, that are used to bend and form sheet metal.

branch   1. Any part of a piping system other than a main, riser, or stack that is used to service individual fixtures or fixture groups. See Fig. D-14. 2. In an extended plenum system, the single ducts that extend off the main duct to service individual rooms.

branch circuit   A single electrical circuit having its own fuse or circuit breaker extending out from the service panel to distribute electricity to one or more outlets, fixtures, or other electrical equipment. Common types of branch circuits follow.

APPLIANCE A circuit, usually 20 amps, that serves a kitchen, laundry room, or other area of high electrical usage.

GENERAL PURPOSE A circuit that supplies power for a number of lighting outlets and receptacles.

INDIVIDUAL A circuit designed to serve an individual piece of equipment that draws a large amount of power, such as a stove. Also called a dedicated circuit.

branch vent   A plumbing vent that connects one or more individual vents with a soil stack or vent stack. See Fig. D-14.

brass A metal allow of copper and zinc, typically in a ratio of approximately 2:1.

brass-plated A thin layer of brass over a steel base. Brass-plated steel is stronger than solid brass, but it has less corrosion and weather resistance.

braze To join metal parts with a metallic mixture, usually zinc and copper, that has a melting point in excess of 800° F.

brazed joint A joint between two metal parts that is made with alloys, called brazing rods, which melt at temperatures above 800° F.

breaching Access that is cut into a main chimney at a right angle to it, used for the installation of additional chimney connectors.

breadboard screw See *hanger bolt.*

breaking radius The maximum radius to which a piece of lumber or plywood can be bent before breaking.

break iron A curved plate attached to the iron of a plane and designed to cause the wood shavings to curl and break off.

break line In drafting, a standard line and symbol used to indicate a break in the drawing. Used singularly or in pairs, the break lines indicate that the object being drawn continues in the same manner after or between the break lines. See Fig. G-2.

break-off strip A small metal tab connecting the hot terminals on the side of an electrical duplex receptacle. Breaking off the tab separates the top and bottom receptacles, allowing them to be placed on different circuits. See Fig. G-6.

breezeway An open-roofed passage between two buildings or between two parts of a single building.

brick A solid rectangular masonry unit, formed from sunbaked or kiln-fired clay or shale. Bricks are available in many forms for different applications.

ACID RESISTANT BRICK Having additives to make it impervious to chemicals. Used with an acid resistant grout.

ADOBE BRICK Oversize clay bricks, usually tan or brownish in color, that are roughly shaped and sun-dried.

ANGLE BRICK Any brick having an end that has been shaped to an angle other than 90 degrees, designed for use on outside corners.

ARCH BRICK 1. A brick that has been formed or cut in a wedge shape for use in an arch or other circular shape; also called a compass or featheredge brick. 2. A brick that has been overburned in the kiln.

ASHLAR BRICK A brick that has been manufactured or altered to resemble stone.

BUILDING BRICK The standard, general-purpose brick for construction; also called a common brick.

COMPASS BRICK See *arch brick.*

CUTTER BRICK A type of soft brick that can be easily cut to shape with a trowel or other tool.

ECONOMY BRICK A building brick designed for use in standardized 4-inch modular units. Nominal size is 4 x 4 x 8 inches: actual size is 3-1/2 x 3-1/2 x 7-1/2 inches.

ENGINEERED BRICK A brick having nominal dimensions of 3-1/5 x 4 x 8 inches.

FACING BRICK A type of brick, either full size or thinly sliced, used specifically as a decorative facing. Available in a variety of surface textures and colors.

FEATHEREDGE BRICK See *arch brick.*

FIRE BRICK A brick made fire resistant through the use of special high-temperature ceramics; used in fireplaces, stoves, and other high-temperature areas. See Fig. F-4.

FLOOR BRICK Thick, smooth brick made with a high degree of wear and abrasion resistance for use as a floor covering.

GAUGED BRICK A brick that has been sawn, ground, or otherwise worked to produce specific, accurate dimensions.

GLAZED BRICK A building brick to which glazing materials have been fused, producing a glassy, decorative surface.

INSULATING BRICK A special type of brick that is resistant to heat transfer, used to insulate kilns and furnaces against excessive heat loss.

JUMBO BRICK 1. A brick having nominal dimensions of 3-1/2 x 4 x 12 inches. 2. A brick that is larger than standard, usually made to order; available in a variety of sizes.

NORMAN BRICK A brick having nominal dimensions of 2-2/3 x 4 x 12 inches.

PAVING BRICK A vitrified, highly abrasion-resistant brick for paving; similar to a floor brick, but usually larger and interlocking.

ROMAN BRICK A brick having nominal dimensions of 2 x 4 x 12 inches.

SCR BRICK Abbreviation for *Structural Clay Research*; a brick having nominal dimensions of 2-2/3 x 6 x 12 inches and actual dimensions of 2-1/6 x 5-1/2 x 11-1/2 inches, with ten 1-3/8-inch vertical holes and a 3/4- x -3/4-inch vertical slot on one end.

SEWER BRICK A brick made for use in sewer and drainage systems and having a high resistance to absorption and abrasion.

USED BRICK A brick that has been recycled from the demolition of old buildings; used to achieve an old, weathered look. Some types of new bricks are manufactured to give a used appearance.

brick construction A structure in which the exterior bearing walls are made of brick, as opposed to walls having a nonstructural brick veneer. See Fig. B-10.

brick hammer See *hammer.*

brick molding A wooden molding forming a boundary between bricks or other siding and a window or door. It is sometimes rabbeted to receive a screen or storm door.

brick nogging Bricks used to fill the space between structural members; also called back filling.

brick tie See *wall tie.*

brick veneer A nonstructural facing installed one brick deep over wood framing for decoration. See Fig. B-11.

bridging Pieces of wood or metal installed between the floor joists at mid-span; used to stiffen the joist and help distribute the floor load. Common methods follow.

CROSS BRIDGING Strips of wood or special metal braces fit between the joists, from the top of one to the bottom of the other, and vice versa, forming an X. See Figs. B-2 and P-2.

SOLID BRIDGING Blocks of wood the same width as the joist, installed vertically between the joists. See Fig. P-2.

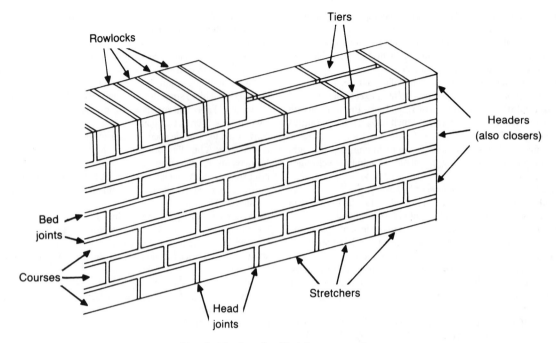

Fig. B-10. Standard brick construction

bright 1. Polished and shiny, as in bright brass. 2. Tools that have been ground and polished, as opposed to black.

bright finished nail See *nails, finishes.*

broached stone A building stone into which wide, decorative grooves have been cut with a punch or chisel.

broadloom Carpet that has been manufactured in widths of 6 to 18 feet.

brocade A hand-applied texture for drywall, created by applying thinned drywall joint compound with a special brush or foamcovered hawk, then lightly troweling over it as it dries. Brocades range from a light, lacy pattern to a heavy, random buildup of peaks and swirls.

broken joints See *staggered joints.*

broom-clean A term often used in construction specifications and contracts to indicate the amount of cleanup to be done upon completion of the project. Broom-clean is generally accepted to mean that the contractor will sweep, remove debris, and perform a light, general cleaning of the structure, or in the case of a remodeling, that portion of the structure that was remodeled.

brown coat In three-coat plaster or stucco applications, the second of the three.

brushability The ease with which a paint or other liquid finish can be applied by brush.

brushed plywood See *plywood siding.*

Btu Abbreviation for *British thermal unit.* The amount of heat necessary to raise the temperature of one pound of water 1° F in an ambient environment of slightly greater than 39° F;

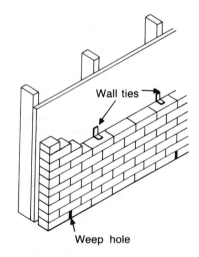

Fig. B-11. Brick veneer over wall sheathing

approximately equal to the heat given off by a common kitchen match.

Btuh Btus per hour.

buck To cut a felled tree into shorter lengths.

buffing compound A very soft abrasive, bonded with wax; used for buffing out a finish. Commonly available in stick form.

builder's hardware See *finish hardware.*

builder's level A device consisting primarily of a small spirit level and a telescope assembly attached to a circular base and usually mounted on a tripod. The telescope can be pivoted horizontally on the base, allowing the accurate sighting of level lines. Sometimes called a dumpy level.

builder's tape A long measuring tape, usually 50 or 100 feet in length, made of steel or fabric and contained in a circular case with a rewinding handle.

building block A hollow, 16-inch-long concrete block, 8 inches high, and available in widths of 2, 3, 4, 6, 8, 10, and 12 inches nominal size (actual size is 3/8 inch less on each face to allow for mortar); used for all types of construction. Common types follow. See Fig. C-11.

BULLNOSE BLOCK A corner block with one or two rounded corners.

CAP BLOCK A flat, rectangular block used to finish off the top of a wall, or as a garden stepping stone.

CORNER BLOCK A block for use at the corner or end of a wall. It is square at one or both ends.

HALF-HEIGHT BLOCK A block having the same width and length as a normal block, but only 4 inches high; used where a wall's height is not a multiple of 8 inches.

HALF-UNIT BLOCK A block that is one-half the length of a normal block.

HEADER BLOCK An L-shaped stretcher block used to facilitate bonding to a brick face.

JAMB BLOCK A rabbeted block used at door and window openings to permit the addition of wood members.

LINTEL BLOCK A U-shaped block designed to be filled with reinforcing bar and concrete; used as a lintel over a door or window opening.

PARTITION BLOCK A block that is smooth on both ends, and half the width of a normal block; used to form a recess in the ducts or pipes or to build non-bearing walls.

SOLID-TOP BLOCK Solid concrete on the top 4 inches, used to support joists and beams, or to cap the top of a wall.

STRETCHER BLOCK The main construction block, having two or three hollow cores, and a half core at each end.

building board A broad, generic category of manufactured boards, usually in 4- x -8-foot sheets, made from a variety of materials and bonding agents for various construction applications. Some of the more common forms of building board follow.

ASBESTOS-CEMENT BUILDING BOARD A rigid board of approximately 85 percent portland cement and 15 percent asbestos fiber, which is steam-cured. Common thicknesses are 1/8, 3/16, 1/4, 3/8, and 1/2 inch, and are either flat or corrugated. Uses include fire-resistant wall sheathing and roof decking.

CEMENT BUILDING BOARD A sheet formed from lightweight cement, designed to be used as a waterproof backing behind ceramic tile. Common sizes are 3 x 4, 3 x 5, and 3 x 6 feet, usually 1/2 inch thick. Also called a tile backing board.

CHIPBOARD A sheet made up of thin wood chips, each about 1-1/2 inches square, that are coated with adhesive, pressed, and cured; uses include underlayment and backing.

CORKBOARD Board formed from pressed cork particles, available in various sizes and colors. It has good acoustical insulating qualities and is often used in offices as a decorative, sound-deadening wall cover.

FIBERBOARD A sheet of wood or asbestos fiber, often with asphalt added as a binding agent; used for insulation panels and underlayment.

HARDBOARD A building board made from wood fiber that is hot pressed into a sheet. It is graded as standard (untempered) or tempered, in which the pressed board is impregnated with drying oils and then baked. Tempered hardboard is harder, heavier, more moisture resistant, and a darker brown in color than standard hardboard. Common thicknesses are 1/8, 3/16, and 1/4 inch. The standard sheet size is 4 x 8 feet, but it is also manufactured in a variety of other sizes for different applications. Hardboard has a number of uses, including siding, prefinished paneling, pegboard, and concrete forms. It is often referred to by the brand name Masonite.

INSULATING ASBESTOS BUILDING BOARD A sheet having asbestos cement faces with a core of fiberglass, particle board, or polystyrene foam. Commonly used as a fire-resistant, insulating backing under siding, roofing, and interior wall coverings.

MINERAL FIBERBOARD Mats of fiberglass, rock wool, or other mineral fiber with a stiff paper or paperboard face; used primarily as roof insulation.

ORIENTED STRAND BOARD A building board made by pressing 2-to-3-inch-long strands of wood into thin sheets, with the strands oriented to run in roughly the same direction; assembling the sheets in layers, with each layer perpendicular to the one before it, like plywood; and bonding the layers together with phenolic resin. The directional orientation of the layers provides greater stiffness and stability than boards manufactured from random strands. Abbreviated OSB.

PAPERBOARD A sheet of pressed paper pulp or corrugated paper, usually 1/8, 3/16, or 1/4 inch thick; used as backing board for insulation and for packing and shipping applications.

PARTICLE BOARD A board made of fine wood particles, larger than sawdust, that are coated with adhesive, then pressed and cured. It has a low resistance to moisture and is not intended for exterior use. It is widely used for floor underlayment, shelving, and as a core for doors and finished panels. Common thicknesses are 3/8, 1/2, 5/8, 3/4, and 1 inch. Standard sheet size is 4 x 8 feet, but other sizes are manufactured for specific applications.

PEGBOARD A sheet of tempered hardboard—usually 2 x 4, 4 x 4, or 4 x 8 feet and 1/8 inch or 1/4 inch thick—containing a series of regularly spaced holes over the panel's entire face. A wide variety of hooks and other accessories are manufactured specifically for use with pegboard to hang tools, create store displays, etc. Pegboard is also used in cabinets, boxes, or other areas where ventilation is required.

STRAWBOARD Compressed straw covered by a kraft paper surface; used for nonstructural panels, plaster base, and roof insulation.

TILE BACKING BOARD See *building board, cement building board.*

WAFERBOARD A sheet formed from randomly oriented wood wafers that are compressed and bonded together with phenolic resin or other adhesives; used primarily for wall, roof, and subfloor sheathing.

building brick See *brick.*

building codes The legal standards and requirements covering the methods and materials used in all phases of construction.

building drain The lowest part of a building's drainage system, receiving the discharge of all the building's soil, waste, and drainpipes. It exits the building for a distance of 3 feet, where it connects to the building's sewer line. Also called a house or main drain. See Figs. D-14 and S-7.

building inspector An official inspector—usually at the city, county, or state level—whose duty it is to inspect residential and commercial construction to ensure compliance with the UBC and local codes. Each phase of the construction process is inspected and must be approved before work may continue. The inspector has the power to stop unsafe or illegal work until necessary corrections are made.

building lines Those lines defined by local codes to keep streets, sidewalks, and structures within uniform limits.

building load coefficient In solar design, the heat loss of a particular structure, except for the passive solar collection areas, expressed in Btus per degree day.

building paper A tough, lightweight, asphalt-impregnated paper; used under finished wood floors, around windows and doors, under siding, and in other areas to reduce air infiltration.

building permit An official permit from the city or county building department allowing construction, alterations, remodeling, or demolition of a building.

building sewer That portion of a building's horizontal drainage piping which extends from the end of the building drain to a public sewer or private septic tank. See Fig. S-7.

building stone In architecture, any stone used in building construction.

building supply The pipe that brings potable water to a building from the water's source; also called a water main or water service.

built-in 1. Of or relating to a bookcase, desk, wardrobe, or similar storage unit that is permanently installed in a building. 2. Of or relating to a permanently installed appliance.

built-up member A single structural member made from two or more smaller members fastened together.

built-up roof A roof covering composed of three to five layers of asphalt felt, laminated with hot tar or asphalt and with rock or gravel embedded in the final tar layer; primarily used on flat or low-pitched roofs.

built-up timber A heavy timber made up of smaller timbers, laid up with the grains parallel, and fastened together with bolts or other mechanical fasteners. See Fig. B-2.

bulk air band A shutterlike assembly on some oil-fired furnaces; used to adjust the amount of combustion air being drawn into the burner.

bulking value The ability of a pigment to add volume to the liquid to which it is being mixed.

bullet connector See *compression ring.*

bull float A long, wide, metal or wooden trowel, usually on a telescoping handle; used for the initial smoothing and leveling of a wet concrete slab. Also called a fresno.

bull header A type of brick that has one rounded corner; commonly used to form the ends of a window sill.

bullnose 1. A heavily rounded-over edge, usually found on stair treads and some types of cabinet doors. 2. A rounded molding, similar to a half round; used to conceal joints or as a decorative edging.

bullnose block See *building block.*

bull's-eye arch An arch constructed so as to form a complete circle.

bungalow A small, 1- or 1-1/2-story house with a wide veranda.

burl An abnormal growth on the side of a tree, often cut into slabs for woodworking because of the unique beauty of its grain structure.

burning in Repairing a finished wood surface by using a hot knife blade or spatula to apply melted stick shellac to the damaged area.

burnisher A hardened steel tool used to polish metalwork by friction.

burr A ragged edge on a piece of metal, usually the result of grinding or cutting.

bus bar, hot A solid conductor, commonly a flat copper bar, located inside a service panel or other main power source to which the service conductors and circuit breakers are connected. See Fig. S-10.

bus bar, neutral A metal bar containing a series of holes and screw terminals, located inside the service panel. It is connected to a ground such as a cold water pipe and serves as the point of attachment for all neutral and ground wires. See Fig. S-10.

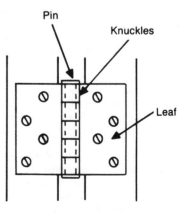

Fig. B-12. Typical butt hinge

butter   To place mortar on a masonry unit with a trowel.

butterfly hinge   A light-duty, decorative hinge, usually brass, having wings that resemble those of a butterfly.

butterfly roof   A type of roof design composed of two shed roofs that both slope inward toward the middle of the building. See Fig. R-8.

butt hinge   A common type of door hinge consisting of two leaves, one of which is mortised into the door jamb and the other into the door edge. Removable pins fit through loops in the leaves, serving to hold the two leaves together while still allowing the door to be removed from the frame if necessary. Also called butts. See Fig. B-12.

butt joint   A joint formed by the meeting of two square-cut pieces of wood.

buttress   A wood or masonry support, usually tapering out from top to bottom, built against a wall to provide lateral support.

butts   See *butt hinge*.

butyl rubber caulk   See *caulking compound*.

BX cable   See *armored cable*.

bypass   1. Of or relating to a system of pipes used to route water or waste around a fixture that is no longer in use. 2. Of or relating to electrical wiring used to bypass a fixture or outlet. 3. Of or relating to a type of door. See *door*.

cab   Abbreviation for *concrete asbestos board;* a fireproof panel used as a substrate for various applications.

cabinet door   A door used in cabinet and furniture construction, generally designed in one of the following styles. See Fig. C-1.

BOARD AND BATT CABINET DOOR Individual boards fastened to cleats to form a solid, surface-mounted door.

FALSE RAISED PANEL CABINET DOOR A solid door, shaped to imitate a raised panel.

RAISED PANEL CABINET DOOR A shaped panel with a raised appearance set into grooves in a framework.

RECESSED PANEL CABINET DOOR A panel door that has a thinner panel than the raised panel door, and no raised effect.

SLAB CABINET DOOR A solid piece of plywood or other building board having edges that are square, rabbeted, or back beveled, often with a routed design on the front.

SLIDING CABINET DOORS Two doors of any style that are set in top and bottom tracks to slide horizontally past each other.

cabinet drawer front   A decorative, finished face on the front of a drawer used in a cabinet or piece of furniture. There are three categories. See Fig. C-1.

FLUSH DRAWER FRONT A drawer front that is flush with the surrounding frame when the drawer is closed.

LIPPED DRAWER FRONT A drawer front that has a projecting lip which contacts the face frame when the drawer is closed.

OVERLAY DRAWER FRONT A combination of the flush and lipped drawer fronts, in which the top edge is flush with the face frame, but the bottom overlaps it.

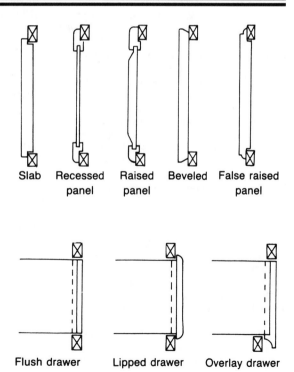

Slab   Recessed panel   Raised panel   Beveled   False raised panel

Flush drawer   Lipped drawer   Overlay drawer

Fig. C-1. Standard cabinet door and drawer styles

cabinet scraper A flat, rectangular steel tool with one sharp edge, used by hand to smooth a piece of wood.

cable Two or more individual, insulated electrical conductors, with or without a bare ground conductor, that are grouped together and wrapped with a common outside covering. The materials used for conductors, insulation, and outer wrapping will vary with the cable's intended application.

cable connector A metal or plastic device that is attached to a metal electrical box by means of a locknut or a snap-in clip. Nonmetallic cable is passed through the connector, which secures the cable and also prevents it from wearing against the box.

cable ripper A U-shaped metal tool with a small triangular blade in the end of one leg. Used for stripping nonmetallic sheathed cable. It is slipped over and squeezed down on the cable, then pulled so as to allow the blade to rip the cable's outer jacket.

cable staple A metal staple with a broad, flat crown, used for securing nonmetallic cable against wood.

CABO Abbreviation for *Council of American Building Officials*. See *Council of American Building Officials*.

cabriole leg A type of gracefully curved furniture leg, often ending in a ball or animal-foot shape.

CAD Abbreviation for *computer-assisted drafting*. Any of a variety of computer software programs designed for architectural or engineering drafting and design. 2. Also an abbreviation for *computer-assisted design*; or as CADD, an abbreviation for *computer-assisted drafting and design*.

cafe door See *door*.

caisson A watertight, pressurized enclosure for performing work under water, such as constructing bridge footings.

calcine To expose a substance to heat so as to make it easy to pulverize or crumble; commonly performed on lime and sometimes called lime burning.

calipers An adjustable measuring tool with two legs, similar to a compass. The tips of the legs are either bent out, for measuring inside areas, or bent in, for measuring outside areas, and are usually referred to as inside calipers and outside calipers, respectively.

calipers, proportional A specialized type of reducing or enlarging calipers, which can be set in any ratio from 1:1 to 1:10.

calorie The amount of heat needed to raise the temperature of 1 gram of water 1°C.

camber See *crown*.

cambium See *wood*.

came A grooved strip of soft metal, usually lead or brass, used to divide and secure individual panes of glass in a window.

candela The light intensity of 1/60 square centimeter of a perfect radiator at the freezing temperature of platinum.

candlepower A measurement of the intensity of a beam of light, expressed in terms of candelas.

cane The dried stems of various plants and grasses, woven for use as chair seats or furniture pieces. Depending on the size and type, the cane may be left whole or split lengthwise before use.

canopy switch Any of a variety of small toggle, button, or chain-operated switches attached to a light fixture and used to control that fixture only.

cant To install or set at an angle.

cantilever To extend one structural member out past the member supporting it, with no other form of support for the extended portion. Typical applications include the overhanging of floor joists over a foundation to support a bay window, or the overhanging of rafters past the wall plates to cover a projection in the building.

cantilever out To project a structure, such as a second-floor or bay window, out past the building sidewalls with no exterior supports. See Fig. D-2.

cant strip A strip of lumber, triangular in section, installed where a flat roof covering meets a vertical wall, or under shingles at a gable end. It prevents both cracking of the roof covering and accumulation of water in corners.

cap 1. A plumbing fitting that fits over the end of a pipe to close it off. See Figs. S-13 and T-4. 2. The upper member of a column, wall, chimney, or other structure. In most cases the cap is only decorative, but in some applications it is structural also, as with a column. 3. To close off an unused pipe, duct, or other opening.

cap block See *building block*.

capillary action A form of surface tension between the molecules of a liquid and a solid that causes the liquid to seep up the solid to a point higher than the level of the surrounding liquid; for example, water which seeps up a piece of paper that is placed into a puddle.

capital The upper member of a pillar or column, often ornately decorated. See Fig. C-8.

cap sheet A layer of mineral-surfaced felt that forms the top layer of a built-up roof.

carbide In common usage, an extremely hard, long-wearing material, typically an alloy of carbide and tungsten, used in the manufacture of cutting tools. Also referred to as tungsten carbide.

carbide-tipped Of or relating to saw blades, router bits, drill bits, and other tools having carbide cutting tips soldered onto a steel core. The hard carbide tip wears considerably longer and needs less sharpening than an ordinary steel tip.

carbon dioxide analyzer An instrument that measures chemicals present in the waste gases which leave a gas or oil furnace through the flue. The readings are used to indicate the efficiency of the burner.

carbon monoxide A dangerous colorless and odorless gas that results from incomplete combustion.

carbon steel See *steel*.

carborundum An abrasive material composed of carbon and silicon, commonly used in sandpaper, cutting wheels, and sharpening stones.

car decking 2- or 4-inch-thick tongue-and-groove lumber used over girders as a subfloor in residential construction. See Fig. P-10.

carpenter's pencil A broad, flat type of pencil with a large, soft lead, traditionally used by carpenters.

carpet roller A type of paint roller covered with short nap carpeting; used to apply paint in a stipple texture.

carport A roofed area open on two or more sides; used to protect an automobile, boat, etc.

carriage The complete framework of wood or steel that supports the steps in a stairway. See Fig. S-18.

carriage bolt A steel bolt for use in wood, having a round domed head and a square shoulder, which locks into the wood and prevents the bolt from turning.

case See *box.*

cased opening An opening in an interior wall, similar to a doorway, that is finished off with jambs and casing, but does not have a door.

case hammer See *hammer.*

case hardening Hardening the surface of certain types of metal alloys so that the case, or outer surface, is made substantially harder than the interior; used primarily on metal parts being subjected to high wear.

casement window See *windows.*

case opener See *hammer: case hammer.*

casing A molding used around doors and windows to cover the area between the wall and the edge of the jamb. Available in a variety of sizes, shapes, and materials. See Figs. B-4, D-6, and W-3.

casing nail See *nails, common types.*

caster A small wheel mounted in a stationary or pivoting frame; used under furniture, appliances, and similar objects to make them easier to move.

casting An object produced by pouring molten metal into a mold.

cast in place Concrete that is poured into forms at the spot where it will be used, as opposed to pre-cast.

cast iron Iron that has a high carbon content—from 2 to 4.5 percent. It is very hard, brittle, and not malleable.

cast-iron pipe Plumbing pipe formed from cast iron; widely used in drain, waste, and vent (DWV) systems.

catalyst A substance that causes an increase in the rate of a chemical reaction.

catch basin A reservoir used to regulate flow in a drainage system or to trap foreign objects before their entry into the sewer.

cat's paw A steel nail puller having a curved head with a tapered, pointed, V-shaped groove. The groove is driven into the wood so it straddles the nail's head, then is levered back to pull the nail up.

caul A metal plate used to support the sheets of wood particles being pressed into particle board.

Caulder coupling The trade name for a type of band clamp used to join soil pipes of various dissimilar materials.

caulk 1. To apply a caulking compound or sealant. 2. A term commonly referring to a variety of caulking compounds and sealants. See *caulking compound* and *sealant.*

caulking compound Any of a variety of compounds used to seal joints in wood, metal, masonry, and other materials. Caulking compounds are generally less expensive than sealants, are less flexible, and have a shorter life span. Some common types of caulking compounds follow.

ACRYLIC LATEX CAULKING COMPOUND One of the most commonly used compounds. It offers better adhesion, less shrinkage, and easier application than oil-based compounds. Elasticity and weather-resistance are relatively low, and it will not bleed through most paints. Its life expectancy is about 10 years.

ADHESIVE CAULKING COMPOUND A compound that acts as both a caulk for sealing joints and as an adhesive for a variety of jobs, such as resetting loose ceramic tile or repairing cracked porcelain.

BUTYL RUBBER CAULKING COMPOUND A rubber-based caulk having good weather resistance and fair elasticity and shrinkage. It works well between many dissimilar materials and has a life expectancy of up to 20 years. A less expensive form that contains styrene-butadiene rubber does not age as well, but is suitable for interior use.

OIL-BASED COMPOUND The least expensive and the most difficult to work with of the caulking compounds. It will fill most small cracks, but has low elasticity and a short life span, usually 1 to 5 years, and can bleed through some paints if not sealed.

PAINTER'S CAULKING COMPOUND A fairly flexible compound used to seal joints and to repair cracks and holes in various materials prior to painting. Compounds of different materials may generically be referred to as painter's caulks.

SILICONE ACRYLIC CAULKING COMPOUND A compound similar to acrylic latex, but with silicone added for increased flexibility. It is easy to work with, is compatible with most materials, and can be painted over.

caulking gun A gunlike tool for applying caulking and other materials from a tube. It uses a trigger-operated plunger to force the material from the tube.

cavetto A molding having a right-angled back and a concave, quarter-round front; also called a cove molding.

cavity wall A hollow wall consisting of two brick faces connected with brick or metal ties. The dead air between the two faces increases the wall's thermal resistance. Also called hollow wall.

C-clamp A C-shaped steel clamp with a threaded bar; used for clamping by applying pressure between the screw and the top of the C.

cedar, red See *softwood.*

ceiling amount See *guaranteed maximum cost.*

ceiling box See *electrical box.*

ceiling height The clear distance, measured vertically, from the finished floor to the finished ceiling. See Sig. C-15.

ceiling insulation See *insulation, thermal.*

ceiling joist One of the series of structural members running across the top plates of a wall; used to carry the finished ceiling. See Figs. C-4, P-3, and R-5.

ceiling panel A panel made of pressed wood fibers, usually 2 x 2 or 2 x 4 feet, and available in a variety of colors and patterns; intended for use in suspending ceiling grids.

ceiling tiles Interlocking tiles manufactured from pressed wood fibers, usually 1 foot square. Most are factory-painted white and come in a variety of surface textures. They are applied by gluing to an existing, solid ceiling, or by stapling to furring strips.

cellar An enclosed area beneath a building, partly or completely underground, that is used for storage.

cellulose 1. In energy conservation, see *insulation, thermal.* 2. In forestry, see *wood.*

cement 1. Any of various liquid or plastic materials that harden and become strongly adhesive after application; available in a variety of forms for use with hundreds of different materials. 2. A mixture of burned limestone and clay, often with other additives; see *portland cement.* 3. To attach one object to another using an adhesive.

cement-based paint A dry powder paint comprised mainly of portland cement, lime, and pigment, mixed with water for application; is primarily used over masonry.

cement board See *building board.*

cement-coated nails See *nails, finishes.*

cement colors Dry mineral or manufactured pigments that are added to cement to produce various colors in the finished surface.

cement joggle Mortar that is poured into equal slots cut in the faces of two adjacent stones to prevent movement between the stones.

cement plaster See *plaster: unfiltered gypsum.*

center-hung sash A window sash hung on its center points so that it swings on a horizontal axis.

centerline A line that denotes the center of an object. Abbreviated CL. See Fig. C-2.

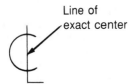

Line of exact center

Fig. C-2. Standard centerline symbol

center punch A marking tool, essentially a steel rod with one pointed end; used to mark a center or starting point on metal.

center square A V-shaped measuring tool; often used in conjunction with a combination square to find the center of round or cylindrical objects.

center to center The distance from the center of one object to the center of another; also referred to as on center.

center to end See *end to center.*

centimeter A metric system measurement of length, equal to 1/100 meter or .3937 inch.

central air conditioning An air-cooling system that may be added to a forced-air heating system, allowing use of the same blower and ducts, or installed independently with its own ductwork. The system consists primarily of an outdoor condensing unit (which also houses the compressor and a cooling fan), an indoor evaporator coil (also called an A-coil) inside the plenum, liquid and suction lines charged with refrigerant (which connect the two units), and a special heating/cooling thermostat. Also see *cooling system.*

ceramic tile Flat pieces of fired, glazed or unglazed clay, in a variety of shapes, sizes, and colors; used as wall, floor, and countertop coverings.

Certificate of Occupancy A written notice issued by the building department or other municipal agency, stating and certifying that the building is ready to be occupied.

Certificate of Substantial Completion A notice issued by an architect to a building owner, certifying that the building is essentially finished and that the contractor is entitled to payment for work completed to that point.

cesspool A storage tank used to collect the sewage from a building. It differs from a septic tank in two ways: having no leach field, liquid waste seeps out in a concentrated area around the tank; there is also no provision for the breakdown of solid waste, but it is simply stored for later removal. Cesspools are considered highly unsanitary, and are now prohibited by most codes.

cfm See *cubic feet per minute.*

chain cutter A tool used for cutting cast-iron pipe that consists of a long, ratcheting handle and a length of special chain with embedded cutting disks. The chain is wrapped around the pipe, then tightened until the pipe snaps. Also called a soil pipe cutter.

chair rail A decorative molding applied horizontally to a wall around the inside of a room at a height of about 3 feet from the floor; used to prevent chairs from marring the wall surface.

chalk box A metal or plastic box containing powdered chalk and a chalk line in a spool with a rewind handle; often just referred to as a chalk line.

chalking Paint film that has decomposed into a loose powder. Mild chalking can be washed off with special solutions, while heavier chalking exposes the surface below, and must be removed and repainted.

chalk line 1. A chalk-covered string used for marking a straight line between two points. The string is held at the two points, then raised and snapped, transferring the chalk to the surface. Some lines are rubbed with solid chalk, but most often chalk boxes are used, which coat the line in powdered chalk automatically as the line is rewound. 2. The line created by the snapping of a chalk-covered string. 3. A chalk box.

chamfer 1. A 45-degree bevel cut along the edge of a board. See Fig. D-1. 2. To cut the bevel.

chamfer plane See *plane.*

change order A written document between the building

owner, contractor, and, if applicable, the architect, stating and verifying a change in the original construction contract or the project specifications, as well as any resulting increase or decrease in the overall cost of the project.

channel A groove or trough that is cut or formed into a material for decoration, water runoff, or as a point of attachment for another object.

channel groove plywood See *plywood siding.*

channel iron A metal bar having two upturned sides, forming a U-shape with a flat bottom.

channel siding A type of wood siding similar to shiplap, but having different-sized rabbets. It forms wide grooves when assembled and resembles a batten-and-board pattern.

charge 1. The amount of electricity stored in an electrical storage battery. 2. To pressurize an air-conditioning or refrigeration system or a heat pump with refrigerant. 3. To put solar heat into storage through radiant absorption or convective transfer.

chase 1. A groove or recess that is built or cut into a wall to receive pipes, ductwork, etc. 2. An enclosure constructed around a chimney or flue pipe to conceal it. See Fig. F-5.

chase cap A sheet metal cap that overlaps the top of a chase to seal and protect the chase from weather. See Fig. F-5.

check See *lumber defect.*

checkered A hammer or other tool having a face that is scored with straight or diagonal lines or grooves running at right angles to one another; designed to reduce the chance of the hammer slipping off the nail head. See Fig. H-1.

checkered nail See *nails, head types.*

checking Small cracks and fissures that appear in a coat of paint; similar to alligatoring, but less severe.

check rails The meeting rails of a double-hung window that are usually beveled and of a thicker material then the sash sides in order to fill the opening between the top and bottom sash. See Fig. W-3.

check valve A water valve with a hinged, one-way metal flap; used in a plumbing system where protection from backflow is needed.

cherry See *hardwood.*

chevron A V-shaped or zigzag design consisting of angles that meet in a point.

chimney A structure that combines one or more individual flues and is enclosed with a surrounding masonry support work.

chimney block See *flue lining.*

chimney capacity The maximum capacity for a metal or masonry chimney, related to the total flow of exhaust gases from all appliances connected to that chimney.

chimney chase See *chase.*

chimney connector A pipe or fitting used to connect a fuel-burning appliance to the chimney that serves it.

chimney effect The tendency of air in a vertical passage or other enclosed area to rise when heated, pulling colder air in behind it. In a building, for example, unheated air that enters from outside will displace heated air from inside.

chimney fire See *flue fire.*

chimney lining See *flue lining.*

chimneypiece See *mantelpiece.*

chink 1. A small crack, fissure, or similar opening. 2. To fill such cracks.

chinking Mortar, caulk, oakum, or any of a variety of other materials used to fill the spaces between exterior members, particularly between the logs in a log building.

chip A thin, uniform sliver of wood approximately 5/8 inch wide and 1 inch long; used in making chipboard and other building boards, and also processed into wood pulp.

chipboard See *building board.*

chippendale Of or relating to a type of furniture design originated in the eighteenth century by furniture maker Thomas Chippendale and characterized by ornate decoration and flowing lines.

chipper A machine that uses rotating steel knives to reduce wood stock to chips.

chisel Any of a variety of hand-held woodworking tools used for mortising, shaping, and other wood-cutting and trimming operations. It consists of a flat steel blade ground to an accurate bevel on one end to form the cutting edge, and a handle on the other end.

chisel point nail See *nails, point types.*

chord 1. One of the main horizontal structural members of a truss. See Fig. T-7. A straight line connecting the end points of an arc. See Fig. C-3.

chroma The purity of a color. The word *chrome,* derived from chroma and used as a prefix to describe certain colors, differentiates pure, intense color from colors that contain whites or grays.

chromated copper arsenate See *wood preservative.*

chrome-plated A coating of chrome over a base of steel, plastic, or other material, designed to combine the strength of the underlying material with the appearance and anti-corrosion properties of the outer chrome coating; common for bathroom fixtures that will be subjected to high humidity levels.

chrome steel A very hard steel that has been alloyed with chromium.

chrome-vanadium steel See *steel.*

chromium A grayish white, very hard metallic element; used in electroplating and in the making of alloys and pigments.

chuck The jaw mechanism on a brace or drill motor that holds the bit.

chute A trough or passage of wood, metal, or other material; used for pouring concrete into forms, removing debris, and many other applications.

cinder block A hollow block made up of cement and crushed cinders; used for foundations, walls, chimneys, etc.

circle cutter 1. Any of a variety of tools or attachments used for cutting circles or round holes in wood, metal, and other materials. 2. See *fly cutter.*

circuit A continuous loop of electrical current, originating at a source (such as the service panel), extending out to the items being supplied, then returning to the source.

circuit breaker   A safety device installed at the beginning of a branch circuit designed to automatically stop the flow of electricity to the circuit if the current passing through the circuit breaker exceeds a preset limit. Circuit breakers can be reset once the excessive demand is eliminated. See Fig. S-10.

DOUBLE-POLE CIRCUIT BREAKER   A circuit breaker that clips to both hot bus bars, providing 240 volts.

SINGLE-POLE CIRCUIT BREAKER   A circuit breaker that clips to one hot bus bar, providing a circuit voltage of 120 volts.

circuit, three wire   An electrical circuit that contains three wires: two 120-volt hot wires, usually color coded black and red, and a neutral wire, usually white. A green or bare wire is often included with the other three for use as a ground, but is not included in the wire count. Three-wire circuits have two primary applications. By connecting the two hot wires and the neutral to an appliance, 240 volts is provided. Both hot wires extend from separate circuit breakers and run to one receptacle. By removing the break-off strip and connecting the hot wires to separate hot terminals, the top half of the receptacle will be on one circuit, while the bottom half will be on a different circuit. Also called split-circuit wiring.

circuit, two wire   A standard 120-volt electrical circuit containing one hot wire, that is black or any color other than white or green, and one neutral wire that is white. A bare or green ground wire is usually included in the circuit but is not figured into the wire count.

circular measurement   A method of measuring a circle or portion of a circle by dividing it into degrees. See Fig. C-3.

circular saw   A hand-held motorized saw with a rotating circular blade. Blade sizes range from 3 to 14 inches, in a variety of types for cutting wood, metal, stucco, etc.

circular stairs   See *stairs: winding stairs.*

circulating fireplace   A type of metal or masonry fireplace surrounded by air chambers. Fan-forced or naturally flowing air, heating in the chambers, is directed into the room to increase the fireplace's ability to heat the room.

circumference   The line that bounds the perimeter of a circle. It is equal to pi times the diameter. See Fig. C-3.

circumscribe   1. To lay out boundaries or limits. 2. To draw a line around an object so as to enclose it completely.

cistern   A watertight tank, usually of concrete, where water is collected and stored for later use.

CL   Abbreviation for *centerline.* Typically shown as an L superimposed over a C, with the vertical leg of the L denoting the exact centerline. See Fig. C-2.

cladding   A material used to cover the exterior of an object for protection, such as aluminum placed over the outside of a wood sash.

clad window   See *window.*

clapboard   1. A 6-x-8-inch board having a wedge-shaped section 1/2 to 3/4 inch thick at the bottom and tapering to almost nothing at the top, applied horizontally as exterior

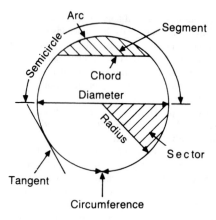

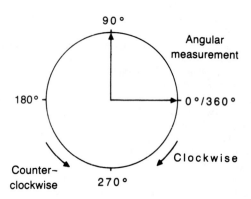

Fig. C-3. Terms associated with circles and circular measurement

siding. 2. Flat, untapered boards applied in overlapping horizontal rows as siding.

classical   Of or relating to a style of architecture that adheres to the principles of ancient Greece and Rome at the period of their highest cultures.

claw   Any of a variety of tools having a forked and curved head used for prying or removing nails. See fig. H-1.

claw hammer   See *hammer.*

cleanout   An opening with a threaded plug that provides access into a waste line, septic tank, or trap for cleaning. See Figs. B-1 and S-8.

clear   1. Of or relating to lumber that is free of knots, checks, and other imperfections. 2. Of or relating to a finish that has no color and is designed to protect the underlying surface while not obscuring it.

clear span   That portion of a building's interior that is free from posts, columns, walls, or other supporting structures. See Fig. C-15.

cleat 1. A strip of wood attached to a wall to support a shelf or other object. 2. A board that is nailed or otherwise fastened across the face of other boards, used to secure those boards together. See Fig. B-6.

clerestory A wall supporting a roof that is higher than the adjacent wall, typically done to allow for the placement of windows or ventilation. See Fig. R-8.

clerestory window A fixed or operable window placed in a clerestory.

clinch To nail in such a way that the nail is driven through both boards, then the protruding end is bent back over into the wood in the direction of the grain.

clip A piece of brick that has been cut to length.

cloak rail A finished board fastened to a wall, to which coat hooks are attached for hanging clothes.

clock thermostat A clock-controlled thermostat that permits the setting of different temperatures for day and night; also called a setback or night-setback thermostat.

clockwise Of or relating to a rotation from right to left, in the same direction as the hands of a clock; indicating the direc-

tion of rotational movement. See Fig. C-3.

close cornice A cornice in which there is no rafter projection beyond the wall.

closed cell A material containing millions of tiny pockets, or cells, of air or gas, each of which has a completely closed perimeter, trapping and holding the air inside. The opposite of open cell.

closed circuit An electrical circuit in which all switches are closed and the electricity has a continuous path in which to flow.

closed-coat paper See *sandpaper*.

closed cornice A cornice in which all overhanging structural members are covered, usually with plywood panels; also called a box cornice. See Fig. C-4.

closed loop A liquid solar heating system in which both ends of the system of pipes are connected to a heat exchanger within the storage tank, forming a complete, continuous loop. The heat exchanger removes the heat from the liquid as it passes through, giving it off to water in the tank for storage and distribution. See Fig. C-5.

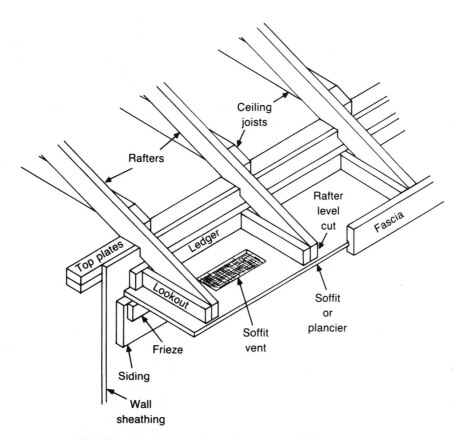

Fig. C-4. Components that make up a typical closed cornice

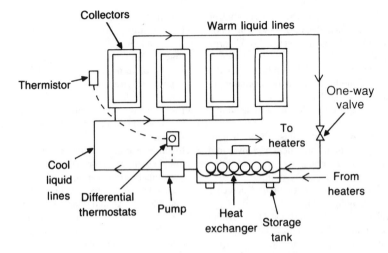

Fig. C-5. Closed loop solar heating system

closed stairs See *stairs.*

close grain See *grain.*

close nipple A short piece of pipe that is 1-1/4 to 1-1/2 inches long and is threaded along its entire length.

closer The last brick or portion of a brick laid in a course; used to finish the row or close up the bond. See Fig. B-10.

closet auger See *toilet auger.*

closet bend A short-radius, 90-degree pipe fitting that connects the toilet flange to the soil pipe.

closet flange A flat, round fitting with bolt holes or slots in the outer rim, which is attached to the closet bend and usually screwed down to the subfloor, providing a base to which the toilet is bolted.

closet pole A round molding that is usually 1 1/4 inches in diameter and is supported by brackets in a closet; used to accommodate clothes hangers.

co-al Abbreviation for *copper-aluminum*; of or relating to electrical devices that are approved for use with both copper and aluminum wires.

coal tar A thick, black liquid that is formed during the distillation of coal; used in a variety of building materials.

coarse grain See *grain: open grain.*

coat A single layer of paint, varnish, or other finishing material that is applied over all surfaces at one time and allowed to dry and harden.

coaxial cable A cable consisting of two concentric conductors—an inner wire and an outer, braided sleeve—that are insulated from each other by a plastic coating. Commonly used for transmission lines, such as for television. Often abbreviated coax.

cob A small masonry unit composed of unburned clay mixed with straw or other materials.

cobblestone A fieldstone rounded by the actions of water and friction; often used in road paving.

code See *building code.*

coefficient of performance The ratio of an electrical appliance's output of energy to the input energy needed to power it; a rating of electrical efficiency. Abbreviated cop.

coffer A decorative recessed panel in a ceiling or dome.

cofferdam 1. A temporary watertight structure of masonry, metal, and/or other materials used for underwater construction. 2. A watertight enclosure temporarily attached to a ship's side in order to make repairs below the water line.

cold chisel A hardened, tempered, steel chisel that is sufficiently strong to cut cold metal.

cold colors Colors in the blue and green ranges, so named for the impression of coolness they give to a room or building.

cold form The bending, cutting, or other working of sheet metal at room temperature.

cold-weather concrete Concrete containing various additives that allow it to be poured at temperatures of 40° F or less without freezing.

collapsible anchor See *anchor, hollow wall.*

collar beam A board installed horizontally between two opposing rafters, usually about one third of the way down from the ridge, to prevent the rafters from spreading away from each other; also called a collar tie or tie beam. See Figs. C-15 and R-5.

collar tie See *collar beam.*

collector Any of a wide variety of devices that are designed to collect and absorb radiation from the sun and convert it to heat. See Figs. C-5, C-6, G-4, O-3, and T-2.

collector aperture The glazed area of a solar collector through which sunlight is admitted.

collector efficiency In solar heating, a rating of the effectiveness of a collector; the ratio of the amount of solar radiation absorbed by a surface to the amount of solar radiation that strikes it.

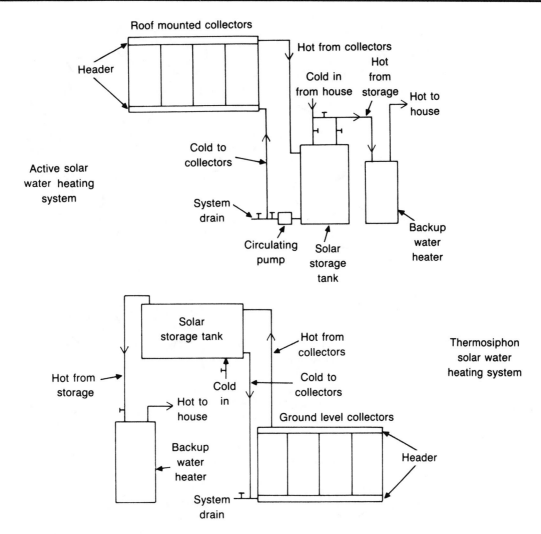

Fig. C-6. Solar collectors used in domestic hot water applications: an active system (top) and a thermosiphon system (bottom)

colloidal suspension Fine particles of a substance that remain in permanent suspension in a liquid.

colonial Of or relating to a style of architectural and furniture design characteristics in use around the time of the American Revolution (1775-83).

colonnade A parallel series of equally spaced columns, used to support a roof structure or series of arches. 2. In classical architecture, the supporting structure of an entablature.

color coat The third and final coat of stucco to which powdered colors are added when the stucco is mixed, producing a finished wallcovering that does not require painting.

color coding In electrical wiring, using different-colored insulating jackets for conductors as a means of identification.

colored nail See *nails, finishes.*

color retention The ability of a paint to hold its color quality and intensity when exposed to the elements.

color wheel A color spectrum arranged in a circular pattern; used to determine various combinations of colors for painting and decorating. See Fig. C-7.

column A vertical member, often slightly tapered, slender in relation to its height and typically round in section, used to support a horizontal load. A column consists of three principal parts: the base, the shaft, and the capital. See Fig. C-8. In classical architecture, columns are divided into three main categories, called orders, based on the designs of traditional Greek architecture.

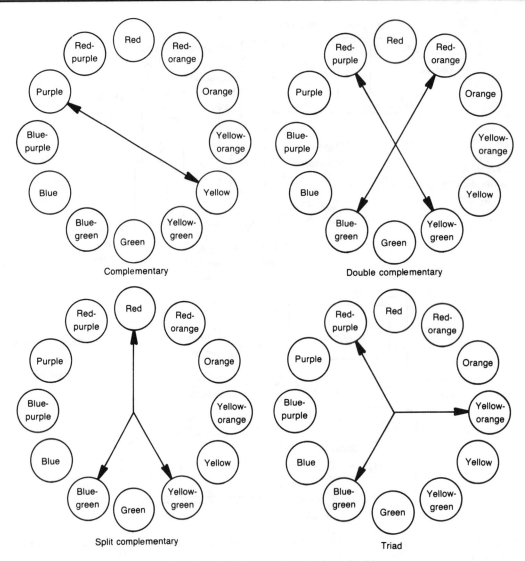

Fig. C-7. Color wheel, with examples of color wheel harmony

CORINTHIAN The most ornate of the three orders, having a slender fluted shaft and a capital decorated with a leaf pattern.

DORIC A simple style of column, having a heavy fluted shaft and
a plain, unornamented capital.

IONIC A type of column typically having a fluted shaft and a
capital decorated with ornamental scrolls.

comb See *drag.*

comb board See *saddle board.*

combed Of or relating to a surface that has been raked or
otherwise treated to produce a surface texture of small
ridges and valleys, often used on siding shingles.

combination blade A circular saw blade having a combination of rip and crosscut teeth; for general-purpose use in
table or portable power saws.

combination fitting A fitting for waste and soil pipe systems,
resembling a Y-branch with a 45-degree bend attached to the
side outlet.

combination plane See *plane.*

combination square An adjustable tool that combines inside
and outside try square, miter square, and a depth gauge. See
Fig. C-9.

combination tools Any of a variety of tools having two or
more uses.

combination window See *window.*

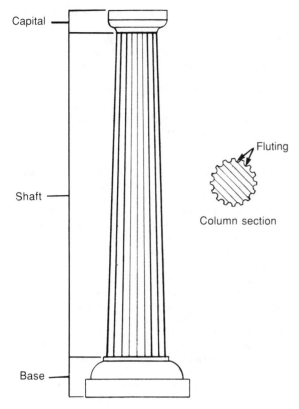

Fig. C-8. Column

combination wrench See *adjustable wrench.*

combined moisture Water that is naturally contained in the cell walls of wood.

combustible Any material that will ignite and burn.

combustion The act or process of burning; rapid oxidation accompanied by heat and usually light.

PRIMARY COMBUSTION The main or principal burning of fuel in a combustion appliance, which occurs within the combustion chamber.

SECONDARY COMBUSTION The additional burning of gases and other secondary materials not burned during primary combustion. Some combustion appliances have a specific combustion chamber designed for secondary combustion in order to utilize otherwise wasted combustion capacity.

combustion air The total amount of air being supplied to the area containing a fuel-burning appliance, combining the air circulating around and into the appliance, the air actually needed for combustion, and the air exhausted up the flue.

combustion appliance Any furnace, heater, stove, or other device that utilizes the combustion process to produce and in some way utilize heat.

combustion chamber The sealed area of a furnace or other device where combustion takes place.

combustion-chamber liner A liner of flexible, heat-resistant fibers; used to line the combustion chamber of an oil-fired furnace to stop leaks and seal crumbled firebrick.

combustion efficiency See *heating efficiency.*

combustion heat See *combustion system.*

combustion system A heating system in which air and a fossil fuel are combined and burned to produce heat, which is then distributed to the building through ducts or other means.

comfort zone The range of temperature and humidity in which the average person is comfortable.

common brick See *brick: building brick.*

common lime See *lime.*

common nail See *nails, common types.*

common rafter See *rafter.*

common rafter table A table stamped or printed on the face of a framing square; used to determine the length of common rafters for various slopes.

common stock A lowgrade lumber containing knots and defects; used for general-purpose work, such as sheathing and temporary braces.

compass 1. An instrument used to determine and indicate geographical direction. 2. An adjustable, two-legged instrument having a metal point at the end of one leg and a pencil or pen tip at the other; used for drawing circles.

compass brick See *brick: arch brick.*

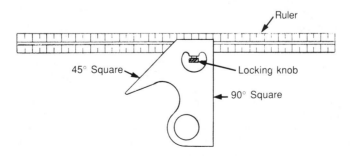

Fig. C-9. Combination square

compass saw See *handsaw.*

complementary colors Two colors directly opposite each other on the color wheel, such as blue and orange. See Fig. C-7.

Completion Bond A bond posted by a contractor prior to commencement of a construction project, intended to ensure the contractor's completion of the project in accordance with the terms and specifications of the contract.

component construction Structures made up of prefabricated modules that are assembled on the job site.

composition shingle A roofing shingle manufactured from asbestos fibers and asphalt impregnated paper, usually with a surface coating of crushed mineral in various colors; available in various shapes and sizes.

compound arch An arch constructed by placing a succession of smaller, concentric arches within and behind each other.

compound miter A cut containing two angles. One is a miter cut across the face of the workpiece, and the second is a bevel cut across the end. The two cuts are usually made simultaneously by tilting the saw blade.

Compreg The trade name for a plywoodlike product formed by soaking layers of wood in resin, then curing them under great pressure. The resulting compressed wood is hard and stable, but retains its natural appearance and beauty.

compression Stress imposed upon an object by forces which attempt to compact the object; the opposite of tension. For example, when a post is used to support a beam, the post is in compression. See Fig. S-22.

compression fitting A brass or plastic plumbing fitting consisting of a body, a compression nut, and a compression ring. Tightening the nut squeezes the ring down around the pipe, forming a seal. See Fig. C-10.

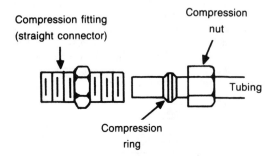

Fig. C-10. Compression fitting used on copper tubing

compression joint Any joint that supports a downward force, such as the joints in a brick wall.

compression ring 1. A short, tapered copper ring, used in conjunction with a compression fitting. See Fig. C-10. 2. A short copper sleeve that is crimped down onto the ends of two or more electrical wires being joined. The sleeve can be covered with an insulating cap, or left bare if used on ground wires. Also called a bullet connector.

compression test A test for the compressive strength of the concrete being used on a particular job. Samples of the concrete are taken at the site as it is poured, then cured and tested in a laboratory.

compression wood See *lumber defect.*

compressive strength The maximum amount of compression that an object can withstand before it fails.

compressor See *cooling system.*

concave Curving inward, like the inside of a bowl; the opposite of convex. See Fig. P-8.

concentrated load A weight that is directly on, and carried by, a beam, girder, or other structural member.

concentrating collector A type of solar collector for use with a liquid system that consists of a series of black painted pipes, each of which passes through a troughlike polished aluminum device called a parabolic reflector. Machinery continuously pivots the reflectors to allow them to follow the sun's path.

concentration The amount of one substance that is present in another substance, such as a pollutant dispersed in air.

concentric Of or relating to two or more objects, especially circles, that have the same center; the opposite of eccentric.

concrete A building material made from a combination of portland cement, fine aggregate, coarse aggregate, and water. Ratios of the dry ingredients are usually 1:2:3, 1:2:4, or 1:3:5, depending on the application, with fresh water added at 1 to 1.5 times the volume of the cement. It is available in bags of dry, premixed material; individual dry materials in bulk; or premixed in wet form.

concrete and masonry sealant See *sealant: silicone sealant.*

concrete asbestos board A fireproof panel used as a substrate for various applications; commonly abbreviated cab.

concrete block A hollow or solid block made from lightweight or small aggregate concrete; available in various standard shapes and sizes. Some of the more common types follow. See Fig. C-11.

BUILDING BLOCK See *building block.*

FACE BLOCK A block to which a glazed or stone face has been bonded; often used in areas where sanitation is important, because the smooth face permits easy cleaning.

SCORED BLOCK A stretcher or corner block having an etched, horizontal groove that resembles a mortar joint. When these blocks are installed in a block wall with unscored blocks, the appearance of different-sized units is created.

SCREEN BLOCK An open, decoratively patterned block used for sun and privacy screens. The common size is 4 x 12 x 12 inches.

SLUMP STONE A long narrow block, often colored to resemble adobe, used primarily in decorative wall and fireplace construction; also called a slump block.

SPLIT BLOCK A concrete block with a roughened face that resembles stone.

TEXTURED BLOCK A block having a raised pattern on the face to

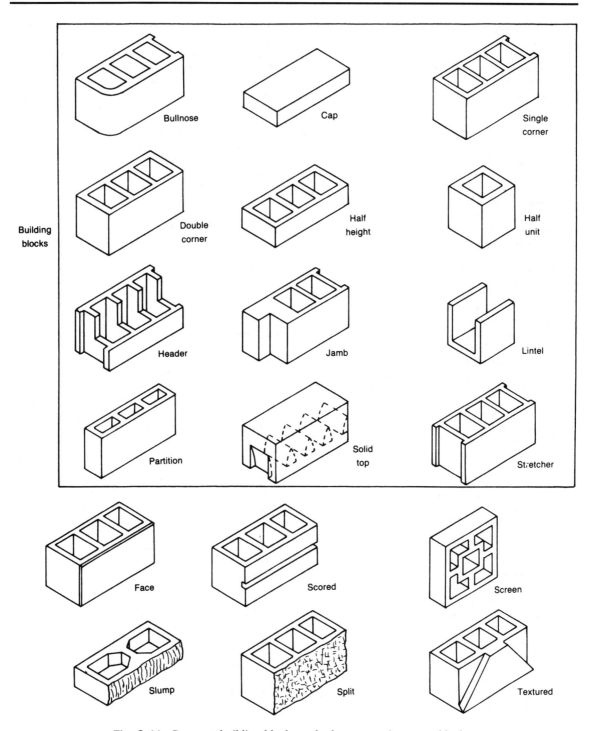

Fig. C-11. Common building blocks and other types of concrete blocks create a variety of geometrical designs and shadows when installed.

concrete form A boxlike enclosure, usually made from plywood or wood planks, into which concrete is poured and left to harden; used in pouring foundations, piers, and other structural components.

concrete nail See *nails, common types.*

concrete pumper Any of a variety of machines, usually truck mounted, that are used for pumping wet concrete into hard to reach areas.

concrete slab A floor or other flat, weight-bearing area constructed of poured concrete, with or without additional reinforcement. See *slab on grade.*

condensate That liquid which separates from air or other gases due to a reduction in temperature.

condensate drain A small pipe or tube used to carry off condensed moisture from an air conditioning unit or air-to-air heat exchanger.

condensation Drops of water or frost that accumulate on windows and the inside of exterior wall and roof coverings when warm, moisture-laden inside air contacts a cold surface. The cold surface extracts the heat from the air, and the water vapor contained in the air changes to a liquid.

condenser See *cooling system.*

conditional use The use, usually temporarily, of a structure or parcel of land that is not consistent with normal, approved usage; typically regulated by the municipality having jurisdiction over that location and requiring special approval.

conditional use permit A special permit granting the use of a structure or parcel of land, subject to specific conditions. See *conditional use.*

conditioned space Any area within a residence or building that is served by a heating or cooling system; technically, a space that is capable of being heated to 68° F at a location of 3 feet above the floor, under winter design conditions.

conduction See *heat transfer.*

conductivity A measure of a material's ability to permit conductive heat flow through it.

conductor 1. A wire used to carry electricity. 2. A material or medium that permits the easy flow of electricity or heat.

conductor pipe See *downspout.*

conduit bender A tool having a curved head with a long, straight handle that is used to form smooth curves in EMT conduit and some types of pipe; also called a hickey.

conduit connector A fitting designed to connect a piece of conduit to an electrical box or fixture.

conduit coupling A fitting designed to join two pieces of conduit.

conduit, flexible A type of conduit that is easily bent and shaped. It is composed of a continuous wrapping of interlocking metal, and is intended for interior electrical installations. It's sometimes referred to by the trade name Greenfield.

conduit, rigid Any of a variety of metal or plastic tubes used to house and protect electrical wiring.

conifer A family of evergreen trees and shrubs including fir,

pine, spruce, and juniper and characterized by needles and cones; a softwood.

consistency 1. The thickness or viscosity of a fluid. 2. A measure of how well different lots or batches of the same product, such as cans of paint or rolls of carpet, will match when intermixed.

console An ornate bracket used to support a cornice or other ornamental fixture.

construction adhesive See *adhesive.*

Construction Lien See *mechanic's lien.*

contact cement See *adhesive.*

contaminant Any substance that contaminates or pollutes another substance.

continuous beam A beam that lays across three or more columns or other supports.

continuous duty Of or relating to machinery designed for operation under a constant load for an indefinite period of time.

continuous header Structural lumber of 2 x 6 inches or larger that is placed on edge on top of or in place of the top plate and encircles the entire building. It acts as a header over all the wall openings, thus eliminating the need for individual headers and support studs.

continuous load Of or relating to an electrical device designed to handle maximum current for three or more hours.

continuous vent See *individual vent.*

continuous waste Of or relating to a drainpipe that connects a common set of fixture drains, such as those from each compartment of a two- or three-compartment sink, with the fixture's trap.

contour gauge A marking tool consisting of several movable "fingers" in a metal frame, which, when pressed against an object, slide back to conform to the object's contours; used to copy curves, molding, and other irregular shapes.

contour line A line on a map or drawing that shows a contour and slope of ground on the same elevation.

control joint A groove formed in concrete flatwork as it dries, usually 1/5 to 1/4 of the slab's depth. If the concrete cracks as it dries, it will happen along this predetermined groove.

convection See *heat transfer.*

convection current See *heat transfer: convection.*

convection grate A fireplace grate made of C-shaped hollow metal tubes. Convection causes cool air to be drawn into the lower ends of the tubes, where it is warmed by the fire before exiting into the room through the upper ends. An electric blower can be attached to further improve the circulation.

convection oven A type of oven design that uses the principle of convection to circulate heated air around the food being cooked. This circulation reduces stratification in the oven and cooks the food more evenly.

convective loop See *heat transfer: convection.*

convector A pipe with an attached series of vertical plates, called fins, that is used in a hot water or steam heating system. As hot water or steam in the pipe heats the fins, warm air is given off into the room by convection.

convenience outlet An electrical outlet that is part of a general-purpose branch circuit; not intended for use with equipment that draws a large amount of power.

convex Curving outward, like the surface of a globe; the opposite of concave. See Fig. P-8.

cooling system A system of machinery, tubing, and refrigerant used to provide cool air, as in an air conditioner or refrigerator. The main components of a cooling system follow.

COMPRESSOR A device that compresses the refrigerant, causing it to liquefy and release heat.

CONDENSER A heat-transfer device that distributes the heat released from the refrigerant to the surrounding air.

EVAPORATOR A second heat-transfer surface that vaporizes the liquid refrigerant back into a gas as it passes through the surface. The gas draws heat out of the surrounding air, which, now cooled, is pumped into the building. The refrigerant, now a gas, passes back into the compressor to begin the cycle again.

REFRIGERANT A gas that circulates through the system, of which Freon is probably the best known brand.

cool white See *fluorescent lamp*.

cooper A carpenter or woodworker skilled in the making of barrels or casks.

cop See *coefficient of performance*.

coped joint A method of joining two pieces of molding by cutting the end of one piece with a coping saw to fit over the contours of the adjoining piece. See Fig. C-12.

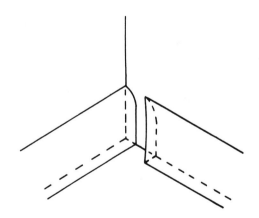

Fig. C-12. Coped joint on a piece of baseboard

coping The top course or projecting finish of a building's exterior wall; usually sloped to shed water.

coping saw See *handsaw*.

copolymerization The interaction of two or more different molecules to form a new compound having different physical properties.

copper-clad aluminum conductor An electrical wire consisting of an aluminum core and a copper covering. The copper protects against corrosion and abrasion and must, by code, make up a minimum of 10 percent of the wire's cross-sectional area.

copper naphthenate See *wood preservative*.

copper pipe, hard supply A type of rigid thin-walled pipe used extensively for water supply systems. It is assembled by soldering, using slip, compatible fittings. Hard copper pipe is broken into four categories, each bearing a letter designation and color-coded stripe for identification.

TYPE DWV HARD-SUPPLY COPPER PIPE Pipe used for drain, waste, and vent systems, although cost may be prohibitive. Diameters range from 1 1/4 to 2 inches. It is color-coded yellow.

TYPE K HARD-SUPPLY COPPER PIPE Pipe with a thick wall; used primarily for underground installations. Diameters usually range from 1/2 to 1 1/4 inches. It is color-coded green.

TYPE L HARD-SUPPLY COPPER PIPE Pipe with a medium wall; used in some underground installations and in above-ground commercial work where higher water pressures are encountered. Diameters range from 1/4 to 1 inch. It is color-coded blue.

TYPE M HARD-SUPPLY COPPER PIPE Pipe with a thin wall; used above ground only and limited to water pressures of 100 pounds or less. Diameters range from 1/4 to 3/4 inch. It is color-coded red.

copper pipe, soft supply A soft-tempered, flexible copper pipe, also called copper tubing; usually available in 60 to 100 foot rolls. It is more expensive than hard-supply copper pipe, but can be installed with fewer fittings. There are two types of soft copper with diameters ranging from 1/4 to 1 inch, sometimes more.

TYPE K SOFT-SUPPLY COPPER PIPE Pipe with a thick wall; used for all above-ground and some below-ground installations. It is color-coded green.

TYPE L SOFT-SUPPLY COPPER PIPE Pipe with a medium wall; for above-ground use only. It is color-coded blue.

corbel 1. A decorative wooden bracket used to support a projecting weight, such as a shelf or countertop. 2. Any weight-supporting structure projecting horizontally out from a vertical surface.

corbel-out To progressively build out one or more courses of a masonry wall in order to form a supporting ledge.

cord 1. A unit of measurement for firewood equal to a stacked pile 4 feet high, 4 feet deep, and 8 feet long (128 cubic feet). 2. Flexible cable containing two or more stranded conductors with a common outer wrapping, designed for specific uses.

APPLIANCE CORD For use with heating appliances, such as irons, toasters, portable heaters, etc., it contains a wrapping of asbestos in addition to its other insulation; also called a heater cord.

LAMP CORD A light-duty, small-gauge cord with a thermoplastic outer jacket; used for lamps and other small appliances.

POWER CORD A heavy-duty, neoprene-jacketed cord designed for power tools, large-gauge extension cords, etc.

core 1. The center veneer in a sheet of plywood. See Fig. P-5. 2. The inside area of a flush door or other panel between the two face veneers. See Fig. D-3.

corkboard See *building board.*

cork tile A type of floor tile produced by blending a resin with cork granules and then baking. It is resilient and non-skid, but is not extremely durable in heavy traffic areas.

corner accessories cabinet See *appliance garage.*

corner bead 1. A strip of formed sheet metal used under plaster or plasterboard cement to reinforce and protect an outside corner. 2. A surface-mounted wooden molding that serves the same purpose as 1; also called an angle bead.

corner block See *building blocks.*

corner boards Two boards installed vertically with overlapping edges that cover and protect the outside corners of exterior siding.

corner brace A diagonal brace at the corner of a wood-frame building used to stiffen the wall. Most commonly found as one of two types.

BLOCKED CORNER BRACE Angled pieces of lumber cut and nailed into each successive stud space; also called cut-in. See Fig. F-10.

LET-IN CORNER BRACE A continuous length of lumber nailed into notches cut in the edge of the studs, so as to be flush with the outside of the wall surface. See Figs. B-2, F-10 and P-3.

corner clamp A type of woodworking tool used to clamp two boards at right angles to each other.

cornering tool A small, flat metal tool with curved ends and a small, sharp, oval-shaped hole in each end, used for shaving off and rounding over the edge of a piece of wood. Each end typically has a different-sized hole for making edges with a different radius of curve.

corner trowel A drywall trowel having a flat metal blade bent at an approximate right angle, with an attached wooden handle. Used for smoothing joint compound in corners.

cornice 1. The overhanging area of a structure at the point where the roof and wall meet, including all appropriate boards, panels, and trim. See Fig. C-4. 2. In a structure having no roof overhang, the exterior trim at the meeting of roof and wall. 3. The upper area of a sidewall when it extends above the roof level. 4. Any of a variety of decorative molded pieces or projections at the top of a wall.

cornice lighting See *soffit lighting.*

cornice return The portion of a cornice that wraps around a corner onto a gable end wall. See Fig. C-13.

corona The portion of a cornice that projects out past the face; designed to protect the walls by throwing off rainwater.

corridor kitchen A kitchen in which the cabinets and appliances are arranged along two opposite walls, with a doorway at one or both ends. See Fig. K-2.

corrosion inhibitor A black, pastelike material used for coating the bare ends of aluminum electrical wires to prevent corrosion.

corrugated Of or relating to any material or surface that is formed into a regular series of symmetrical concave and convex curves.

corrugated fastener A small piece of sheet metal formed into a series of inside and outside curves and sharpened along

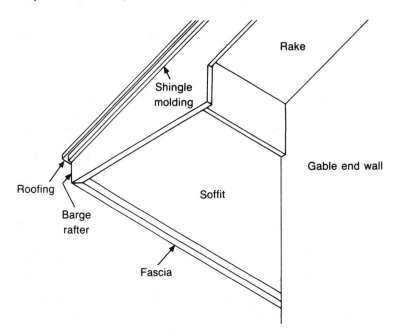

Fig. C-13. Typical cornice return on a gable end

one edge; used to join and reinforce butt joints in wood. Also called a wiggle nail. See Fig. W-6.

corrugated plywood   See *plywood siding.*

corundum   An aluminum oxide used for its abrasive properties; second only to the diamond in hardness.

cost ceiling   See *guaranteed maximum cost.*

coulomb   A unit of electrical current equal to the amount of electricity flowing through a current of one ampere in one second.

Council of American Building Officials   Organization of building officials established in 1972. Authors of *CABO One and Two Family Dwelling Code* and other building-code-related books that compile and standardize national construction codes and requirements. Abbreviated CABO.

counterbore   A larger-diameter hole drilled over another, smaller hole and having the same center, done for the purpose of recessing a screw or bolt head. See Fig. C-14.

counterbracing   Diagonal bracing that acts in the opposite direction from the main bracing, thus helping to balance the stresses acting on the structure.

counterclockwise   Of or relating to a left to right rotation, in the opposite direction as the hands of a clock; indicating rotational movement. See Fig. C-3.

counterflashing   A sheet-metal flashing set between courses of brick, as on a chimney at the roof line, and bent down to cover the exposed edge of the shingle flashing. See Fig. F-7.

countersink   A hole that is beveled inward from top to bottom; designed to receive a flat-headed screw, leaving the top of the screw head flush with the surrounding surface. See Fig. C-14.

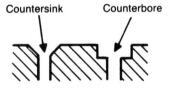

Fig. C-14. Cross-section view of countersunk and counterbored holes

countersink bit   A boring tool used to create a cone-shaped hole to receive the head of a flat-headed screw. It can be hand operated or used with a portable drill or drill press.

coupling   A plumbing fitting used to connect two lengths of like-sized pipe in a straight run. See Figs. S-13 and T-4.

course   A horizontal row or layer, such as bricks, stones, or shingles. See Fig. B-10.

coursed ashlar   A type of masonry construction in which random blocks of stone are grouped and laid up in rows of matching height.

cove   A molding or surface that forms a concave curve.

cove base   A rubber or vinyl strip with a curved bottom that is installed in place of a baseboard over the finished flooring to create a coved corner where the floor meets the wall.

coved ceiling   A ceiling that meets the sidewalls with a con-

cave curve instead of a 90-degree corner.

coved linoleum   Linoleum that is curved up onto the side walls for a short distance, usually 4 inches, for easier cleaning in the corners.

cove molding   A wooden molding having a rounded, concave face; primarily used to cover the joint where two boards meet at a 90-degree angle.

coverage   The amount of area that will be covered by a particular product at a given quantity and thickness.

cover strip   See *batten.*

cps   See *cycles per second.*

cpvc   See *plastic pipe.*

cramp   1. A portable tool having a metal bar with ends bent at right angles, used to lift masonry or heavy timbers. Also called a cramp iron. 2. Any of a variety of portable clamps used primarily for woodworking.

cramp iron   See *cramp.*

crawl   For paint or varnish, to gather into globules, instead of covering a surface evenly; caused by poor adhesion of the paint or varnish.

crawl hole   An opening with a removable cover; used to provide access to an otherwise enclosed area, such as a crawl space.

crawl-hole vent   A sheet-metal frame, usually 24 x 18 inches, with a screened, lift-out, sheet-metal door. Set in the foundation wall, it allows for both crawl-space ventilation and access.

crawl space   The area beneath a house enclosed by the foundation walls; used for access and inspection of framing, plumbing, etc. Entry is by an inside trap door in the floor or by an outside door set in the foundation wall. See Fig. C-15.

crazing   1. A fine, netlike pattern of tiny cracks that often appears on aged finishes. 2. Numerous fine cracks that appear in freshly dried concrete as a result of surface shrinkage.

cream   See *laitance.*

creep   The slow, permanent increase in an object's deformation under the stress of a permanent load.

creosote   1. A type of wood preservative. See *wood preservative.* 2. The liquid component of smoke, primarily tar and soot mixed with water vapor, that collects and hardens inside a chimney or flue.

cresting   A decoration or series of decorations, usually of metal but sometimes of stone or other material, placed along the top of a wall or roof.

cribbing   1. Heavy, crossed timbers that are stacked to form supports for a building being raised or moved. 2. Stacked or cross-piled lumber used to line a shaft or retaining wall.

cricket   See *saddle.*

crimped   1. Of or relating to the end of a round sheet-metal pipe or duct that has been formed into a series of accordion folds. These folds reduce its diameter to allow it to slip into the straight end of the next pipe. 2. Of or relating to a pinch or fold placed through one object into another, usually with

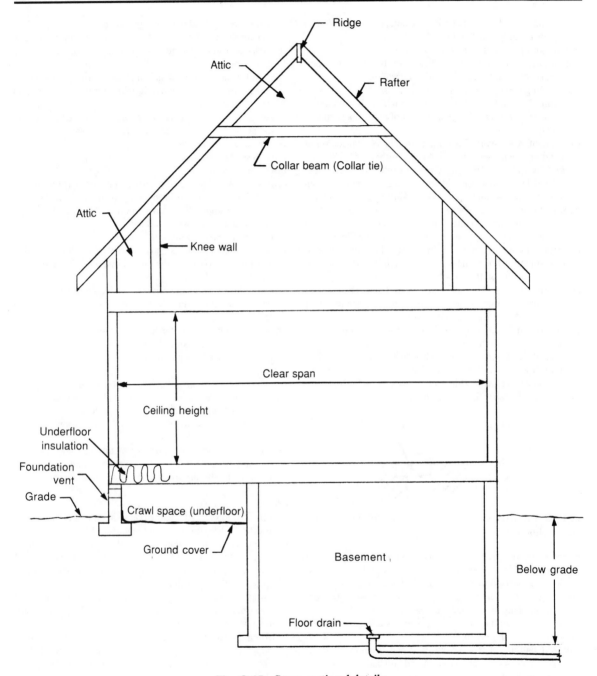

Fig. C-15. Cross-sectional details

pliers or a crimping tool, to hold the two pieces together. 3. Of or relating to a sharp crease in a tube, pipe, conduit, or other similar object, caused by too sharp of a bend, that reduces the flow through the pipe.

crimper 1. A type of pliers used to crimp electrical connectors onto wires. 2. A type of pliers used to crimp the end of a sheet-metal pipe to allow it to enter another pipe.

cripple A framing member cut shorter than the average surrounding members, such as one of the pieces under a window that support the sill. See Fig. F-10.

critical distance The maximum length of drainpipe allowed between a plumbing trap and the vent or soil stack. See Fig B-5.

critical level A mark stamped on a vacuum breaker or backflow protection device showing the minimum elevation above the fixture being protected to which the device should be installed; abbreviated C-L.

crook See *lumber defect: warp.*

cross A plumbing fitting having four outlets, each of which is at a 90-degree angle to the ones on either side of it.

crossband In plywood of five or more plies, those layers that lie between the core and the two faces. See Fig. P-5.

cross bridging See *bridging.*

cross-buck door See *door.*

cross connection An incorrect plumbing arrangement in which contaminated water becomes mixed with potable water.

crosscut To cut across the grain of a piece of wood; the opposite of rip.

crosscut saw See *handsaw.*

cross fence Fences used to subdivide a parcel of land into smaller areas, as for livestock.

cross furring Smaller furring members fastened perpendicular to the main furring or framing members.

cross grain See *grain.*

crosshatching In drawing, parallel diagonal lines used for shading or for indicating a cut section.

cross lap A woodworking joint for connecting two pieces that cross each other, in which half of the board's thickness is removed from each piece at the area of intersection, allowing the faces of the two pieces to be flush.

cross-peen hammer See *hammer: peen hammer.*

cross section 1. A section cut from an object at right angles to its length. 2. A detailed drawing that represents the appearance of an object after a similar, imaginary cut. See Fig. C-15.

crotch 1. The fork formed in a tree when two branches angle up and away from each other. 2. A forked pole or similar device used as a support.

crotch veneer Veneer cut from the crotch, or fork, of a tree, which forms an unusual decorative pattern.

crow bar See *wrecking bar.*

crown 1. A slight arch in a beam or other horizontal structural member. When the beam is installed, the crown is placed face up, helping to prevent the beam from bending downward under vertical load. Also called camber. 2. A slight

convex curve on the face of a hammer or other striking tool. See Fig. H-1.

crown molding A decorative molding used primarily to cover the area where a wall meets a ceiling.

cubic content The total space contained within a building or other enclosed space, equal to the height times the width times the depth. See Fig. C-16.

cubic feet per minute A measure of the number of cubic feet of a substance flowing past a given point in one minute; abbreviated cfm.

cubic foot A measurement of volume equal to an area 1 foot long, 1 foot wide, and 1 foot deep, or 1,728 cubic inches. See Fig. C-16.

cubic footage The size of a room, building, or other area in cubic feet. See Fig. C-16.

cubic inch A measurement of volume equal to an area 1 inch long, 1 inch wide, and 1 inch deep. See Fig. C-16.

cubic measurement A measurement of volume, equal to an object's width times its length times its depth. See Fig. C-16.

cubic yard A measurement of volume equal to a space 1 yard long, 1 yard wide, and 1 yard deep, or 27 cubic feet. The cubic yard is a standard measuring increment for such bulk items as concrete, dirt, and rock, and when used in this application is commonly referred to simply as a yard. See Fig. C-16.

cup See *lumber defect: warp.*

cup-head nail See *nails, head types.*

cup hook A thin, J-shaped steel or brass rod having a wood thread at one end; used for hanging cups and other light-weight objects.

cupola A small, roofed structure installed on the ridge of a roof. Usually four- or eight-sided, it might be vented to provide an escape for hot attic air or be purely decorative.

curb roof See *mansard roof.*

curing Of or relating to the process during which concrete, adhesive, paint, or some other material dries and reaches its final degree of hardness.

current The total flow of electrons passing through a circuit, measured in amperes.

current electricity See *static electricity.*

curtain board A decorative box fastened to the wall above a window to house the curtain rod, usually painted or covered with fabric.

curtain wall 1. A non-bearing interior wall used to divide space. 2. Any non-bearing wall.

cut To dilute the strength or thin the consistency of a material by adding water or a suitable solvent or thinner.

cut-in 1. To use a brush to paint corners, the areas around trim, and other hard-to-reach areas preparatory to painting the rest of the surface with a roller or spray. 2. To recess an object into an existing surface; for example, a new electrical outlet box into a plasterboard wall.

cut-in box An electrical box with side wings or toggles that is inserted into a precut hole and held in place by a screw, which is turned to draw the wings up against the back of the surface. Available as a wall box or or round ceiling box for fixtures.

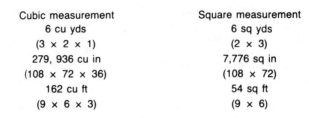

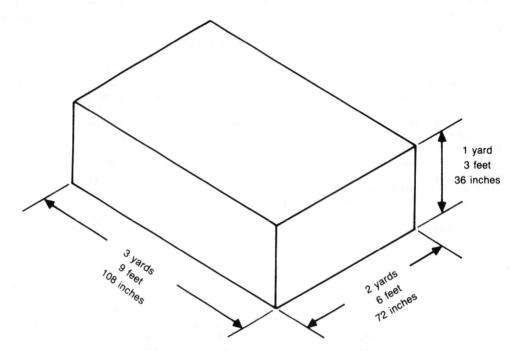

Fig. C-16. Cubic and square measurement

cut nail An older style of nail cut by machine from metal stock into a square taper, as opposed to today's wire nails. Also called a square nail.

cutoff wheel A thin, circular, abrasive blade mounted on a portable or stationary power saw and used for cutting metal.

cutter brick See *brick*.

cutter head The group of rotating blades on a planer or molding cutter that contacts and cuts the wood being fed past it.

cutting oil A type of oil used as a lubricant while cutting, threading, milling, or performing other operations on metal.

C-value See *heat resistance*.

cyanoacrylate adhesive See *adhesive*.

cycle In electricity, two reversals of an alternating current's direction.

cycles per second A measurement of electrical and sound frequency equal to the number of complete cycles that occur in one second.

cylinder The portion of a lockset that contains the lock tumblers and the mechanism for moving the latch.

cylinder guard A tapered steel ring used with a dead bolt and designed to rotate freely around the cylinder to prevent the cylinder from being twisted off with pliers.

cylinder ring A steel ring designed to fit behind the cylinder to allow the use of a standard lockset on a thin door or an oversized hole.

cylindrical lock See *lock*.

cyma A type of commonly used architectural molding, having a cross section of one concave and one convex curve.

CYMA RECTA A molding having the concave curve nearest the top of the molding.

CYMA REVERSA A molding having the concave part nearest the bottom.

cypress See *softwood*.

D

d See *penny*.

dado A rectangular slot cut across the grain of a piece of wood. See Fig. D-1.

dado blade A circular saw blade designed to cut dados, grooves, rabbets, etc. There are two basic types.

ADJUSTABLE DADO BLADE A blade having a set of attached, beveled washers that can be dialed to the desired width of cut, causing the blade to spin in an inclined circle.

DADO SET A pair of outside, conventional blades combined with a set of inside, Z-shaped blades, called chippers, to create a single unit of the desired width.

damper A metal disk or plate inside a chimney or duct used to close off, regulate, or direct the amount of air flowing past it. See Fig. F-4.

damp location See *location*.

damp-proofing Materials applied to a wall to make it impervious to moisture penetration.

d & m Abbreviation for dressed and matched. See *tongue and groove*.

danish oil A penetrating, clear or pigmented, resin-based oil used to seal, preserve, and finish wood.

darby A long, narrow steel trowel with two handles; used primarily for plastering.

dash A type of texture applied to an exterior stucco wall by splattering it with small lumps of wet stucco thrown from a heavy brush.

date nail A nail with a stamped date on its head, driven into a structure as a permanent indicator of when construction was completed.

datum line 1. In surveying, the base line from which all levels and measurements are taken. 2. On a drawing, the line from which all heights and depths are measured.

daylight basement A basement having at least one wall exposed to the outside, typically cut into a hillside. See Fig. D-2.

daylight factor The ratio of the amount of sunlight transmitted through a window to the amount of sunlight falling on the window's outside surface; also called visible transmittance.

daylighting See *daylight factor*.

dc See *direct current*.

dead air Air that does not circulate, especially the air trapped between two structural surfaces for thermal insulation.

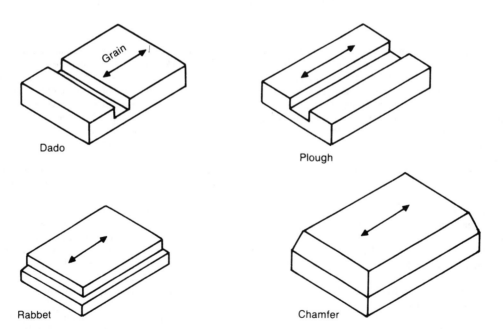

Fig. D-1. Commonly used cuts in wood; arrow indicates the direction of the wood's grain

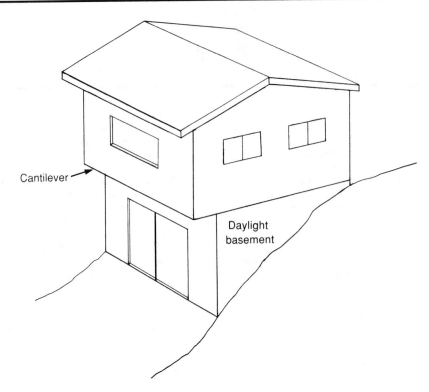

Fig. D-2. Typical daylight basement

dead bolt   1. The long, tempered-steel bolt used with a deadlock. 2. The common term used to describe a deadlock.

dead end   Any branch pipe leading off a DWV system and having a developed length of 2 feet or more. Used to provide a point of attachment for future expansion of the system.

deadening   Of or relating to sound-absorbing materials applied to walls, floors, and other areas to stop the transmission of sound waves.

dead front   Of or relating to a machine or appliance having no live electrical parts exposed to the person on the operating side of the equipment.

dead level   Absolutely level or horizontal.

dead load   Any fixed, stationary load in a building, including the weight of the building itself, framing, doors, windows, etc.

deadlock   A security door lock that is key or latch operated and features a dead bolt, which is much longer and heavier than a normal latch. Common types follow.

DOUBLE-CYLINDER DEADLOCK A deadlock that is key operated from both inside and outside.

ONE-SIDED DEADLOCK A deadlock that is latch operated from inside only, with no exterior access.

ONE-WAY DEADLOCK A deadlock that is key operated from outside only, with no interior access.

SINGLE-CYLINDER DEADLOCK A deadlock that is latch operated from inside and key operated from outside.

dead man   A heavy block, usually concrete, with an eyebolt attached. The block is buried in the ground and guy wires are attached to the eyebolt to secure a pole, tower, tree, etc.

decal   A painted or printed decoration on a paper-backed, clear plastic film that is designed for transference, usually by wetting, onto a permanent surface.

decay   See *lumber defect.*

decibel   The common unit of sound intensity, equal to 1/10 of a bel. Decibels range from 0 (silence) to over 120 (deafening).

deciduous   A tree or shrub that produces and then sheds leaves during specific seasons, as opposed to an evergreen.

decimal equivalent   Of or relating to a fractional number that is expressed as a decimal, which is derived by dividing the numerator (upper number) by the denominator (lower number); for example 1/4 = .25.

deck   Any flat, floorlike roof or platform generally built for storage or outdoor living.

decking   1.2-inch or 4-inch thick lumber, usually side matched, that is used between roof beams or rafters as a base for the roofing materials. The underside is often finished and left exposed. Also called roof decking. See Figs. O-1 and P-10.

2. Lumber or plywood secured to the top of base cabinets as a supporting surface for the installation of ceramic tile. Also called tile decking.

deck paint An enamel with a very high resistance to wear; designed for use on walkways, decks, and other high-traffic areas.

declination 1. The deviation between true north and the magnetic north shown by a compass needle. 2. The position of the sun at solar noon, given as an angle between the equator and the sun, which varies with the time of year; used in the calculation and design of solar heating systems.

dedicated circuit See *branch circuit.*

deep well See *well.*

deflection The amount of bending or sagging of a beam or other structural member under a load.

deformation The alteration in the shape of a structure or structural member when it is subjected to a force or a load.

deformed Of or relating to the intentional cutting, grooving, threading, or other roughening that is subjected to the surface of the shank of a nail or steel reinforcing bar to give it greater holding power.

degree 1. 1/360 of a circle. 2. A division on a scale used for measuring, as a thermometer.

degree day A measurement of the difference between the average outside temperature and a fixed standard of 65° F; used in calculating the demands on a heating or cooling system for a particular location. One degree of difference equals one degree day; for example, an average outside temperature of 50° F for one day would equal 15 degree days.

delamination The separation of two pieces that have been glued together as a result of failure of the adhesive.

demand factor A ratio of the maximum demand placed on an electrical system at a given time, to the system's total connected load.

demand load 1. The amount of water required to properly operate a particular fixture. 2. The maximum amount of electrical current drawn by a particular piece of equipment. 3. The total water- or space-heating needs for a building that the building's energy supplies must meet.

demarcation A fixed line or mark describing a boundary or other limit.

density The mass of a substance per unit of its volume.

dentils Small blocks of wood that are regularly spaced in rows like teeth; used as trim.

design temperature 1. A percent, usually 97.5, of the average yearly low temperatures for a given area; used in estimating heating-system needs. It is an important factor in designing a system that will be adequate to maintain a comfortable indoor temperature, even during the area's coldest weather. 2. A percent of an area's average yearly high temperature; used when calculating cooling-system needs.

detached building A building that is not physically attached to the house, but is still considered part of it, such as a garage.

detail drawing A larger-scale drawing that illustrates either a particular component or a section of a larger area shown in another drawing.

developed length The length along the centerline of a run of pipe, including the fittings attached to it.

device Any component of an electrical system, such as a switch or receptacle that carries but does not use electricity.

dew point The temperature at which water vapor in the atmosphere begins to deposit as a liquid. It varies with the humidity and pressure of the air.

dhw See *domestic hot water.*

diagonal Of or relating to a straight line between two nonadjacent corners of a square, rectangle, or other figure. See Fig. P-8.

diagonal cutting pliers A type of cutting pliers having rounded, slightly angled cutting jaws that are in line with the handles; also called dikes.

diagonal sheathing Boards applied diagonally to an exterior wall, usually at a 45-degree angle, as a base for the finish wall covering. Application in this manner causes each board to act as a brace, greatly strengthening the structure.

diameter A straight line that passes through the center of a circle and connects opposite points on the circle's circumference; the "width" of a circle, equal to twice the radius. See Fig. C-3.

diamond point nail See *nails, point types.*

diaphragm tank A steel tank containing a heavy vinyl or rubber bag; designed to ensure constant water pressure in hot water heating systems and in water systems that are supplied by wells. Water enters the bag, compressing the air in the top of the tank. The air presses down on the bag, forcing water out into the plumbing or heating system.

diazo print A method of reproducing drawings. An original, drawn on tracing paper, is placed on light-sensitive diazo compound paper, then run through a blueprint machine. The paper is developed dry in ammonia fumes, and depending on the paper used, produces black, dark blue, or maroon lines on a white or very light blue background. Also called whiteprint.

die 1. A tool for cutting outside threads on a rod, bolt, dowel, or other cylindrical object. 2. A mold or other device used in various metal-forming processes to give the desired shape to the object being produced.

die casting A method of making metal objects in which molten metal is injected into a die under pressure.

dielectric union A special union fitting for joining galvanized and copper pipe. It is similar to a regular union, except that it contains an insulating washer and sleeve which keeps the two dissimilar metals from contacting each other, thus preventing electrolysis, a corrosive chemical reaction.

differential thermostat A device used with an active solar system that senses temperature differences between the various components of the system and regulates them accordingly. See Figs. C-5 and O-3.

diffuser A device such as a translucent plastic sheet that is used with a light fixture to direct, scatter, or soften the light being emitted.

diffuse radiation See *solar radiation.*

diffusion The spontaneous movement of particles and molecules from an area of high concentration to an area of low concentration.

dikes See *diagonal cutting pliers.*

dimension A measurement or specification of size.

dimensional stability The ability of a material to resist change in its dimensions from moisture, temperature, or stress.

dimension lines The lines on a drawing that indicate the actual size of an object.

dimension lumber See *lumber.*

dimmer switch A type of rheostat used to control the brightness of the light emitted from a fixture.

dimple A small concave area, formed with the head of a hammer when a nail is being set in plasterboard, that provides a recess which, when covered with joint cement, conceals the nail head.

ding A small nick or gouge, especially in wood trim or other finished surfaces, caused by impact during construction.

dinger A small, easily constructed house with few amenities and little ornamentation; a slang term.

dip tube The long, cold-water inlet pipe inside a water heater tank. It extends down to near the tank's bottom so that incoming cold water forces heated water out the tank's top. See Fig. W-2.

direct current Electrical current that flows in one direction only, such as the type generated by a battery. Abbreviated dc.

direct gain Of or relating to solar energy that enters a building directly through the glazing and is then absorbed and stored.

direct-gain system A type of passive solar system in which the building is constructed and oriented so as to be warmed by direct penetration of the sun through windows, skylights, greenhouses, etc.

direct lighting Light from a fixture that is cast directly onto an object without being reflected off any other surface.

direct-nail See *face-nail.*

direct radiation See *solar radiation.*

direct-vent heater An upright gas furnace attached to the inside of an exterior wall that vents directly out from the back, through the wall, to the outside.

disappearing stairs A flight of open stairs attached to a ceiling-mounted, spring-loaded frame. The stairs fold down from the ceiling for attic access and are concealed behind a ceiling panel when not in use.

discharge To remove heat from solar storage.

disconnect Any switch, fuse, circuit breaker, or other device that provides a means for disconnecting an electrical conductor from its source of supply.

dishwasher front A wood or painted-metal panel used to cover the front of a dishwasher for decorative effect.

dishwasher panel A wood panel with one faced edge and a toe-kick cutout; designed to be incorporated into a run of base cabinets to conceal and protect one or both sides of a dishwasher.

disk sander A portable or stationary power tool having a rotating disk to which a round sheet of sandpaper is affixed; used for the rapid sanding of wood or metal.

dispersal field See *disposal field.*

disposal field An area of buried, gravel-lined trenches that receives the effluent from the drainage lines in a septic system; also called a leach field or dispersal field. See Fig. S-7.

distillate The condensed liquid that results from the distillation process.

distillation The process of heating a substance until its more volatile parts separate into a vapor, then condensing the vapor into a liquid by cooling it.

distressing Selectively damaging the face of a piece of wood to give the appearance of age. A wide variety of methods are used, including striking the wood with a chain or rock, rubbing it with abrasives, or spattering on dark paint or stain to give the illusion of small holes.

distribution box A buried receptacle that receives liquid waste from a septic tank through a single inlet, then distributes it to several drainage lines through a series of outlets. See Fig. S-7.

distribution medium Air, water, or steam that is heated by a combustion system and then carries the heat into the building through the distribution system.

distribution panel See *service panel.*

distribution system The series of ducts, pipes, registers, radiators, or other devices through which heated air, water, or steam is moved to warm a building.

diurnal Occurring each day. Certain diurnal events, such as daily high morning humidity, are factors in solar design.

diverter valve A type of mixing valve for use in a bath/shower combination. A separate handle is provided to divert water up to the shower head or down to the tub spout.

dividers A tool similar to a compass, but having two metal points; used for dividing or laying off distances or for scribing parallel lines.

DIY See *do-it-yourself.*

dog 1. A metal device used in conjunction with a woodworking vise for clamping large work pieces. It is slipped into predrilled holes in the workbench and serves as a stop for the item being clamped. Also called a bench dog. 2. A short metal cleat with two tapered, pointed legs; designed to be driven into the butt joint of two pieces of wood to draw the wood together and secure it. See Fig. W-6.

dogear To trim the top two corners off a board, usually at 45 degrees, for decorative effect; most commonly found on fence boards.

do-it-yourself Of or relating to construction methods and materials specifically designed or adapted for installation by a nonprofessional.

Dolly Varden siding A type of bevel siding having a rabbeted bottom edge.

domestic hot water Heated water circulated within a house for such purposes as washing and bathing.

domestic sewage The liquid and solid wastes normally generated within a home that are free from industrial wastes and are able to be disposed of in a public sewer or private septic tank with no special treatment.

door A barrier to an entry that usually swings or slides to open and close the entry. Available in a wide variety of sizes, styles, and types, depending on the intended application. Some of the most common styles, sizes and specifications follow. See Fig. D-4.

ACCORDION DOOR A door made up of a number of narrow vertical panels, usually 6 inches wide, alternately hinged to fold up like an accordion.

ACTIVE DOOR In a pair of doors, the door that is opened first.

ATRIUM DOOR A type of double-door configuration in which one door is fixed and inoperable, and the second door is hinged to the edge of the first so as to close against the jamb.

BI-FOLD DOORS Narrow doors, usually 1 to 1-1/2 feet in width, hinged to fold against each other and flat against the jamb. An overhead track, attached to the head jamb, keeps the doors in line. Bi-folds can be set up to open completely to one side or from the center to both sides. Common sizes are 1-0, 1-3, 1-6, 1-9, and 2-0 per panel, assembled into units of the desired width; 6-8 or 7-0 in height; and 1 to 1-3/4 inches thick.

BY-PASS DOORS Two or more doors set on rollers that slide in staggered top and bottom tracks. They open by sliding to the side in front of or behind the other door. Also called sliding doors.

CAFE DOORS Short, decorative doors that are hinged as a pair to swing open in either direction.

CROSSBUCK DOOR A door made up of two stiles and three rails, with two diagonal rails forming an X in the lower half. Four triangular panels fit between the lower stiles and rails, and glass is used in the upper half.

DOUBLE-ACTING DOOR A door set on top and bottom pivots on one side and installed in a frame with no stop, allowing it to swing open in either direction; also called a swinging door.

DUMMY DOOR A false, unhinged door that cannot open; used with an active door as decoration to give the appearance of double doors.

DUTCH DOOR A door cut in half across the width, with the two halves separately hinged to open independently of each other. A shelf is often set on the top of the lower half for use when the upper half is open.

EXTERIOR DOOR A door that is assembled using waterproof glue to prevent delamination of the face panels when exposed to weather.

FALSE LOUVER DOOR A less expensive version of the louvered door, having a panel with cuts that resemble louvers, but that do not go all the way through the door.

FIRE DOOR A solid-core door that is either cased in steel, has a steel lining, or is in some other way enhanced to prevent the spread of flames. The frame used with a fire door is weather-stripped well to prevent the spread of smoke and fumes.

FLUSH DOOR A door having two flat, paneled faces glued over a

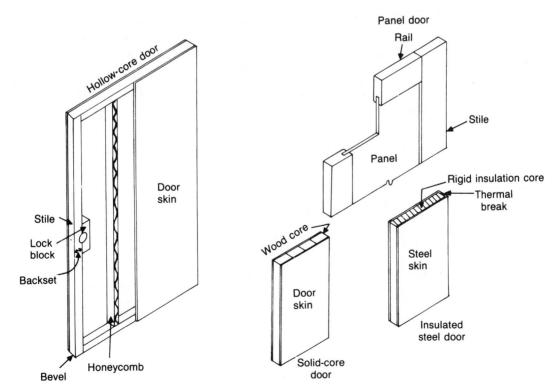

Fig. D-3. Components of various types of doors

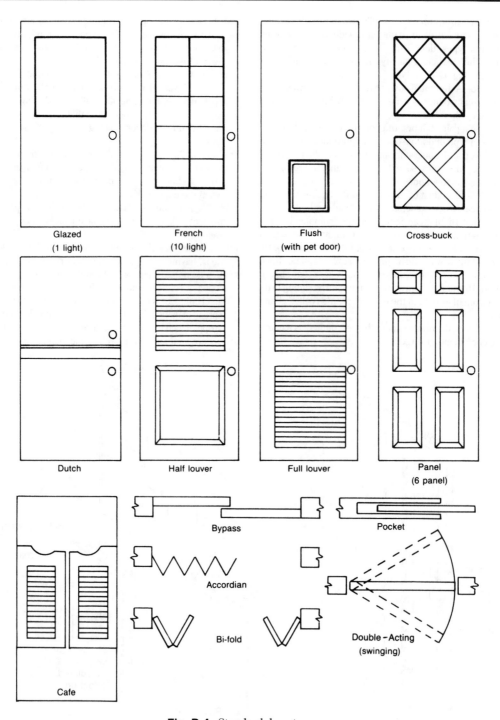

Fig. D-4. Standard door types

wood frame, as opposed to a paneled door. The core may be hollow or solid, and the face is often decorated with moldings in various patterns.

FRENCH DOOR A door having a four-piece frame with glass in the middle. The door is specified by the number of panes of glass (lights) it contains: a one-light door has one solid pane; a nine-light door has three rows of three, etc. Also called a french window.

GARAGE DOOR A wide, spring-balanced or track-mounted door designed for use in residential garages. Common sizes are 7-0, 7-6, and 8-0 (single car) or 16-0 (double car) in width; 7-0, 7-6, and 8-0 in height; varying thicknesses.

GLAZED DOOR A door having one or more panes of glass.

HALF-LOUVER DOOR A door having louvers on the top half and a solid or paneled bottom.

HOLLOW-CORED DOOR A door constructed of four wooden strips that form a rectangular frame, which is covered on both sides by a plain or decorative veneer. The veneer is often reinforced by the addition of corrugated paper or wood strips within the framework. See Fig. D-3.

INACTIVE DOOR In a pair of doors, the door that is opened last. It is usually latched to the head jamb and floor or threshold with barrel bolts.

INSULATED STEEL DOOR A four-piece, thermally broken steel frame that is covered with decorative steel panels over a core of foam insulation. It offers good thermal and security properties. Some types are decorated with a covering that resembles wood. See Fig. D-3.

LEAD-LINED DOOR A solid-core door with a thin lead sheet in the middle: used to block radiation for use in labs, X-ray rooms, etc.

LEFT-HAND DOOR A door having the hinges on the left side of the door, which opens away from you; abbreviated LH. See Fig. D-5.

LEFT-HAND REVERSE DOOR A door having the hinges on the left side of the door, which opens toward you; a right-hand door. Abbreviated LHR. See Fig. D-5.

LOUVERED DOOR A door having a four-piece frame, with horizontal slats set at an angle between the two stiles for ventilation.

PANEL DOOR A door having a number of stiles and rails with wood panels in between, as opposed to a flush door. The door is specified by the number of panels it contains. See Fig. D-3

PATIO DOOR See *door: sliding glass door.*

PERSONAL DOOR A door that is intended for use by a person, as opposed to a garage, pet, or other special-purpose door. Common sizes are 2-0, 2-2, 2-4, 2-6, 2-8, 2-10, 3-0, and 3-6 in width; 6-8 and 7-0 in height; and 1-3/8 inches (most interior uses) or 1-3/4 inches (most exterior uses) in thickness.

PET DOOR A small swinging door, usually a rubber flap in a metal frame; designed to be cut into a regular door for use by a pet.

POCKET DOOR A door that rolls on an overhead track into a frame or pocket hidden in the wall.

PRE-HUNG DOOR A door that has been factory set up for quick installation. It consists of an assembled frame with the door already hinged to the jamb and the holes bored for the locks. Precut casing pieces are often included.

RIGHT-HAND DOOR A door having the hinges on the right of the door, which opens away from you; abbreviated RH. See Fig. D-5.

RIGHT-HAND REVERSE DOOR A door having the hinges on the right side of the door, which opens toward you; a left-hand door. Abbreviated RHR. See Fig. D-5.

ROLL-UP DOOR A door consisting of hinged horizontal steel or wood panels that ride on rollers in side and overhead tracks.

SLIDING DOOR See *door: by-pass door.*

SLIDING GLASS DOOR A door having a wood or aluminum frame fitted with one fixed panel and one sliding panel of glass. Common sizes are 5-0, 6-0, and 7-0 (total of both panels) in width; 6-8 and 7-0 in height; varying thicknesses. Often called a patio door.

SOLID-CORE DOOR A door constructed of a four-piece framework that surrounds a solid core of wood strips or blocks and is covered by plain or decorative veneer. See Fig. D-3.

SWINGING DOOR See *door: double-acting door.*

WARDROBE DOOR A wide by-pass door designed for access to closets or other storage areas. Wardrobe doors may be wood or metal, solid or louvered, or covered entirely with mirrors.

doorbell transformer An electrical step-down transformer used to reduce 120-volt house current to a lower voltage, usually 25 volts, to power a doorbell.

door bevel A bevel placed on the edge of the lock side of a door.

door-boring jig A jig that clamps to the edge of a door and contains two guide holes: one on the edge for drilling the latch hole and one on the face for drilling the cylinder hole. It can be adjusted for different door thicknesses and different lock backsets, and allows for rapid, accurate drilling of lockset holes.

door buck A rough wooden framework built into a door opening in a masonry wall to which the door jambs are secured.

door casing See *casing.*

door check See *door closer.*

door closer A pneumatically operated device that pulls a door closed; also called a door check.

door frame The combination of parts that enclose and support a door, including two side jambs, a head jamb, and three door stops. See Fig. D-6.

doorjamb See *jamb.*

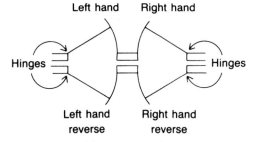

Fig. D-5. Door swing designations

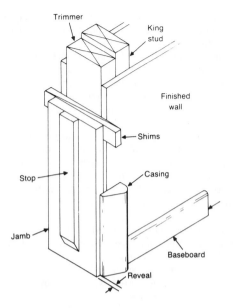

Fig. D-6. Parts of a typical door frame

door shoe   A weather-stripping device, usually metal with a rubber or vinyl strip, that is mounted at the bottom of the door and seals against the threshold to block air infiltration.

door skin   Veneer or plywood panels, usually 1/8 inch thick, used as the face material for flush doors. Skins may be smooth or textured hardboard, hardwood veneer (usually birch, mahogany, or oak) or a variety of other materials. See Fig. D-3.

doorstone   In masonry construction, a stone that is used to form the threshold of a door.

doorstop   1. A wooden molding attached to the face of a jamb to prevent the door from swinging the wrong way. 2. A rubber-tipped metal rod or spring attached to a door, wall, or hinge to stop the door as it opens, preventing contact with other objects.

door sweep   Any of a variety of weather-stripping devices that is attached to the base of a door and sweeps along the floor as the door opens and closes; designed to block air infiltration beneath the door.

door viewer   A small, tubular device that fits in a door at eye level, permitting a one-way view, from the inside, of any person at the door.

dormer   A framed opening that projects out from a sloped roof and forms a vertical wall to allow for a window. See Fig. R-8.

double-acting door   See *door: double-acting door.*

double-acting hinge   A type of door hinge for swinging doors, that has three leaves and two sets of knuckles and pins, allowing it to move in either direction.

double-complementary colors   Two sets of complementary colors obtained from the color wheel; used in selecting a four-color decorating scheme. See Fig. C-7.

double-cut file   A metal cutting file having rows of teeth or ridges that cross each other at an angle to the length of the file.

double-dipped galvanized   See *nails, finishes: hot-dipped galvanized.*

double header   A header comprised of two pieces of lumber of the same width and thickness that are secured on edge alongside each other. See Fig. W-3.

double-hung window   See *window.*

double joist   Two floor joists nailed together to provide extra support under bearing walls.

double-pane window   See *window.*

double-pitch skylight   A V-shaped skylight that slopes in two directions; designed for use at the peak of a building's roof.

double plate   See *plate: top plate.*

double-pole circuit breaker   See *circuit breaker.*

double roll   Of or relating to a roll of wallpaper containing 72 square feet, the most common type of wallpaper. See *single roll.*

double trimmer   Similar to a double joist, but used specifically around a stairwell, hearth, or other opening cut in a floor.

double wall   1. Having two walls or surfaces separated by an airspace; commonly used with flue pipes to prevent heat buildup. 2. In construction, a second wall built inside the exterior wall to deepen the wall cavity and allow for increased amounts of wall insulation.

doubling   Using two structural framing members, such as studs or joists, which are nailed together to provide greater strength.

Douglas fir   See *softwood.*

dovetail jig   An adjustable template device used with a router to guide the cutting of dovetails.

dovetail joint   A woodworking joint in which a flared tenon interlocks with a similarly flared mortise, preventing the joint from pulling apart. See Fig. D-7.

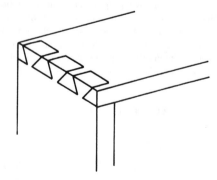

Fig. D-7. Dovetail joint

dovetail saw   See *handsaw.*

dowel center   A small marking tool for wood joints, consisting of a short aluminum or steel rod machined with a thin overhanging lip and a short, sharp pin in the exact center of one face. The tool is placed in a predrilled dowel hole in one piece of wood, then pressed against an adjoining piece so that the pin marks the second point, indicating the exact location for drilling the matching dowel hole. See Fig. D-8.

Fig. D-8. Dowel center

doweling jig   A metal block with guide holes and a screw clamp that when clamped to a board, provides an accurate means for guiding the drilling of vertical holes for dowel pins.

dowel pin   A short length of dowel rod, beveled at each end and having longitudinal or spiral grooves for holding glue; used to pin together and strengthen various joints in woodworking. Sizes vary from 1/4 to 1/2 inch in diameter and 1-1/2 to 2-1/2 inches in length. See Fig. D-9.

Dowel pin

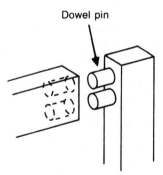

Fig. D-9. Doweled joint

dowel rod   A round length of hardwood, usually birch, available in a variety of diameters from 1/8 to 1 inch or more and commonly in lengths of 3 or 4 feet.

downdraft   A flow of air coming down a chimney, usually caused by windy conditions or a chimney that is too short.

down-feed system   A water supply system used in some buildings over 60 feet in height in which water is pumped to storage tanks on the building's roof, then fed downward by gravity to service the fixtures.

downlight   A light fixture designed to direct all its light downward.

downspout   A wood, metal, or plastic pipe descending from a gutter to allow accumulated water to drain to the ground; also called a leader.

draft   1. The flow of air and other gases being pulled over a flame and up the flue in a fireplace, furnace, or other such device. 2. Outside air leaking into a building through cracks, holes, or other points of entry.

draft curtain   A fire-resistant dividing wall or other partition, designed to retard the spread of fire. Most commonly used in attics between attached dwellings.

draft diverter   A flared hood or similar device that sits on top of a gas-fired furnace, water heater, or other appliance to allow any downdrafts in the flue to be diverted into the room rather than going directly into the combustion chamber. See Fig. W-2.

draft dodger   A sand- or fiber-filled tube used at the base of a door or window to block air infiltration.

draft gauge   An instrument used to measure the draft in the firebox and flue of a gas- or oil-fired furnace.

draft hood   A device on a gas-fired appliance that mixes secondary air with the combustion gases being vented out of the unit, allowing a smooth, continuous flow of exhaust gases out of the vent. On oil-fired appliances, the device is usually known as a draft regulator.

drafting   The preparation of detailed and accurate drawings used in manufacturing and construction.

draft regulator   See *draft hood*.

draft stop   See *fire block*.

drag   A steel-toothed masonry tool used to dress the surface of a piece of stone; also called a comb.

drain   Any pipe within a building's drainage system that carries liquid and solid waste material.

drainage   Pipes, gravel, grade slopes, or other methods used to channel water from one location to another.

drainage fitting   See *sanitary fitting*.

drainage lines   Perforated or open-joint pipes that extend out from the distribution box in a septic system, and are used to distribute the effluent from the system throughout the disposal field. See Fig. S-7.

drain auger   A springlike, hand- or electric-operated tool that has a cutting head at one end, which is forced into waste pipes to clear obstructions; sometimes called a snake.

drain-down   Of or relating to a liquid solar heating system having a thermostat that signals the circulating pump to drain all the water in the system down to the storage tank if the outside temperature approaches 32° F so that the water does not freeze in the system.

drain field   That area on a building site where the drain lines for the septic system are buried, and which receives the effluent from the septic tank. Also called a disposal field or leach field. See Fig. P-4.

drainpipe   The pipe that connects a fixture trap with a waste pipe; intended to carry liquid waste only. See Fig. B-5.

drain tile   1. Round sections of tile or perforated pipe that are laid in gravel-lined trenches around the foundation of a building; designed to collect and remove excess ground water. See Fig. D-10. 2. Any round tile or pipe that is used to direct ground water or rainwater away from a building.

drain valve   A valve located at the bottom of a water heater to allow the tank to be drained for servicing or removal of sediment.

drawband   A flexible collar of galvanized metal connected with screws and nuts; used in some applications to join sections of sheet-metal pipe.

drawboard joint   A mortise-and-tenon joint in which a hole is

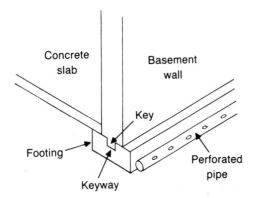

Fig. D-10. Typical basement wall construction, with drain tile

bored through the joint, and the joint is then secured with a pin. See Fig. D-11.

drawer pull A handle attached to the front of a drawer to facilitate its opening and closing.

drawer slide A unit consisting of metal tracks and rollers, available in various sizes and styles and used to support a drawer and provide smooth operation.

drawknife A long, flat blade that is beveled to a chisellike cutting edge on one side and has a raised handle at right angles to the blade on each end. It is pulled across a log or piece of wood for rough shaping.

dress 1. To sand, plane, shape, or otherwise finish a piece of wood. 2. To smooth and square a piece of stone.

dressed and matched See *tongue and groove.*

Dresser coupling The trade name for a type of plumbing fitting used to join two pieces of galvanized pipe that have not been threaded.

drier See *paint.*

drill A hand or portable power tool used with a variety of bits and other accessories for drilling holes, driving screws, sanding, and performing a number of other operations.

drill guide Any of a variety of devices used with hand or portable power drills to guide the accurate drilling of holes, either straight or at an angle.

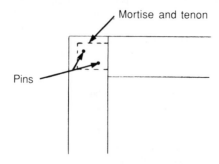

Fig. D-11. Drawboard joint

drill press A stationary power tool consisting of a motor and a handle-controlled, vertically moving chuck mounted on a pole; used to accurately bore holes and perform a variety of other jobs. A table attached to the pole below the chuck holds the workpiece and can be tilted to permit the boring of holes at an angle.

drill stop 1. A plastic or metal collar attached to a drill bit to limit its penetration into the material being drilled. 2. A device used on a drill press to limit the vertical travel of the drill bit and control the depth of the hole.

drip 1. Any exterior horizontal course or molding that projects over a wall or other surface to throw off water. 2. A small groove on the underside of a drip cap or windowsill to prevent water from running back under the cap or window. See Fig. W-3.

drip cap 1. A molding or flashing commonly installed over windows and doors to direct water away from the building in order to prevent seepage; also called a drip molding. See Fig. W-3. 2. A rounded or beveled metal strip attached to the bottom of an exterior door to prevent water from draining or blowing under the door.

drip edge An L-shaped sheet-metal flashing, usually 1 to 3 inches in width on each leg, that is installed over the ends of composition roofing at the roof edge to protect the edge of the roof sheathing and to cause water to drip free of the underlying cornice.

drip leg A capped pipe extending vertically downward from a T-fitting in a run of gas pipe and usually placed next to the shutoff valve; used to catch any condensation in the line before it can enter the appliance.

drip loop One of a series of short, downward curves formed in electrical service conductors at the point where they attach to the service drop to prevent rainwater that is flowing along the service drop wires from entering the weather head. See Fig. S-9.

drip molding See *drip cap.*

drip pan A tray or other receptacle located under an evaporator coil to catch dripping condensate.

dripstone In masonry construction, a stone or brick placed over a door or window to throw off water.

driptight Of or relating to electrical panel boxes that have an overlapping top to protect the box against water from above, but that are not considered completely watertight.

drive anchor See *anchor, hollow wall.*

drive home To hammer or screw a nail, rod, pin, screw, or other similar object into its final position.

drive screw See *screw nail.*

drop chute A funnel-like device used for placing wet concrete into narrow forms.

drop cloth A large piece of cloth, canvas, plastic, or other material used to cover and protect furniture, rugs, etc. from being damaged by work being performed in the area.

drop forging A method of metal forming in which a piece of metal is compressed into a die under pressure or impact to produce the desired shape.

drop leaf An extension on the side of a table that is hinged to the table on one end. It hangs down alongside the table and can be swung up and locked in place when extra table area is needed.

drop siding A type of siding, usually 4 to 8 inches wide and 3/4 inch thick, with tongue-and-groove or shiplap edges; commonly used without sheathing.

drop window See *window.*

drum sander Rubber cylinders in a variety of diameters, over which a sandpaper sleeve is attached; used in a portable power drill or drill press primarily for sanding curves.

drum trap A type of trap for tubs and showers that consists of a drum with inlet and outlet pipes at different levels.

drum wall See *water wall.*

dry bulb temperature The temperature of air or other gases, indicated by an unwetted thermometer. It does not take evaporation into consideration.

dry cell A type of battery that utilizes dry chemicals and is spillproof, such as a flashlight battery.

dryer outlet See *range outlet.*

dry location See *location.*

dry masonry A type of masonry construction in which the masonry units are laid up without mortar.

dry rot A general term applied to moisture-generated decay in wood; commonly characterized in early stages by wood that is spongy and soft, and in advanced stages by wood that can easily be crushed into a dry powder.

dry-to-touch Of or relating to a product, such as a paint film, that has dried sufficiently to be lightly handled, but has not yet dried to complete hardness.

dry vent A vent that does not carry water.

drywall 1. Gypsum, sometimes with fiber or other additives, made up into paper-covered sheets for use as an interior wall covering. Common sheet sizes are 4 x 8 feet and 4 x 12 feet, in thicknesses of 3/8, 1/2, and 5/8 inches. In common usage, the term is interchangeable with plasterboard, wallboard, gypsum board, and gypsum wallboard. 2. Interior wall covering materials that do not require mixing with water or other additives. The term can apply to a variety of materials, such as plywood and paneling, but most often refers to plasterboard.

drywall clip One of a number of small metal clips used at various framing corners to back-up drywall sheets. The clips

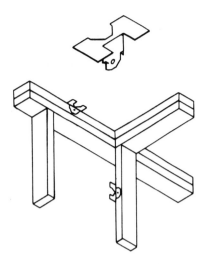

Fig. D-12. Drywall clips

eliminate the need for additional wood blocking or other supports. Also called drywall stops. See Fig. D-12.

drywall jack A portable, hand-operated lifting machine designed to raise and position a sheet of drywall and hold it in contact with the wall or ceiling until it is fastened.

drywall nails See *nails, common types.*

drywall saw See *handsaw.*

drywall square A T-shaped tool used as a guide for the measuring, marking, and cutting of drywall sheets. Typically made of aluminum or steel, with an 18- to 24-inch head and a 48-inch blade. Some types also feature an adjustable head for laying out and cutting angles. See Fig. D-13.

drywall stop See *drywall clip.*

dry well An underground pit, usually lined with concrete blocks and having a gravel bottom, that is connected to a building's gutter system to serve as a receptacle for rainwater runoff, which eventually drains out of the well and is absorbed into the ground.

dual-voltage motor An electric motor designed to be run at lower voltage, usually 110 to 120 volts. By a predesigned

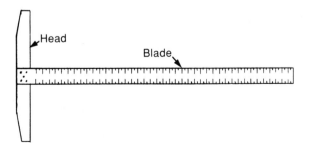

Fig. D-13. Drywall square

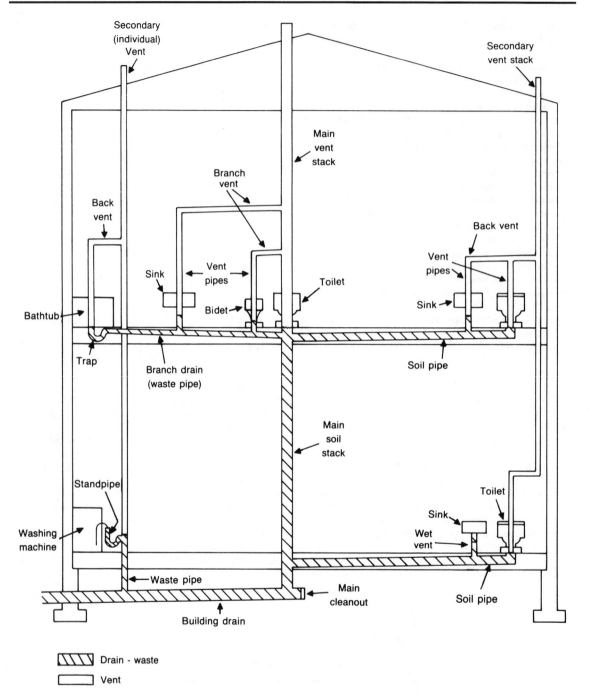

Secondary
(individual)
Vent

Secondary
vent stack

Main
vent
stack

Branch
vent

Back
vent

Sink

Vent
pipes

Toilet

Back vent

Vent
pipes

Bidet

Sink

Bathtub

Trap

Branch drain
(waste pipe)

Soil pipe

Main
soil
stack

Standpipe

Toilet

Sink

Wet
vent

Washing
machine

Waste pipe

Main
cleanout

Soil pipe

Building drain

Drain - waste

Vent

Fig. D-14. Common configurations used in a residential drain-waste-vent (DWV) system

switching of wires inside the motor, it will run at higher voltage, usually 220 to 240 volts.

duct A round or rectangular pipe, usually sheet metal, used with a heating and/or air conditioning system to carry and distribute warm and cool air throughout the building. Round ducts are available rigid or flexible; rectangular ducts are made in rigid form only.

ducted hood See *hood.*

ductile Of or relating to metal or other materials that can be stretched, hammered, or otherwise substantially elongated without breaking.

ductile iron pipe A type of cast-iron plumbing pipe that is somewhat more flexible and less brittle than conventional cast-iron pipe; widely used for water mains.

duct insulation See *insulation, thermal.*

ductless hood See *hood.*

duct system The continuous system of ducts through which conditioned or return air circulates within a structure. The duct system is usually considered to include the plenum, duct connections, boots, registers, fans, and all other air-handling and air-moving equipment, but not the furnace itself.

duct tape A tough, metallic tape used to seal seams and joints in sheet-metal pipe.

dummy door See *door.*

dummy knob A doorknob with no cylinder or latch; used as decoration on the inactive door in a set of double doors.

dumpy level See *builder's level.*

Duncan Phyfe Of or relating to a furniture style originated by American furniture maker Duncan Phyfe during the late 1700s and early 1800s.

duplex Two individual houses sharing at least one common wall and located on common property.

duplex nail See *nails, head types.*

duplex receptacle A standard electrical receptacle with outlets for two plugs. See Fig. G-6.

durham system A system of threaded or other type of rigid soil or waste pipes, that use sanitary or drainage fittings corresponding to the type of pipe.

dust-free Of or relating to the point at which a paint film or other finish has dried sufficiently that dust will not adhere to it.

dusting A fine white or gray powder that sometimes appears on the surface of new concrete as it cures and hardens.

Dutch door See *door.*

Dutch hip roof A combination of the hip and gable roof styles, having a modified hip on two ends that terminate in a shortened gable end. See Fig. R-8.

dutchman A nickname for any odd piece of wood, stone, or other material used to fill an opening or hide a defect.

Dutch metal Thin leaves of bright brass or bronze used to simulate gold leaf.

dwelling unit As regards the building codes, a dwelling unit is defined as a single unit designed to provide living facilities, including specific areas and equipment for sleeping, cooking, and sanitation.

DWV 1. Abbreviation for drain-waste-vent, a plumbing system that removes liquid and solid waste, drains contaminated water, vents gas and odors, and maintains correct atmospheric pressure in the system. See Fig. D-14. 2. A designation on a pipe, such as ABS-DWV, showing that the pipe is suitable for use in such a system.

dye See *paint.*

E

E In electrical calculations, the symbol for voltage.

earlywood See *wood: springwood.*

eased edge A corner or edge that has been slightly rounded.

easement 1. Any legal right or agreed permission to use another person's property. Easements may be given to a municipal government, such as for sidewalks and power lines, or to an individual for access to his property across another's, for irrigation canals, etc. 2. A curved molding used to ease a transition from one material or surface to another.

eastern framing See *braced framing.*

eaves The area of the roof that projects out over the exterior wall.

eaves trough See *gutter.*

ebonize To apply black stain to any of a variety of straightgrained woods in order to achieve a finish resembling real ebony.

eccentric Of or relating to two or more objects that do not have common centers; commonly used to convert a rotary motion into a reciprocating motion. Opposite of concentric.

economy brick See *brick.*

edge banding Thin strips of wood, wood veneer, or plastic

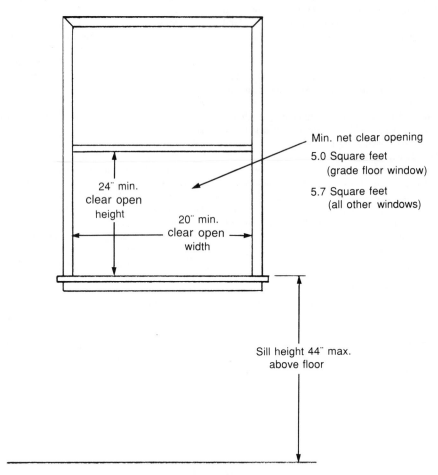

Fig. E-1. Typical egress window specifications

laminate affixed to the edge of plywood or other materials to provide a finished edge.

edge grain See *grain.*

edge joint A butt, doweled, or splined joint between two pieces of lumber that meet edge to edge.

edge-nail See *blind-nail.*

edge plane See *plane.*

edger A type of steel trowel having one downward curving edge; used to form a rounded edge on new concrete.

edging Finishing the outside edge of concrete flatwork into a convex curve, to give a good appearance and to lessen chipping.

EER See *energy efficiency ratio.*

efficiency The ability to produce a desired effect, condition, product, etc. with a minimum of waste, expense, or effort.

efflorescence A white crust or powder that forms on the surface of masonry; caused by the evaporation of water during the drying process which leaves behind deposits of water-soluble salts.

effluent An outflow; usually the liquid waste flowing out of a septic tank. See Fig. S-7.

eg See *electroplated galvanized.*

egg and dart A classical form of molding, having an egg-shaped form alternating at regular intervals with the form of an arrow, anchor, or other pointed shape. Also called egg and anchor.

egress 1. A place of exit, as from a building. 2. The right to go out, such as a variance granted for an exit road. 3. A window large enough for emergency egress. See *egress window.*

egress window Any type of operable window that, by virtue of its location and the size of its operable sash, is large enough to provide an alternate exit from a building in an emergency. The building codes establish which rooms, typically bedrooms, require egress windows, and set the minimum standards for their size. See Fig. E-1.

1/8 bend A plumbing elbow having a radius of turn equal to one-eighth of a circle, or 45 degrees.

elasticity The ability of a material that has been stretched, compressed, or otherwise distorted to regain its shape once the distorting force is removed.

elastic limit The maximum amount of elastic stress that an object can withstand before it fails.

elastomer Any rubber or rubber-based material that is capable of extreme elasticity.

elbow A fitting used for making a turn in a run of pipe or ducting; designated either by the degree of the turn, as a 90-degree elbow or 45-degree elbow, or by that portion of a full circle which the elbow's turn represents, such as a 1/4 bend or an 1/8 bend. See Figs. B-5, S-13, and T-4.

elbow catch A cabinet door latch used on the inside of the inactive or left-hand door of a pair of doors.

electrical box Any of a variety of boxes constructed of metal or various nonmetal materials and used as a place for joining wires or attaching devices or fixtures.

CEILING BOX A round box mounted in a ceiling to receive a light fixture. It may have flanges for nailing directly to a joist or a hanger bar for positioning between joists.

EXTENDER RING A square box with an open back; attached to the front of a square junction box to increase its depth and capacity.

JUNCTION BOX A square box designed for use with a blank cover as an enclosure for splicing wires.

WALL BOX A square or rectangular box that holds switches or receptacles; also called a switch box. See Fig. E-2.

electrical device See *device*.

electrical symbol Any of various standardized drawings and letters used on building plans to designate the location and use of various types of electrical components.

electrical system A complete installation of electrical conductors, devices, and other components; designed to serve a specific purpose.

electrical tape A type of nonconducting black vinyl tape used to wrap electrical connectors; also called insulating tape.

electric ignition See *pilotless ignition*.

electric meter A glass-enclosed electrical device installed in a building's electrical system at a point before the incoming electricity reaches the service panel; used to measure and record the building's electrical usage in kilowatt hours. See Figs. S-9 and S-10.

electric moisture meter A meter that uses electrical resistance to measure the amount of moisture present in a piece of lumber. The resistance varies with changes in the wood's moisture content.

electronic air cleaner A type of air-filtering device, used alone or in combination with a forced-air heating system, that places an electrical charge on dust and other airborne particles, which are then trapped against plates holding an opposite charge and removed from the air.

electroplated galvanized See *nails, finishes*.

electroplating A metal coating applied to the surface of a piece of metal using an electrical current. Various coating materials and thicknesses can be applied in order to reduce wear or corrosion or to provide a decorative finish.

elevation 1. The front, back, and side views of a building, room, machine, etc., especially as depicted on a drawing. 2. The height above sea level.

ell 1. That part of a building constructed at right angles to the front of the building, forming a shape similar to the letter L. 2. An L-shaped addition or other structure. 3. Any elbow fitting.

ellipse A curved shape that has the appearance of an elongated circle such that the sum of the distances from any point of the curve to two fixed points within the curve remains constant.

elliptical arch An arch formed in an elliptical curve, rather than a circular one.

elm See *hardwood*.

embossed A surface that is raised above the surrounding surfaces for decoration.

emery A very hard, black variety of corundum mixed with other minerals for use as an abrasive.

emery cloth A tough cloth material to which powdered emery has been cemented; used primarily for polishing and rubbing down metal.

emery paper Paper coated with powdered emery that is used for sanding and polishing and is similar to emery cloth, but usually less flexible and more abrasive.

emery wheel A wheel made of compressed powdered emery that is used to grind metal.

eminent domain The government's right to take private

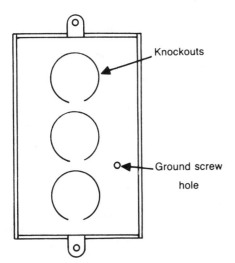

Knockouts

Ground screw hole

Fig. E-2. Typical metal electrical box (wall or switch)

property for public use, usually after the payment of just compensation to the owner.

emission   The release or discharge of pollutants into the atmosphere.

emittance   The ability of a material to give off or reflect thermal radiation; used in measuring the effectiveness of a solar collector.

emt   Abbreviation for *electro-mechanical tubing,* a thin-walled, nonthreaded galvanized metal conduit that is easily cut and bent. It is usually sold in 10 foot lengths and is joined by special fittings that use either a set screw for inside use or a weathertight locknut and compression ring for exterior applications. Also called thinwall.

emulsion   A suspension of fine particles of one liquid within another liquid.

enamel   A type of paint coating produced by blending pigments with varnish.

encased knot   See *lumber defect.*

encaustic   Of or relating to materials, such as tile or porcelain, that are colored by the burning in of pigments.

enclosure   Any structure, such as a fence, building, etc., that surrounds and closes in an area or object on all sides.

encroachment   A building, fence, road, or other structure or improvement on one parcel that extends across a property line and onto the adjacent parcel.

encumbrance   See *incumbrance.*

end check   See *lumber defect.*

end cut   A square or beveled cut across the end of a piece of lumber.

end grain   See *grain.*

endiron   See *andiron.*

end joint   A butt or scarf joint of two pieces of lumber that causes the lumber pieces to meet end to end.

end-matched   Of or relating to boards having a tongue at one end and a groove at the other, permitting them to be joined without support directly under the joint.

end skin   A thin sheet of veneer used to face the unfinished side of a cabinet.

end to center   The distance from the end of one object to the center of another, a common measurement found on construction drawings, showing distances from the end or corner of a building to the center of cross walls, windows, doors, etc.

energize   To initiate a flow of electrical current into a circuit or fixture.

energy   The capacity to do work and overcome resistance; potential force and the capacity for action; such power in action. There are two forms of energy.

KINETIC ENERGY Energy that is in motion, such as electricity, moving water, etc.

POTENTIAL ENERGY Energy that is stored and has the capacity for action, such as coal, gasoline, etc.

energy audit   The evaluation of a building to determine its heat loss and energy use together with the specific recommendations made concerning those areas where heat loss can be lessened and energy can be saved.

energy-efficiency ratio   The ratio of the usable output of power to the input of energy; commonly used as a rating of how well an electrical appliance or component conserves energy. Abbreviated EER.

energy efficient   A generic term generally applied to any structure, appliance, or device designed to minimize the use of whatever energy source or material (electricity, gas, etc.) is supplied to it.

energy system   The total, interrelated machinery and processes in a building that use energy, including space heating and cooling, water heating, etc.

engineered brick   See *brick.*

engrailed   Of or relating to a surface that is decorated with indented lines or shapes.

entablature   A horizontal structure resting upon and supported by columns. It consists of the architrave, frieze, and cornice.

entrance elbow   A 90-degree elbow fitting for use with electrical conduit, having a removable plate that allows the wires to be easily bent around the corner.

entrance head   See *weather head.*

entrance lock   See *lockset.*

envelope   The protective physical shell of a building that separates the inside living space from the outside environment.

EPA   Abbreviation for *Environmental Protection Agency,* an independent agency of the United States Government's executive branch that regulates pollution, radiation, and the handling and disposal of toxic substances.

epoxy paint   Paint made with epoxy resins, either premixed or in two components, which employ a catalyst to dry the paint.

epoxy resin   See *paint resin.*

epoxy resin glue   See *adhesive.*

equilateral triangle   See *triangle.*

equilibrium   A state of balance between the forces acting on an object.

equilibrium moisture content   The point at which wood neither gains nor loses moisture when surrounded by air at a given relative humidity and temperature.

equivalent length   A method for compensating for the loss of air pressure in a duct fitting by obtaining from charts the number of feet of straight ducting that would equal the loss for the one fitting, and then calculating the air pressure in a given heating or cooling system using the figures in the charts.

ergonomics   The study of the functions and techniques of humans and machines and how they relate. Research in this field is instrumental in designing machinery, equipment, furniture, etc., that is more comfortable and efficient for human use.

escutcheon   1. A piece of decorative builder's hardware consisting of a plate with a keyhole, usually held in place with small nails called escutcheon pins. 2. In plumbing, a decorative plate that slides over a pipe and seals against the wall. See Fig. B-5. 3. Any of a wide variety of decorative plates and flanges.

eutectic salt Phase-changing salts used in place of rocks or water in storage units for air-type solar collection systems. The salts melt at temperatures in the 80° F to 100° F range, allowing them to absorb much greater amounts of heat per pound, which they release as they cool and resolidify. Glauber's salt, the most commonly used, holds as much heat in 2-1/4 pounds as 132 pounds of water or 690 pounds of rock, permitting a much smaller storage area.

E value See *modulus of elasticity.*

evaporative cooler A large window- or roof-mounted cooling unit in which water is dripped over thick fiber pads by a float-controlled pump and outside air is drawn across the pads by a blower. The water absorbs the heat from the air; so cooler air enters the building.

evaporator See *cooling system.*

evergreen Of or relating to a tree or shrub having foliage that remains green until the formation of new foliage, as opposed to deciduous.

excavation A hole or trench dug in the ground to receive footings, piers, pipe, etc.

excavation line A string stretched between batter boards to indicate one of the edges of an excavation.

excess air In heating or combustion appliances, combustion air being supplied that is in excess of the amount needed by the appliance to operate efficiently.

existing In reference to the building codes, any structure that was completed prior to the adoption of a new or revised building code requirement.

exotic wood Any of a variety of wood types not native to a particular area, especially those from another country.

expanded metal See *metal lath.*

expanded polystyrene See *insulation, thermal: rigid boards.*

expansion anchor See *anchor, masonry.*

expansion chamber See *expansion tank, 2.*

expansion coefficient A rating of how much a particular material expands when heated, measured in inches of expansion per degree F of heat applied.

expansion joint A bituminous fiber or wood strip used to separate large areas of concrete; designed to prevent cracking from expansion and contraction during temperature changes.

expansion tank 1. A steel tank used with a well, in which a layer of compressed air is used as a cushion over a layer of water, exerting pressure on the water and helping to maintain an even water pressure in the system. 2. In a hot-water space-heating system, a tank that allows for changes in water volume brought about by temperature differences; also called an expansion chamber.

expansive bit A type of drill bit that can be adjusted to bore a variety of different-diameter holes.

exposure 1. That portion of a shingle or horizontal siding board which is exposed to the weather. It is the distance from the butt of one shingle or the edge of one board to the butt or edge of the one immediately above or below it. 2. The direction a building faces. A building facing north has a northern exposure.

extended plenum system A system of forced-air heating ducts in which one or more main ducts extend from the plenum and branch ducts run from the mains to individual rooms.

extender See *paint.*

extender ring See *electrical box.*

extension box An open-ended sheet-metal box used to extend a new branch off the main duct of an extended plenum system into the floor or ceiling directly above or below the main duct.

extension cord An electrical cord having a male plug at one end and one or more female outlets at the other; used to extend the operating range of an electrical device.

extension ladder A straight ladder having two or more sections of which the upper section slides over the lower section to extend the ladder's height, then locks in position.

extension plank A scaffolding plank, usually made up of 1 x 3s on edge and bound by metal bands, in which the boards can slide side by side and be adjusted to different lengths.

exterior chimney That portion of a masonry chimney or metal flue on the outside of a building and exposed to the elements.

exterior door See *door.*

exterior plywood See *plywood.*

external 1. Of or relating to the outer edge, face, or surface. 2. Of or relating to that portion of an object which is visible or exposed to the elements.

externally operable Of or relating to electrically powered equipment that may be used without exposing the operator to contact with live parts.

extra A general term applied to additional labor or material on a construction project, beyond the original contracted price for the job.

extrados In architecture, the exterior curve of an arch. See Fig. A-7.

extruded aluminum Aluminum that is forced through a die under pressure. Aluminum extrusions are relatively inexpensive and are produced in a number of shapes for the construction industry, including thresholds, channels, tubes, conduit, and much more.

extruded polystyrene See *insulation, thermal: rigid boards.*

eyebolt A bolt having a machine thread and nut at one end and a loop at the other end.

eyebrow window A type of small dormer window, constructed by extending a short section of roof up into a raised curve.

F

facade The front or principal face of a building.

face 1. The exposed or visible part of a wall. 2. That side of a board, panel, masonry unit, or other building material designated to be exposed. 3. The working surface of a tool. See Fig. H-1. 4. To overlay an underlying surface with a decorative covering.

face block See *concrete block*.

faced wall See *veneered wall*.

faceframe The finished stile and rail assembly of a cabinet, which is fastened to the front of the box and provides a point of attachment for the door hinges.

face-nail To nail directly through the face of a board, perpendicular to the board's surface.

face of an arch The exterior or exposed vertical surface of an arch.

facia Variant of *fascia*.

facing Any molding, trim, or veneer used to cover an exposed edge or surface.

facing brick See *brick*.

facing hammer See *hammer*.

facing tile See *structural clay tile*.

factor of safety The difference between the maximum stress acting on an object and the maximum amount of stress the object can withstand before failing.

factory and shop lumber See *lumber*.

factory built Building components constructed in whole or in part in a factory. See *prefabricated*.

faience Pieces of decorated, glazed ceramic used as facing for buildings, fireplaces, etc.

fall 1. A downward slope that facilitates drainage to a pipe, trench, driveway, etc. See Fig. S-2. 2. The pulling rope in a block and tackle.

false beam A decorative, nonsupporting beam, usually made up of a wooden nailer attached directly to the wall or ceiling and covered by three pieces of decorative wood.

false louver door See *door*.

false rafter A short rafter that extends out off of a main rafter to provide an extension or a change in pitch.

false tread A stair tread that is not full width. Typically used to cover the ends of the rough treads, leaving a space in the center of the treads for carpeting.

false work Temporary scaffolding, walkways, covers, etc., used to protect or brace a building during construction.

fan-coil heater A small, thermostatically controlled heater that pipes hot water from the home's existing water system through coils within the unit and uses an electric fan to force air over the coils and out through grills in the unit's face; designed to be recessed between wall studs.

fan light 1. A ceiling-mounted unit containing an exhaust fan and a light; commonly used in bathrooms and laundry rooms. 2. A wide-bladed, ceiling-mounted paddle fan having a light fixture below the fan motor.

fanlight A semicircular window having a sash containing bars that radiate out from the center of the base. See Fig. F-1.

FAS See *firsts and seconds*.

fascia A wood or plywood strip nailed to the overhanging ends of rafters, forming the outer face of the cornice; also spelled facia. See Figs. C-4, C-13, and O-2.

fast track A method of construction in which work begins on the structure before all of the design details are finalized.

fat lime See *rich lime*.

F-chart A computer program developed at the University of Wisconsin that calculates the solar fraction supplied by a solar heating system and evaluates its economics.

feather To smooth and taper one surface onto another, leaving no visible transition, such as applying joint compound on plasterboard or patching a hole in plaster.

featheredge A tapered edge on a piece of lumber, particularly siding, that has been milled to a thin line.

featheredge brick See *brick: arch brick*.

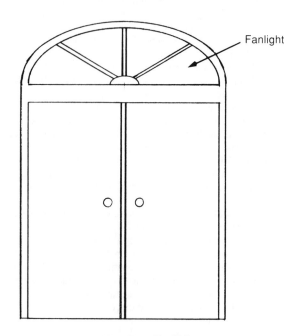

Fanlight

Fig. F-1. Fanlight

feeder An electrical conductor or conductor pair used between service equipment and overcurrent protection devices for branch circuits, between a meter and a service panel, or between a service panel and a subpanel. See Fig. S-9.

feed-through-switch A small electric switch that can be attached directly to a fixture's cord to control the operation of that fixture.

fell To cut down a tree.

felt paper A thick, strong paper made from long-fibered wood pulp saturated with molten asphalt. It is sold in rolls, usually 36 inches wide, and is specified by the weight per square (100 square feet), the most common being 15 pound and 30 pound. The higher the weight, the thicker the paper. Primarily used under roofing and siding material for protection against moisture.

female See *male/female.*

fence 1. A flat, adjustable bar or other straightedge, such as that found on a table saw; used to provide a true edge against which the workpiece can be guided. 2. A structure, usually of wood, metal, or wire, used to enclose an area or parcel of land or to show its boundaries.

fencing tool Any of a variety of combination tools used for installing and repairing fencing. The typical fencing tool has a cutting plier, a hammer face, and a small slotted claw.

fender A short metal screen or plate, usually brass, used in front of a fireplace to prevent logs or coal from rolling out onto the hearth.

fenestration The arrangement, proportioning, and design of a building's windows and exterior doors. See Fig. F-2.

ferroconcrete Small aggregate concrete work, often in curves and other patterns, that is reinforced with embedded wire mesh or bars.

ferrous Of or relating to any of a group of different metals in which iron is the major ingredient.

ferrule A metal ring or tube used for protection and strength. Common ferrule applications are as a support for gutters, whereby the gutter hanger nail is driven through the tube to prevent collapse of the gutter, and as a metal ring on some hand tools placed where the blade enters the handle to prevent damage to the handle.

festoon An architectural or woodworking decoration consisting of a series of carved flowers and leaves linked together in a garland.

fiber See *wood.*

fiberboard See *building board.*

fibered gypsum plaster See *plaster.*

fiberglass 1. Long filaments of spun glass, chopped and blended with resin, then molded and cured in a variety of shapes for use in many different products. 2. See *insulation, thermal.*

fiberglass fabric Any of a variety of fabrics made from

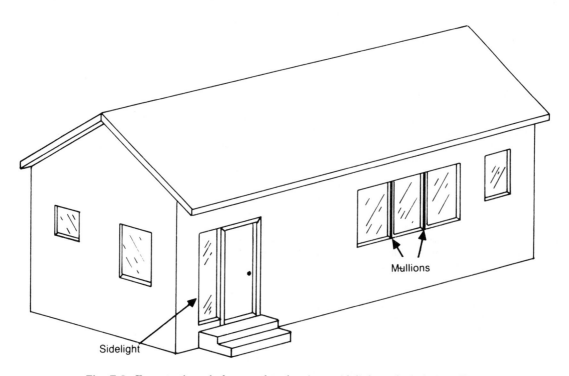

Fig. F-2. Fenestration of a house, also showing a sidelight and window mullions

bunched, pressed, or woven glass fibers, called rovings. The fabric is designed for use with a liquid resin in the construction of fiberglass items.

fiberglass shingle A type of fire-resistant, long-lasting roofing shingle manufactured from a fiberglass compound, usually in conjunction with asphalt, and topped with a mineral surface.

fiber pipe Plumbing pipe, usually manufactured from a coal-tar formula; primarily used in the exterior portion of a sewage or septic system. Orangeburg is a commonly used trade name for fiber pipe.

fiber saturation point The point at which all free moisture in wood has evaporated, leaving only combined moisture, which is about 30 percent water by weight. Any further evaporation from this point will cause the wood to shrink.

fiber stress In engineering, a calculation of how much bending stress can be placed on a piece of lumber before permanent damage is done to the wood's fibers.

fibrotile Large, thick, corrugated tiles made from asbestos cement; used for fireproofing.

fieldstone Natural, smooth or irregular stone in sizes suitable for rubble masonry construction.

field tile 1. Porous tile placed around a building's foundation to absorb groundwater. 2. Ceramic tile having no special edge treatment that makes up the main part, or field, of a tile installation, as opposed to tile used specifically for the edges. 3. Any tile other than an edge or trim tile.

file card A brush fitted with short, slender, stiff wires and used for cleaning files.

fill Loose gravel or dirt thrown back into an excavation.

filled ground A grade level that has been achieved by the addition of dirt, or dirt over debris to even out low spots.

filler See *wood filler*.

filler strip A strip of wood or other material used to fill a gap or to add length, as in the area between a cabinet and a wall or between two cabinets.

fillet A narrow molding used as an accent to larger moldings or to cover a corner where two surfaces meet.

fillet weld A weld, roughly triangular in section, that fills the corner created by two overlapping steel parts.

fillister 1. A type of narrow-bladed plane used for making grooves or rabbets. 2. The rabbet on a window sash that receives the glass and glazing compound. 3. A type of cap screw with a rounded head.

fin One of a series of metal plates attached to the elements in a resistance heater to increase the total amount of heated surface that comes into contact with the air, thus providing better heat transfer.

final inspection 1. The last inspection performed on a building by the building officials, prior to issuing a certificate of completion. 2. The last inspection performed by an architect, prior to authorizing final payment to the contractor.

finals See *final inspection*.

fine-line plywood See *plywood siding*.

fines 1. Monetary assessments for the violation of building codes or other infractions. 2. Small pieces of coal that chop

off during processing and handling. 3. Any smaller than normal piece of material, occurring naturally or as a result of handling.

finger joint A joint in which two short lengths of wood are joined by a series of glued, interlocking "fingers"; commonly used for producing moldings and door cores. See Fig. F-3.

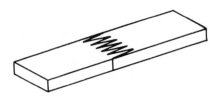

Fig. F-3. Finger joint

finial A decorative object that adorns the top of a spire, gable, post, or pinnacle.

finish 1. The material used to color or protect a surface, such as paint, stain, varnish, etc. 2. The surface or appearance achieved by the application of such material.

finish carpenter A person skilled in the installation of finish materials.

finish carpentry A branch of carpentry dealing with the installation of the interior trim or other visible components of a building, including doors, baseboards, shelving, etc.

finish coat See *top coat*.

finished floor The final floor-covering material, such as linoleum, tile, hardwood strips, etc., that is laid over the subfloor or underlayment.

finished grade The final ground level surrounding a structure following completion of all excavating, backfilling, etc. See Fig. P-3.

finished string The finished, visible stair stringer fastened to the rough carriage.

finish hardware The visible hardware used on a building's interior, including locks, hinges, handles, etc.; also called builder's hardware.

finishing nails See *nails, common types*.

finish plaster See *plaster*.

fink truss The common roof truss used in residential and light commercial applications for relatively short spans; also called a W-truss because of its shape. See Fig. T-7.

fir, Douglas See *softwood*.

fireback A metal plate bolted to the firebricks on the back wall of a fireplace. The fire heats the plate, causing more warm air to be radiated into the room.

fire block 1. One of a series of blocks of wood nailed horizontally between wall studs to prevent the spread of flame and smoke in the airspace between those studs. See Figs. B-2, F-10, and P-3. 2. Any concealed, tightly enclosed space that serves such a purpose; also referred to as a fire-stop or draft stop.

firebrick See *brick*.

fireclay A type of clay that can withstand high, concentrated

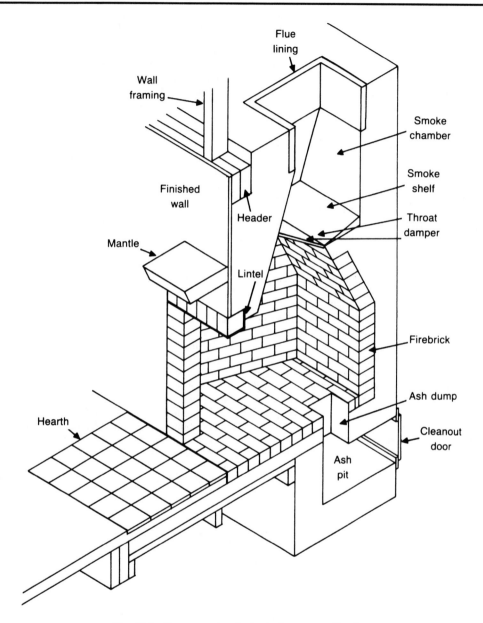

Fig. F-4. Components of a typical masonry fireplace

temperatures; used primarily in making up mortar for setting firebricks.

fire cut   One of a series of beveled cuts on the end of floor joists that are installed in a building having masonry walls. If the joist burns through in the event of a fire, the cut causes it to fall inward without disturbing the exterior walls.

firedog   See *andiron.*

fire door   See *door.*

fireplace   A framed opening in a chimney, made to hold a fire. See Fig. F-4.

fireplace insert   A type of air-tight, steel wood stove designed to be retrofit into an existing masonry fireplace, thus providing more efficient operation.

fireproofing   Chemicals or other coatings that will allow a material to withstand flames and high heat. Some materials, such as wood, cannot be made literally fireproof; instead fireproofing makes wood hard to ignite and less able to sustain combustion.

fireproofing tile   See *structural clay tile.*

fireproof plaster   See *plaster.*

fire-resistant   In general terms, of or relating to any material that will not support combustion and that can withstand a fire for one hour without suffering serious damage. Several agencies offer specific criteria on fire resistance of a wide variety of materials.

fire-retardant chemical A chemical used to reduce flammability or to retard the spread of flames.

fire-stop See *fire block.*

fire wall A fireproof or fire resistant wall extending to the roof, which subdivides a building to prevent the spread of fire.

firsts and seconds A grade for hardwoods, indicating lumber of top quality for use in cabinets, furniture, and similar finish applications; abbreviated fas.

fish   To work a wire, pipe, or other object through a concealed and otherwise inaccessible area.

fished joint A butt joint between two structural members that is secured and reinforced with a fish plate.

fish glue   See *adhesive: animal and fish adhesive.*

fish plate   A wooden or metal plate, used alone or as a pair, installed with nails or bolts to fasten two members together at a butt joint.

fish scales   Exterior siding shingles having a rounded bottom, most often used as gable end ornamentation. See Fig. F-5.

fish tape   A length or roll of flexible spring steel used to pull electrical wires through conduit or other inaccessible areas.

fitting   Any device or piece of equipment used to join pipes, conduit, ducts, fixtures, etc.

fitting brush   A small wire brush used for cleaning male and female copper pipe fittings prior to soldering; available in various diameters.

fixative   A clear, protective coating, usually applied by spray; used to prevent printing or drawings from being rubbed off the paper.

fixed   Permanent, nonmovable.

fixed appliance   See *appliance.*

fixed window   See *window.*

fixture strap   A slotted metal bracket that is secured across the front of an electrical box and to which a light fixture is attached.

fixture unit   A standard of measurement equal to 7-1/2 gallons (1 cubic foot) of water per minute; used to determine pipe size, septic tank capacity, and other plumbing calculations.

flagstone   A flat stone from 1 to 4 inches thick; used for rustic walkways, floors, etc. Also called flagging or flag.

flame hardening   A method of hardening the surface of a metal part without affecting its internal structure by rapidly heating the metal with a flame or electrical current, then quenching it in a liquid such as water or oil.

flame spread rating   A rating of how rapidly flame will spread over a given material, as tested in laboratory conditions. The higher the rating number, the faster the flame spreads.

flange   A projecting rim, lip, or collar; designed to add stiffness or to provide a means of attachment.

flank   The side of an arch.

flapper   See *stopper.*

flare fitting   A brass, copper, or plastic plumbing fitting having a beveled end designed to mate with a corresponding flare on the end of a pipe. The pipe and fitting are secured to each other with a locknut. See Fig. F-6.

flaring tool   An adjustable, two-piece tool used to flare the end of a pipe for use with a flared fitting.

flashed joint   A joint or meeting between two surfaces, for example a roof valley, that is covered with sheet metal or other material to repel water.

flashing   Any solid, waterproof material, most commonly sheet metal, used over the joints in wall and roof construction to

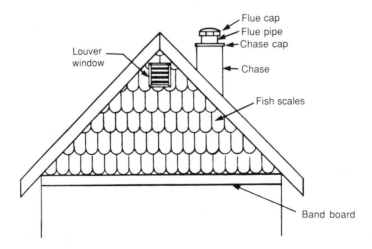

Fig. F-5. Fish scales

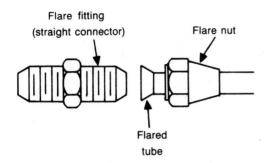

Fig. F-6. Flare fitting used on copper tubing

prevent the entry of moisture. See Fig. F-7.

flash point The lowest temperature at which the vapors of a combustible liquid will ignite.

flat 1. The sides of a bolt head or nut. 2. Two opposing sides of an object such as a hexagon or octagon.

flat corner iron A predrilled, L-shaped, flat metal plate used for securing two pieces of wood or other material at a right angle joint. See Fig. F-8.

flat countersunk nail See *nails, head types.*

flat grain See *grain.*

flat paint A type of paint containing a high degree of pigment that dries to a flat or lusterless finish.

flat plate collector A basic style of solar collection device consisting of a black, heat-absorbing plate contained horizontally within a boxlike enclosure below which insulation is packed and above which one or two layers of transparent glass or plastic are located.

flat roof Any roof that has no pitch or inclination; also a roof that is pitched slightly, usually no more than 1/2 inch per foot, to allow for water runoff. See Fig. R-8.

flat sawn See *grain: flat grain.*

flatwork Concrete poured flat and directly on grade; used for driveways, patios, walks, etc.

flexible connector A tube or pipe that can easily be bent as needed; used to connect a water or gas shutoff valve to a fixture or appliance.

flexible duct 1. A type of soft duct pipe that is easily curved and bent without fittings, consisting of a long wire spring wrapped with insulation, an inside and outside vapor barrier, and a conventional sheet-metal collar or clamp at each end for connection to other pipes or fittings. 2. A corrugated, noninsulated sheet-metal pipe; commonly used for venting bath fans and similar installations.

flexible rule See *rule.*

flier One of the uniform, parallel steps in a straight stairway.

flight of stairs A series of steps, with or without intermediate landings, that connect two floors of a building.

flint paper See *sandpaper.*

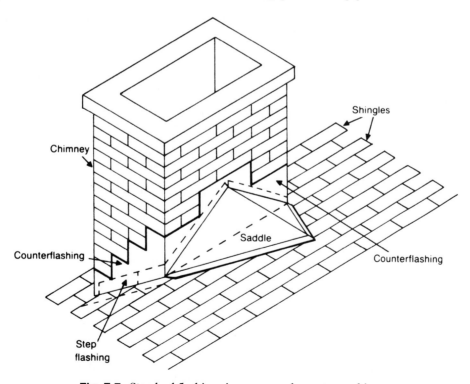

Fig. F-7. Standard flashings in use around a masonry chimney

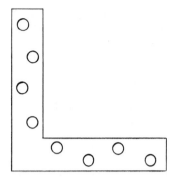

Fig. F-8. Flat corner iron

flitch   A section of log that has been cut lengthwise so as to be flat on two or four sides, and that is ready for further cutting in the manufacture of lumber or plywood.

flitch beam   A heavy beam constructed by sandwiching one or more thick steel plates between two or more wooden joists and binding them together with bolts.

float   1. A wooden trowel used to level and finish the surface of concrete flatwork. Wood floats produce a relatively rough surface, compared to that obtained from a steel trowel. 2. To level and smooth fresh concrete. 3. Any of a variety of buoyant balls, discs, or other shapes, linked to a pump, valve, or other device. The float rests on top of a liquid and follows its movement, activating or shutting the device it's linked to. 4. A type of coarse woodworking file with large, parallel teeth.

float arm   A threaded brass rod used in a toilet tank to connect the float ball to the ball cock. See Fig. F-9.

float ball   A large, buoyant ball made of copper, plastic, or foam that floats on the surface of the water in a toilet tank. It rises and falls with the water level, thus opening or closing the ball cock to control the water coming into the tank. See Fig. F-9.

floating   Spreading a uniform layer of plaster, concrete, or stucco onto a wall, ceiling, or slab.

floating foundation   Raftlike pieces of reinforced concrete used as a foundation for a building constructed on marshy or unstable soil.

flocked   Of or relating to a decorative finish obtained by spraying or sprinkling short fibers of wool, silk, or other materials onto paint, glue, or other wet surfaces.

flocoat   To coat an object by pouring on the coating material and allowing the excess to drain off, usually into a collection tank.

flood coat   The last hot-mopped coat of tar or asphalt applied to a built-up roof, which is then covered with a cap sheet or spread with mineral aggregate.

floodlight   An incandescent bulb designed to provide a wide beam of light; commonly used to light large exterior areas.

floor   See *story*.

floor brick   See *brick*.

floor drain   A plumbing fitting having a slotted strainer top, recessed into the floor and designed to receive waste water from floor cleaning and spillage. See Fig. C-15.

floor flange   A metal disk with female pipe threads in the center and a series of countersunk holes around the outside that enables a piece of standard, threaded, steel pipe to be attached to a floor or other solid surface.

floor furnace   A small gas- or oil-burning furnace set in a metal box in the floor of a house and covered with a single large grill.

flooring nails   See *nails, common types*.

floor insulation   See *insulation, thermal*.

floor joist   One of the structural members in a building that supports the subfloor, usually running across the girders for the floor of the first story and across the lower story's wall plates for the floors of upper stories. See Figs. B-2, P-2, and P-3.

floor load   The maximum combined live and dead loads a floor is designed to support.

floor plan   A drawing of a building in horizontal section as viewed from above, showing the arrangement of rooms, windows, doors, and other construction details.

floor sander   A heavy, motorized belt sander with a long handle; used for sanding wood floors.

floor tile   1. Any of a variety of materials formed into squares or other shapes and used as a finished floor covering. 2. See *structural clay tile*.

floor truss   A flat truss used as a floor joist to allow a longer span between supports.

flow rate   1. The quantity of water or liquid passing a given point in a given time, usually figured in gallons per minute (gpm) or cubic feet per minute (cfm). 2. The quantity in pounds of heat-transfer fluid or in cubic feet of air passing through a solar collector each hour.

flow restrictors   Any of a variety of devices used on shower heads, faucets, toilets, and other plumbing fixtures to restrict the water's flow for better water conservation.

flue   A metal or masonry pipe or passageway used to exhaust smoke, gas, or fumes from a fireplace, stove, or other device to the outside of a building. See Fig. W-2.

flue brush   A stiff wire or nylon brush, either mounted on a pole or pulled by a rope, that is used to remove combustion deposits from flue pipes.

flue cap   Any of a variety of sheet metal caps, designed to interlock with a flue pipe and serve as weather protection and termination for the flue. See Fig. F-5.

flue fire   A fire inside a flue, chimney, or connector, caused by the ignition of excessive creosote deposits in the flue. Also called a chimney fire.

flue gases   The by-product of the combustion process in a furnace or other appliance, mixed with excess air.

flue heat exchanger   A device that fits into the flue pipe of a furnace, consisting of a series of hollow or water-filled tubes that are heated by the exhaust gases passing them. An attached electric blower, activated by a thermostat, forces air across the tubes and out into a duct.

flue lining A round or square, clay or terra cotta pipe, sold in standard cross-sectional sizes and in 2-foot lengths, that is used for the inner lining of the flue space in a masonry chimney. See Fig. F-4.

flue pipe Any of a variety of special-purpose, interlocking pipes that are designed or approved for use as a flue. See Fig. F-5.

fluorescent lamp A type of lamp consisting of a sealed glass tube coated with phosphorus on the inside and containing mercury vapor and an inert gas, usually argon, as well as cathodes on each end that start and maintain a mercury arc, which is absorbed by the phosphorus and radiated as visible light. Different phosphorus can be used to produce colors of light ranging from white (called cool or daylight white), used for shadowless commercial applications; to slightly red or yellow (called warm white), often used for residential lighting because of the more natural flesh tones it renders.

fluorescent paint Luminous paint that glows under exposure to ultraviolet light.

fluorescent starter An electrical device that supplies the electrons needed to start and maintain the mercury arc in a fluorescent lamp.

flush The positioning of two pieces so that their adjacent surfaces are even or in line with each other.

flush door See *door*.

flushometer valve A device activated by the buildup of direct water pressure that discharges a preset quantity of water for purposes of flushing a toilet; used primarily in commercial applications.

flush tank toilet A toilet having a separate or integral tank for water storage. The stored water, flowing into the bowl by gravity when released, provides the necessary pressure to flush and cleanse the bowl. See Fig. F-9.

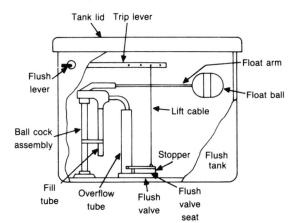

Fig. F-9. Components found within the tank of a typical flush tank toilet

flush valve A device consisting of a seat and a rubber flap or stopper at the bottom of a toilet tank; used to control the flow of water from the tank into the bowl. See Fig. F-9.

flush valve toilet A type of toilet that does not have a tank, but instead uses a high-pressure, 1-inch water line to produce the flushing action; most often used in commercial installations.

flute A decorative concave cut, usually in the form of a round bottom groove; used to decorate a column or furniture leg.

fluted Columns, moldings, furniture legs, or other items decorated with flute cuts. See Fig. C-8.

fluted nail See *nails, shank types*.

fluting and reeding Decorative cuts applied to the side or edge of a piece of wood. The fluted edge contains a series of parallel concave coves, while the reeded edge has a matching series of convex rounded areas.

flux A paste material used when soldering to help distribute the heat evenly in the joint and prevent oxidation of the materials being soldered.

fly ash Small particles of ash or, sometimes, unburnt material, that escape to the outside through the chimney.

fly cutter A tool consisting of a vertical arbor and an adjustable, perpendicular arm with a vertical cutter at the end; used in a drill press to cut large holes and circles in wood. Also called a circle cutter.

flying shore A timber placed horizontally between two vertical forms, intended as a temporary brace for both forms; primarily used during excavations and the pouring of concrete.

fly rafter See *barge rafter*.

foamed-in-place insulation See *insulation, thermal*.

foaming agent A chemical material that is worked into a foam and then added to wet concrete in order to alter or control the concrete's density.

folding rule A type of wooden ruler, commonly 2 or 3 feet in length, that folds down to between 5 and 9 inches for easy transport.

footcandle A unit of measurement of the amount of light falling on a given surface; equal to one lumen per square foot of illuminated surface.

footer See *footing*.

footing A flat masonry section, usually concrete, that is wider than the wall, column, or pier it supports and is used to transfer support for a vertical load over a wide area; also called a footer. See Figs. D-10 and P-2.

footing form A wooden panel used to support soft earth to allow for the pouring of concrete footings.

footlambert A unit of measurement of the brightness of a surface that emits or reflects light, such as a light fixture; equal to one lumen per square foot.

footprint The outline of a building's perimeter, regardless of height, used in determining its overall size and shape, and also the percentage of the building site that it occupies. See Fig. P-4.

forced air system A warm-air space-heating system that uses a thermostatically operated fan to move heated air through a series of ducts into the building and to draw return air back into the furnace for reheating.

forced circulation heater A hot-water space-heating system that employs a pump to circulate the hot water through the distribution system.

fore plane A woodworking plane that is intermediate between a jack plane and a jointer plane in length.

form A temporary wooden or metal mold used to receive and contain wet concrete until it dries.

formaldehyde An organic chemical compound used as a bond material in a variety of products, including some foam insulation, carpet, fabric, plywood, and particle board.

Formica The commonly used brand name for a variety of plastic laminates used in construction and furniture making. See *plastic laminate*.

Forstner bit A type of wood-boring bit with no screw feed; used for boring accurate, flat-bottomed holes.

fossil fuel Fuel from inside the earth, derived from the breaking down of certain plant and animal life over long time periods; gas, oil, and coal are common examples.

foundation The load-bearing structure of a building, usually concrete or concrete blocks, that rests on the ground and upon which the building is erected; commonly both the footings and the foundation walls. See Figs. B-2 and P-3.

foundation forms A temporary wooden mold, braced vertically and diagonally, used to receive and contain the poured concrete for foundation walls until it dries.

foundation plan A drawing of a building in horizontal section as viewed from above, showing the exterior foundation walls, piers, girders, footing specifications, and other details of the building's foundation.

foundation survey A specialized type of survey that accurately locates a building's foundation in relation to the property lines of the land parcel on which it sits; done to legally ensure that no setbacks have been violated.

foundation vent A small, screened metal or plastic vent set into a rim joist or foundation wall, used to provide ventilation to a crawl space in order to prevent moisture buildup. See Fig. C-15.

fourplex Four individual dwelling units sharing common walls, roof, and property.

four-way switch See *switch*.

4 x Pronounced "four by"; any lumber having a nominal thickness of 4 inches, regardless of width or length.

foxtail wedging A method of tightening and strengthening a mortise-and-tenon joint by inserting thin wedges into slots on the bottom of the tenon, then driving the tenon into a wedge-shaped mortise. Contact with the bottom of the mortise forces the wedges to spread the tenon and lock the joint.

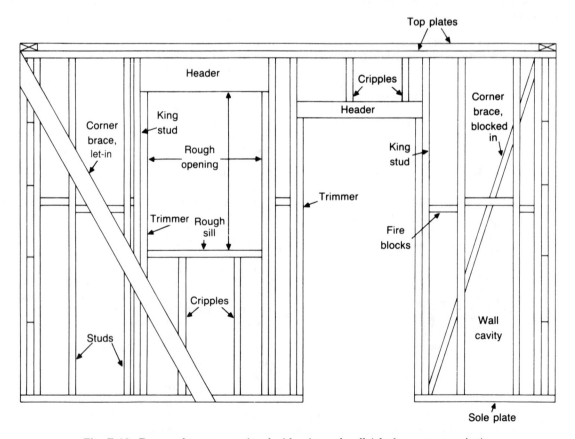

Fig. F-10. Parts and terms associated with a framed wall (platform construction)

fracture A crack or series of cracks in a masonry wall or concrete slab caused by impact or strain.

frame To assemble all the various structural members into a rigid shell, ready to receive siding, roofing, plumbing, interior finish, etc.

frame building A building having all or almost all of its structural components made of wood.

framer See *framing carpenter*.

framing 1. The lumber, including roof sheathing and wall sheathing or siding, that makes up the structural members of a building. See Fig. F-10. 2. Of or relating to the method employed in constructing a frame house.

framing carpenter A carpenter who specializes in assembling a building's rough framework, not including trim, cabinetwork, or other finish carpentry.

framing hammer See *hammer*.

framing hatchet See *hammer*.

framing plan A set of drawings showing details of the framing procedures to be used in a particular building.

framing square A flat, L-shaped, metal square having one 24-inch and one 16-inch leg, and printed with various measurements and tables; used to lay out rafters, stairs, and other framing and to check inside and outside 90-degree corners. Sometimes called a rafter square.

free moisture Water that is naturally contained in wood outside the cell walls.

freestanding fireplace A factory-assembled fireplace, available in a variety of shapes and sizes, designed to be set almost anywhere on a noncombustible base, and vented through the roof with interlocking, insulated pipes. Requires clearances from combustibles of several inches to several feet, depending on the type.

freestone Any of a variety of stone types that can be easily worked into designs with a chisel.

freezeproof faucet A self-draining hose bibb having a long body and valve stem that allows the water flow to be stopped back inside the warmer areas of the house or crawl space, lessening the chance of the hose bibb freezing. See Fig. F-11.

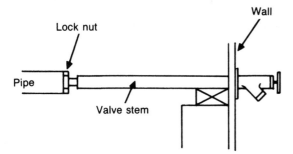

Fig. F-11. Freezeproof faucet

freezing point The temperature at which a liquid freezes into a solid; for water, 32° F, 0° C.

French curve A flat drawing template consisting of a series of concave and convex curves of various sizes; used for laying out curved lines.

French door See *door*.

French drain See *drain tile*.

French polishing A method of finishing wood in which shellac is applied with a cloth bag or pad, using linseed oil as a lubricant.

French window See *door: French door*.

freon See *cooling system: refrigerant*.

frequency 1. In alternating current, the number of cycles per second; often abbreviated Hz, for Hertz, the basic unit of frequency equal to 1 cycle per second. 2. The rate at which sound-activated air molecules vibrate; equal to the number of wavelengths of vibration passing a fixed point in one second. The greater the number of cycles per second, the more high-pitched the sound.

fresno See *bull float*.

fret saw See *handsaw*.

friction catch A small metal device used to latch a cabinet door, usually consisting of a pin or rod on the door that engages rollers or a tube on the cabinet frame; also called a roller catch.

frieze 1. A horizontal trim piece used in closed cornice construction at the angle formed between the soffit and the top of the siding. 2. A series of decorations that form a band around a wall, mantel, or other area, used for ornamentation. 3. The horizontal band between the architrave and cornice of a building, usually decorated and sometimes containing windows. See Fig. C-4.

frieze block A wooden block used between the rafters at the exterior of the plate line in an open cornice to seal off the attic from the outside. See Fig. O-2.

frieze vent A rectangular sheet-metal frame covered with wire mesh used between rafters at the plate line in place of a frieze block to allow for the entrance of outside air into the attic. See Fig. O-2.

froe A wedge-shaped sharpened steel cutting tool, having a wooden handle at right angles to the blade. Used primarily for splitting or shaving wood into smaller or thinner pieces.

frog A depression or groove in the face of a masonry unit; designed as a key to receive mortar or to reduce weight.

frontage A property line that borders a street or other particular feature, such as a stream or lake.

front elevation A drawing showing the front face of a building; one of the elevation drawings in a set of house plans that depict all faces of a building.

front plate The metal plate that screws to the edge of a door to hold the latch unit in place.

frost line The maximum depth to which frost penetrates the soil. This line varies in different parts of the country and is an important consideration in the placement of footings, pipes, and other below-grade installations.

fuel magazine See *hopper*.

full-frame An older method of house framing in which all structural members are joined with mortise-and-tenon joints.

fuming   Of or relating to a method of staining oak lumber by using a special fuming stain, which produces a grayish, antique finish.

fungicide   A chemical that is poisonous to fungi; used in treating wood to eliminate or prevent fungus growth.

fungus, wood   A microscopic plant that exists in damp wood and is responsible for mold, stain, and decay.

funnel-cup plunger   A plunger having on its bottom a long, tapered lip that is specifically designed to seat against the outlet at the bottom of a toilet bowl for clearing clogs.

furnace, combustion   A device that takes in a particular type of fuel—usually gas, oil, coal, or wood—mixes it with air to support combustion, then converts the air/fuel mixture into heat by burning it; it is also responsible for transferring the heat into the distribution system and for getting rid of exhaust gases. Technically referred to as a primary conversion unit.

furnace, electric   A device that uses electric resistance elements to create heat. Air is pulled into the furnace, heated, then fan-forced into the distribution system.

furnace-repair cement   See *adhesive: refractory cement.*

furring   1. Strips of wood or metal used to level a surface and provide a nailing base for another surface. 2. Similar strips used to create a space for the installation of thermal or acoustical insulation.

furring nail   See *nails, common types.*

furring tile   A grooved tile, usually 12 inches square and 1 to 4 inches thick, that is bonded to a masonry wall to create an airspace and to provide a backing for plaster.

fuse   An electrical overcurrent protection device, used instead of circuit breakers in some types of service panels and subpanels and containing a metal strip that is thick enough to withstand the amperage for which it's rated, but that will burn out if subjected to excessive current. Common fuse types follow. See Fig. F-12.

CARTRIDGE FUSE   A tube-shaped fuse used for higher amperages. There are two basic types, both designed to be clipped into a pull-out fuse holder. The *ferrule type* has a metal cap at each end and is usually available in 15- to 60-amp ratings. The *knifeblade type* has a rectangular metal leg at each end and is available in ratings of 60 amps and up.

SCREW-IN FUSE   A type of fuse that screws into a threaded holder

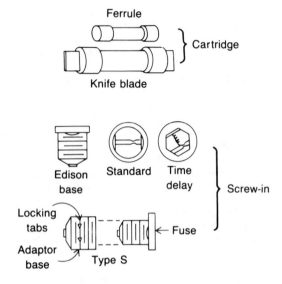

Fig. F-12. Standard types of fuses

and has a small window on top to check the metal fuse link. A fuse burned out by an overload will leave a clear window, while one burned out by a short circuit will cloud up. There are two types of screw-in fuses, both available in sizes from 15 to 30 amps. In the *Edison-based* type, all amperages have the same size of base. The *time-delay* type is designed to permit a brief overload, such as that which occurs when a heavy appliance starts up, without burning out. The *Type S* type is now recommended by most codes instead of the Edison-based type and it has a nonremovable adaptor that is screwed into the fuse socket before the fuse, thereafter permitting the use of only one specific amperage of fuse for that circuit. It is also called a fustat.

fuse box   An older style of electrical service and distribution panel, in which fuses of different amperage ratings are used instead of circuit breakers to protect individual branch circuits.

fustat   See *fuse: screw-in, type S.*

G

gable   In a double-sloped roof, the area of the wall above the plate line and between the two roof slopes. In a single-sloped roof, the area of the wall between the roof slope and a horizontal line projected from the lowest elevation of the roof structure. See Fig. R-8.

gable end   An end wall having a gable. See Figs. C-13, R-3, and R-8.

gable fan   See attic *ventilation*.

gable mold   See *shingle mold*.

gable roof   A roof structure that inclines upward from two sides of a building; as opposed to a hip roof. See Fig. R-8.

gable vent   A sheet-metal or wood frame covered with wire mesh, placed at the top of a gable end to provide an exit for stale attic air.

gage   See *gauge*.

gain   1. A notch cut into a piece of wood for the purpose of receiving a hinge or other piece of hardware. 2. A notch or slot made across the grain of a timber for the purpose of receiving another timber.

gallons per minute   A measurement of the flow rate of a liquid, especially water. Abbreviated gpm.

galvanic action   See *galvanic corrosion*.

galvanic corrosion   A type of corrosive electrical/chemical reaction that occurs between two dissimilar metals in the presence of a liquid.

galvanized   Of or relating to steel that has been dipped or plated in zinc to protect it from atmospheric corrosion.

galvanized nails   See *nails, finishes*.

galvanized steel pipe   Steel plumbing pipe that has been dipped in a galvanizing solution to coat it inside and out for better protection against corrosion.

galvey   Shortened form of galvanized, most commonly used in reference to galvanized nails.

gambrel roof   A modified type of gable roof structure in which the slope is broken into two distinct sections: the upper section slopes gradually from the ridge, while the lower section drops off sharply. Used to provide greater sectional area to a home's second floor. See Fig. R-8.

gang-nail plate   Heavy-gauge sheet-metal plate available in various sizes and shapes and perforated so as to form a series of numerous prongs; applied under pressure to secure the butt joints in a manufactured truss. Also called truss plate. See Fig. T-7.

garage door   See *doors*.

garnet paper   See *sandpaper*.

gas   Any of a variety of gases used for fuel in gas-fired appliances, including natural gas, liquefied petroleum gas (LPG) in a variety of phases, manufactured gas, or a combination of these gases.

gaseous discharge light   A fixture that creates light by passing electricity through a tube containing one of a variety of gases, including fluorescent, neon, sodium, and mercury vapor.

gasket   A sheet of rubber, cork, or other material used at the joining of two parts in order to render the joint leakproof. See Fig. B-5.

gateleg table   A type of table having hinged leaves that hang down next to the table when not in use, and that, when needed, can be swung up and supported by legs which swing out from the main table frame.

gate pier   A reinforced post upon which a gate in a fence is hinged; also called a gate post.

gate post   See *gate pier*.

gate valve   A type of water valve having a tapered wedge at the end of a stem which fits down into a corresponding slot in the valve bottom, allowing water to flow straight through with little obstruction. These valves are designed to be used fully open or fully closed and should not be used to regulate flow. Used primarily for main shutoffs and on machinery and large appliances.

gauge   1. The thickness of sheet metal and certain other materials. 2. The diameter of steel, copper, and other types of wire. 3. A simple carpentry tool consisting of a short bar, a fixed or adjustable stop, and a metal or pencil point; used for marking a line parallel with the edge of a board. See Fig. G-1.

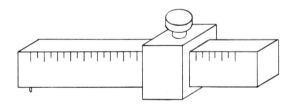

Fig. G-1. Gauge

gauged brick   See *brick*.

gauging   Cutting a stone or other piece of masonry to make it uniform in size.

gazebo   A small structure with open or latticework sides and a slatted or solid roof; designed as an outdoor sitting area with an expansive view of surrounding landscape.

gel-coat   A specially formulated resin used as the final, outer coating on fiberglass. The resin dries smooth and hard, and can be tinted to a variety of colors.

general contractor   An individual engaged in the construction and/or alteration of buildings that house people, animals, or property, particularly when the construction requires the use of three or more unrelated trades. The individual also oversees the work; arranges for materials, subcontractors, permits, and inspections; sets the schedule for the job; and deals directly with the owner.

general engineering contractor A contractor who constructs or alters fixed engineering works such as dams, roads, and pipelines.

general-purpose circuit See *branch circuit.*

Georgian 1. Of or relating to a furniture style popular during the reigns of the first four Georges of England (1714 to 1830). 2. Of or relating to a style popular during the reign of King George V of England (1910 to 1936).

GFCI Abbreviation for *ground fault circuit interrupter,* a special type of electrical device, available as either a receptacle or a circuit breaker, designed to monitor the current entering and leaving the circuit or receptacle it protects. If the device detects a difference in the current, indicating an electrical leak (a ground fault), it will immediately open the circuit. GFCIs are designed to trip in 1/40th second if a ground fault of as little as 0.005 amps occurs.

GFI See *GFCI.*

gimlet A small tool with a screw tip, a spiral cutting edge, and a bar handle; used for boring small holes.

gingerbread Informal term for the detailed ornamentation added to the exterior of some styles of houses, especially Victorian.

girder A large wood or steel beam used to support concentrated loads at various points along its length; in residential construction, a member supporting the floor joists or subfloor. See Figs. B-2, P-2, and P-3.

girder hanger See *timber connector.*

girder truss A type of flat prefabricated truss, typically with wood top and bottom chords and wood or metal inner webbing, used as a girder or floor joist, especially for heavy loads or long clear spans. See Fig. G-2.

girth The distance around something, particularly a round object.

girt strip See *ledger.*

glass The transparent end product of the combining and heating of silica sand, lime, and soda, with small amounts of other ingredients. Glass is broken down into four quality grades.

AA-GRADE GLASS The highest quality of glass; used only for special applications.

A-GRADE GLASS A superior grade of glass.

B-GRADE GLASS A common grade of glass; used for most general glazing.

GREENHOUSE-GRADE GLASS A glass used for nonwindow applications.

glass block A thick, hollow block of opaque glass built into a masonry wall to admit light.

glass cutter A small hand tool with a rotating metal wheel; used to score a line on a piece of glass along which the glass will break cleanly.

Glauber's salt See *eutectic salt.*

glaze A hard, smooth, surface finish that is fused to various clay products under heat.

glazed brick See *brick.*

glazed door See *door.*

glazier The traditional name for a person who fits and installs glass.

glazing 1. Installing glass into window frames and doors. 2. The panes of glass that are installed into the frames and doors. See Fig. W-3.

glazing compound A puttylike plastic substance used to install and seal a pane of glass into a wood or metal frame. See Fig. W-3.

glazing points Small, triangular-shaped metal pieces driven into the sides of a wood-frame window to hold the pane of glass before and during puttying.

glide 1. A thin plate of plastic, nylon, or similar material placed on the faceframe of a cabinet or piece of furniture and over which a drawer slides; used to prevent the drawer from wearing down the wood of the frame. 2. A metal or plastic button attached to the bottom of a furniture leg to enable the furniture to slide easily on the floor.

gliding window See *window: sliding window.*

glitter Small, light-reflecting flakes of a gold or silver metallic material; applied to an acoustic ceiling while still wet for decorative effect.

globe An enclosure used to protect a lamp, and possibly to diffuse, direct, or color the light coming from the fixture.

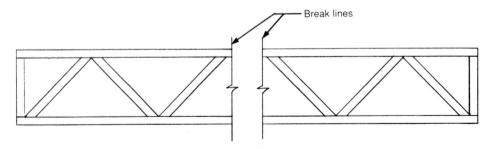

Fig. G-2. Girder truss

globe valve A type of water valve containing a washer at the end of a stem, which closes down onto a corresponding seat in the valve body, and baffles inside the valve, which force the water flowing through the valve to make two sharp turns, thus slowing down the flow. Used primarily as fixture shutoffs and in areas where control of flow is desired.

gloss A surface finish, such as paint or varnish, that reflects a large amount of light.

glossy Shiny.

glue See *adhesive.*

glue block A small wooden block, square or triangular in section, that is glued into a right-angle butt joint for reinforcement; also called a corner block.

glue pot A double pot in which certain types of glue are melted prior to application.

glulam Large straight or curved beams constructed by laminating multiple layers of lumber together under pressure; short for glue laminated.

glyph In architecture, a decorative, vertical groove or channel, such as those in a column.

G-max See *guaranteed maximum cost.*

gold leaf A paper-thin sheet of gold, called a leaf, applied to a surface for decorative effect.

good faith With regard to contracts or construction agreements, a term indicating the absence of premeditated fraudulent intent.

Gothic arch A type of architectural arch that is typically high, narrow, and pointed.

gouge A type of woodworking chisel having a concave cutting edge with an inside or outside bevel.

gpm Abbreviation for *gallons per minute.*

grade 1. The slope or level of the ground surrounding a structure. 2. A designation of the quality of a piece of lumber. 3. To clear, level, or otherwise prepare a piece of land for construction. See Fig. C-15.

grade floor window Any window having a sill height of 44 inches or less above or below the level of the grade adjacent to that window.

grade stamp A stamp of inspection and grade placed on a piece of finished lumber showing the quality, species, and other information about that piece.

graduations Equally spaced marks etched or painted on a scale or measuring gauge.

grain 1. The direction of the cells or fibers in a piece of wood. Structural wood is cut so that the grain runs parallel to the length of the piece. 2. The appearance or pattern of the cut surface of a piece of wood, influenced by how the piece was milled and the size and arrangement of the wood fibers. See Fig. G-3.

ANGLE GRAIN Grain on lumber in which the annual rings are at an angle of approximately 45 degrees to the face.

CLOSE GRAIN Grain on wood having no pores in its cellular structure, commonly softwoods.

CROSS GRAIN Grain on lumber in which the fibers are not parallel to the length of the piece.

EDGE GRAIN Grain produced when the log is sawn in half or in

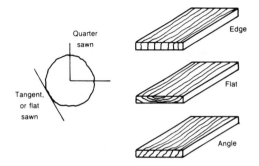

Fig. G-3. Grain types

quarters lengthwise, then the lumber is cut from those exposed faces, causing the rings to form an angle greater than 45 degrees to the face of the board; also referred to as quartersawn. This method of sawing is more expensive and wasteful, but it produces lumber that swells and shrinks less and is not as likely to warp.

END GRAIN Wood fibers that have been cut perpendicular to their length; seen on the end of a piece of lumber, and having little holding power for fasteners.

FLAT GRAIN Grain on lumber that has been sawn tangent to the annual rings so that the rings form an angle of less than 45 degrees to the surface of the board. For hardwoods, often called flat-sawn or plain-sawn. Sawing with this method produces minimum waste and a desirable grain pattern.

OPEN GRAIN Grain on certain types of hardwoods, such as oak, walnut, and ash, that have pores in their cellular structure which appear on the exposed face and usually require additional steps when applying a finish. Also called coarse grain.

graining Using paint or other finishing materials to artificially simulate wood grain on a surface.

graining comb A wooden or metal comb having a number of long, slender, parallel teeth. Used by painters and woodworkers for creating imitation wood grain patterns in painted surfaces.

grain raising Areas of short fibers on the surface of bare wood that swell when contacted by water and other liquids.

gram In the metric system, the basic measuring unit for weight and mass; equal to 15.43 grains, the smallest quantity measure used in the United States.

grandfather An informal or slang term, referring to any structure, component, code, ordinance, or procedure in use prior to the adoption of a current law or building code, and which is therefore allowable for continued use even though it does not comply with current standards. Also grandfathered or grandfathered in.

granulated slag Rapidly cooled blast-furnace slag in small, coarse pellets 1/2 inch or less in diameter; used in various commercial applications.

graph paper Paper that has been ruled into regular squares; used for sketching, doing layouts, and drawing and transferring curves.

grass cloth See *wallpaper.*

grate A heavy, metal frame used inside a coal- or wood-burning appliance, and upon which the fire is built.

gravity circulation Of or relating to a heating system that relies on the natural rising of warm air to bring heated air from a basement or underfloor furnace into a building; also used with certain types of hot-water heating systems. Also known as thermal circulation.

gravity furnace A furnace designed to be placed in a basement, having ducts that slope up to the floor above, allowing warm air to rise naturally into the building.

gray water Nonpotable water.

grease trap A device designed to separate and retain grease from the normal discharge of a building sewage system; sometimes used on the waste line from a kitchen sink to prevent grease from entering a septic tank, where it would then remain without being dissolved.

green Lumber that has been freshly cut and not yet dried; it has a moisture content in excess of 20 percent.

green board See *waterproof plasterboard.*

Greenfield See *conduit, flexible.*

green firewood Relatively wet firewood from a freshly cut tree. Green firewood has less heat capacity than dry firewood, since much of the heat energy produced in combustion is used up in evaporating the moisture out of the wood as it burns.

greenhouse A separate building or a room attached to the side of a building, having walls and roof constructed totally or in large part of glass, fiberglass, or clear plastic, in which sunlight is trapped as a means of providing heat for plants or for circulation into the building.

greenhouse effect The basic principle behind passive solar heating stating that sunlight, in the form of shortwave radiation, will pass through glass to enter a room or collector and once inside will be absorbed, then reradiated as heat, which is longwave radiation and will not pass through glass, thus becoming trapped inside. If thermal mass is provided, that trapped heat can be absorbed and stored in the mass for later use. See Fig. G-4.

green plate See *lumber: pressure treated.*

green terminal A screw terminal on an electrical device that is green in color and intended as the point of connection for a ground wire. See Fig. G-6.

green wire In an electrical circuit, the wire wrapped with a green insulating jacket and universally intended for use as a ground wire.

grill A finned metal faceplate used over the opening of a duct where it enters the room. Grills have no louvers for stopping or regulating air flow and are mostly used over cold-air return ducts.

grip That portion of a tool or other device intended to be held in the hand during use. Tool grips, such as those for a hammer, are often scored, rubber-coated, or otherwise constructed to afford the user a comfortable, nonslip grip. See Fig. H-1.

grommet A metal ring that is placed in the edge of a piece of canvas, leather, or other similar material to reinforce a hole.

groove 1. A rectangular slot cut with or across the grain of a piece of wood. 2. To plough or dado.

groover A type of steel trowel having a formed ridge down the middle; used for creating control joints in concrete.

gross 1. Undiminished by deductions; total, as distinguished from net. 2. A quantity equal to 12 dozen, or 144.

gross floor area The total square footage of a building, calculated by measuring around the building's perimeter at the exterior walls making no deductions for the area taken up by the interior walls.

ground 1. A series of electrical conductors which are designed to connect all conductive metal parts of the electrical system to the earth. All ground wires in the system are routed back to the service panel and terminate at the neutral bus bar, which in turn is attached to the earth through a

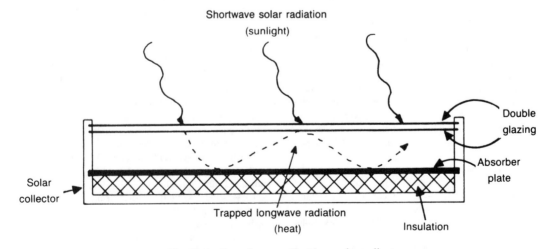

Fig. G-4. Greenhouse effect in a solar collector

coldwater pipe or through a rod buried in the earth or in the building's foundation. The ground wires are not in use during normal operations of the circuit, but when a fault current is present in the system as a result of a malfunction, the current will be carried harmlessly to the earth, preventing injury. 2. See *plaster ground*.

ground clamp A device having a hole in one side with a set screw for attachment of a grounding conductor; designed to be clamped over a water pipe or buried rod.

ground cover Thin sheets of plastic or other material used to cover the ground in a crawl space to help control moisture in that area; also called a soil cover. See Fig. C-15.

grounded Of or relating to any object that carries or uses electricity and is connected to the earth or to a conducting body, such as a water pipe, which serves in place of the earth.

grounded plug An electrical plug with a third leg for grounding. See Fig. G-5.

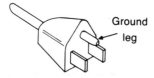

Fig. G-5. Grounded electrical plug

grounded receptacle A single or duplex receptacle having a D-shaped hole, in addition to the two rectangular slots, intended for the insertion of the grounding leg of a grounded plug, and a separate, green screw on the back for the connection of a ground wire. See Fig. G-6.

grounding adapter A device used to adapt a grounded, three-prong plug for use with a nongrounded, two-prong receptacle; also called a three-prong adapter.

grounding bushing A threaded bushing with an attached ground screw for connecting to a ground wire. The bushing is screwed over the end of a threaded metal conduit fitting to provide a means for grounding the conduit.

grounding clip A small, U-shaped metal clip designed to snap over the edge of a metal box to secure a ground wire to it.

grounding electrode An electrical conductor, usually a bare copper wire of #8 gauge or larger, used to connect a service panel to a suitable ground. See Fig. S-10.

grounding jumper A wire used to connect a metal box or device to the other ground wires in the box or device.

ground joint A connection in which two meeting faces are machined to contact each other without a washer, such as a pipe union.

groundwater Water that is present in the earth near the surface. The groundwater level varies with the season and can affect such areas as basements and septic tanks.

ground wire A conductor that does not carry electricity during normal operation, but is used to ground the metal components of an electrical system. Ground wires may be bare or wrapped with insulation that is green, but no other color.

grout Mortar that has been thinned with water to a consistency which will flow into the joints of masonry work and fill them solid; often colored for decorative effect, particularly when used with ceramic tile.

growth ring See *wood*.

guaranteed maximum cost The upper limit amount of cost for a construction project, as agreed to between the owner and the contractor. Abbreviated G-max. Also referred to as a cost ceiling or ceiling amount.

guard Any of a variety of retractable devices used on power-cutting tools to prevent operator contact with the cutter blade during operation.

guard bolt On key-activated locksets, a bolt used next to the latch bolt to prevent tampering.

gun grade Of or relating to the consistency of a material, such as a caulking compound, that is necessary to allow it to move easily through the nozzle of a caulking gun or tube.

Gunite The brand name for a process in which cement, sand, and water are mixed together and applied under pressure by a spray gun; commonly applied over wire mesh in the construction of swimming pools.

gunmetal 1. A type of bronze made from copper, tin, and zinc,

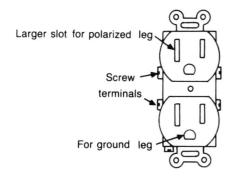

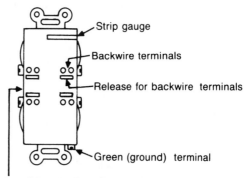

Fig. G-6. Grounded, backwired duplex receptacle

commonly used in the making of gun barrels and ship fittings. 2. Anything of a gunmetal or dark gray color.

gusset　A wood or metal plate placed over the intersection of two or more members, and bolted, glued, or otherwise fastened in order to strengthen the joint; commonly used in truss construction. See Fig. T-7.

gutter　A wood, metal, or plastic channel attached to the fascia to collect water runoff from the roof; also called an eave trough.

gutter-and-lap sealant　See *sealant: silicone sealant.*

gutter hanger　A metal strap designed to be hooked over the front and rear faces of a gutter, then nailed to the roof

sheathing before installation of the roofing material.

gutter spike　A long nail, used in conjunction with a metal ferrule for hanging gutters. The spike is driven through the face of the gutter at the top, through the ferrule, then through the back of the gutter into the fascia.

gypsum　Soft, hydrous calcium sulfate; used in a variety of building materials, such as plasterboard and plaster.

gypsum board　See *drywall.*

gypsum lath　Sheets of perforated gypsum board, usually 3/8 inch thick, applied to interior wood-frame walls as a base for plaster.

gypsum wallboard　See *drywall.*

H

hacksaw　See *handsaw.*

haft　The handle of a tool, particularly a cutting tool such as a file or chisel

hairline　A very fine crack in paint, wood, plaster, or other surfaces

half bat　One half of a building brick.

half bath　A bathroom that contains a sink and a toilet, but no shower or bathtub.

half-height block　See *building block.*

half-lap　Of or relating to a joint used to splice two pieces of lumber by removing half the thickness of each piece, then overlapping the pieces to form a flush joint. See Fig. J-1.

half-louver door　See *door.*

half newel　A newel post that has been sawn in half lengthwise, used against a wall as the termination for a handrail.

half-round　Of or relating to a piece of molding having a half-circular cross-section; primarily used to cover the joints in two adjoining panels. See Fig. M-4.

half section　A type of drawing in which an object is divided in half, with one half depicting the object's exterior, and the other a section view of its interior.

half story　A room with sidewalls approximately 4 feet in height that blend into a sloped ceiling, as in an attic.

half-timbered　Of or relating to a building having a heavy timber framework and the intervening spaces filled in with brick or stone.

half-unit block　See *building block.*

hammer　Any of a variety of hand tools designed to drive nails and other fasteners, remove nails, tear out boards, and perform many other tasks. Hammers differ by their intended use and include the following.

BRICK HAMMER　A tool having a squared hammer face and a long, slender, gently tapered chisel in place of a claw; used for cutting bricks.

CASE HAMMER　A steel tool having a head that combines a hammer face, a hatchet blade, and a curved nail claw. Also called a case opener.

CLAW HAMMER　A type of hammer used for general woodworking and consisting of a slightly crowned face for driving nails and a claw for pulling nails and removing boards. Common weight is 16 ounces. Also called nail hammer. There are two types of claw hammers. The *curved claw hammer* curves down considerably from the head, providing good leverage for pulling nails. The *ripping claw hammer* tapers down and away from the head in an almost straight line and is designed for wedging in between boards to pry them apart. See Fig. H-1.

FACING HAMMER　A special type of beveled, toothed hammer used to dress stone.

FRAMING HAMMER　A heavy, long-handled hammer designed primarily for wood framing and other heavy carpentry and having a milled, checkered face for more positive contact with the nail head, and a long, slightly curved claw. Common weights are from 18 to 32 ounces.

FRAMING HATCHET　A type of framing hammer having a hatchet blade on one end of the head instead of a claw.

LATHING HAMMER　A type of hammer having a nailing face on one side of the head and a small hatchet blade on the other, used for cutting and attaching wood lath prior to plastering.

MALLET　A type of hammer designed primarily for driving another tool, such as a chisel, and having two broad striking faces; usually made of wood, plastic, rawhide, or rubber.

NAIL HAMMER　See *hammer: claw hammer.*

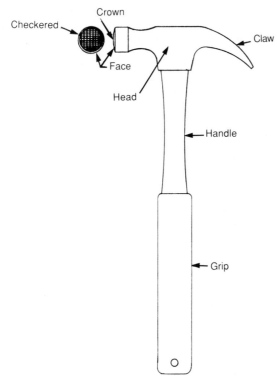

Fig. H-1. Claw hammer

PEEN HAMMER A type of hammer in which the peen, or the end of the hammer's head opposite the striking face, is available in a variety of shapes for hammering indentations in metal. There are three common types. The *ball peen hammer* has a rounded peen similar to half of a ball. The *cross peen hammer* has a wedge-shaped peen that is perpendicular to the hammer's handle. The *straight peen hammer is* similar to a cross peen, except that the wedge is parallel to the handle.

SLEDGE HAMMER A heavy hammer with a broad face on both sides of the head, available with long or short handles and in weights typically ranging from 2 to 20 pounds; used for heavy jobs, such as breaking masonry and driving wedges for splitting wood.

SOFT-FACE HAMMER A hammer having one or two faces made of rawhide, rubber, copper, plastic, or similar materials that are softer than steel; used to assemble and disassemble parts, form soft metals, and perform other tasks where the surrounding surface must be protected.

TACK HAMMER A lightweight hammer with a tapering, rounded head; used for driving tacks, small finish nails, and other light work. Some types are magnetized to hold the tack prior to driving.

hamper cabinet A cabinet that contains a tilt-down or roll-out receptacle for dirty clothes; used in a bathroom, bedroom, or laundry area.

hand The specification of which way a door opens. See *door* and Fig. D-5.

hand breadth An informal measurement using the width of a hand as reference; equal to about 4 inches.

handrail A continuous railing along a flight of stairs provided for the support of any person using the stairs. If used along an open set of stairs, also referred to as a stair rail; if used on a closed set of stairs, also referred to as a wall rail. See Fig. S-18.

handrail bolt A special type of bolt having a wood thread on one end and a machine thread with a locknut on the other and set into a groove on the underside of a handrail to draw together and lock the joint between two sections.

handsaw Any of a variety of saws intended to be operated by hand, with no other power source. Handsaws differ by their intended use, and include the following. See Fig. H-2.

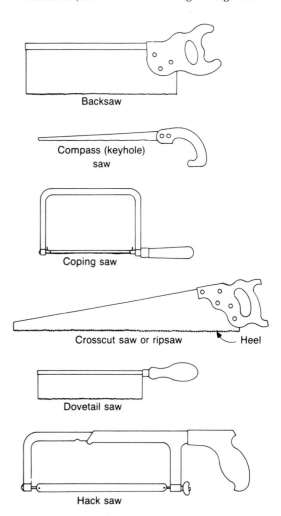

Fig. H-2. Common types of handsaws

BACKSAW A type of handsaw, used primarily for crosscutting fine finish materials, such as wood moldings, having a blade with a metal strip along the back for stiffness, and ranging from 10 to 26 inches in length, with 11 to 16 points per inch. The longer-bladed models are designed for use with a miter box.

BAYONET SAW A type of handsaw having a tapered blade that comes to a point, similar to a keyhole saw. The blade can be thrust through soft materials to start a pocket cut or can be started in a predrilled hole.

COMPASS SAW A handsaw having a blade that tapers to a point; used for cutting holes, openings, and curves, and for general use in confined places. Blade sizes range from 8 to 14 inches in length and have 8 or 10 points per inch.

COPING SAW A type of saw for wood having a deep-throated frame and thin, replaceable blades; used for very fine, accurate cuts and for tight curves and scrollwork. The blades range from 6 to 6-1/2 inches in length, with 10 to 20 points per inch.

CROSSCUT SAW A type of saw designed to cut across (perpendicular to) the grain in a piece of wood. Blade lengths vary from 20 to 26 inches, with 7 to 12 points per inch.

DOVETAIL SAW A type of fine-toothed handsaw designed for cutting dovetail joints and for other fine cuts in wood and having a reinforced blade, like a backsaw, but somewhat smaller.

DRYWALL SAW Any of a variety of saws having large, widely spaced teeth and designed specifically for cutting drywall.

FRET SAW A small, very fine cutting handsaw used for intricate finish work; originally designed for cutting frets in musical instruments.

HACKSAW A type of handsaw used specifically for cutting metal and consisting of a rigid metal frame that is adjustable for different blade lengths and a replaceable saw blade. Blades range from 8 to 12 inches in length, with 14 to 32 points per inch.

KEYHOLE SAW A saw with a small, tapered blade, originally designed to cut keyholes in wood doors; very similar to a compass saw, but slightly smaller.

RIPSAW A handsaw designed for cutting wood parallel to the grain. Most ripsaws have a 26-inch blade, with 5, 5-1/2, or 6 points per inch.

hand screw A clamping device having two threaded wooden jaws and two activating screws, one from each side, which can be adjusted independently, permitting a variety of angles between the jaw faces. Also called an adjustable gluing clamp. See Fig. Fig. H-3.

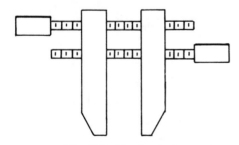

Fig. H-3. Hand screw

hand tool Any tool that is guided and operated by hand, without assistance from any other power source.

hanger bolt A type of metal fastener having a pointed, wood screw thread on one end and a machine screw thread on the other end; also called a breadboard screw.

hardboard See *building board*.

hardboard siding Hardboard sheets or individual strips manufactured to resemble natural wood; used for various exterior siding applications.

hard-burned Of or relating to clay products that have been fired at high temperatures, resulting in a low rate of absorption and high compressive strength.

hardpan A heavily compacted soil, usually containing a large amount of clay, that is almost cementlike in hardness and difficult to dig.

hardwall plaster See *plaster*.

hardware cloth Flexible metal screening, commonly sold in rolls in various mesh sizes; used for covering vents and other applications.

hard water Water that contains calcium, magnesium, iron, or any of a variety of other minerals which do not dissolve, but are instead held in suspension in the water. The hardness of water is often determined by the grains of calcium carbonate in a gallon of water, ranging from less than one (soft) to more than ten (very hard).

hardwired Of or relating to an electrical appliance or machine that is wired directly and permanently into an outlet; for example, an oven.

hardwood A general term referring to lumber produced from broad-leaved, deciduous trees, but having no bearing on the actual hardness of the wood. Some of the more common hardwoods follow.

ASH A hard and heavy wood, with good impact resistance; common in furniture, cabinets, baseball bats, and oars.

BALSA The lightest and softest of all woods; often used in making models.

BASSWOOD A soft, straight-grained, light-colored wood that is easily worked; used for general woodenware, drawing boards, and furniture. Also called linden.

BEECH A wood that is hard, close-grained, heavy, and pale brown in color; used for chairs, trim, woodenware, and plywood.

BIRCH A heavy, strong, hard wood that is fine grained and easy to finish; often used in cabinets, furniture, doors, plywood, and flooring.

CHERRY A close-grained wood that has a rich red to cream color and finishes nicely; used primarily for furniture and trim.

ELM A heavy, tough, hard, and light brown wood that bends well; often used for trim and the curved parts of furniture.

IRONWOOD See *hardwood: lignum vitae*.

LIGNUM VITAE The hardest and heaviest of all wood, milled from a tropical hardwood tree; used for mallets and other tools, bushings, etc. Also called ironwood.

LINDEN See *hardwood. basswood*.

MAHOGANY A rich, reddish-brown wood that is tough and strong, but easily worked, and takes finishes well; extensively used in door, trim, and furniture manufacturing.

MAPLE A light brownish-gray wood that is hard and strong, works well, and is not prone to splits or chips; primarily used for furniture, also for shipbuilding and flooring.

OAK A very tough, hard, strong, and heavy wood that has a pronounced grain, finishes well, but is hard to work and susceptible to shrinking and warping. Common species are red oak and white oak, which appear similar, although red oak has a coarser grain. Common in furniture, flooring, plywood, and shipbuilding.

POPLAR A light, soft, straight-grained and easily worked wood that has a pale, uniform buff color; used for cabinets and plywood veneers.

WALNUT A deep brown, durable, hard, and even-textured wood that works and finishes well; used in fine furniture, flooring, shipbuilding, gunstocks, and plywood.

hardwood floor A finished floor covering made up from side and/or end-matched strips or blocks of hardwood, usually oak, and then varnished or otherwise sealed: also called strip flooring.

hardwood plywood See *plywood.*

harp A brass rod formed into a large loop and attached to a lamp to hold the lamp shade.

hasp A hinged, slotted metal strap that passes over a metal loop and to which a padlock is then secured through the loop; used for locking doors, boxes, etc.

hatchet Any of a variety of short, light axes. There are dozens of variations, designed for specific applications. Common hatchet types include the following.

DRYWALL A wood- or metal-handled hatchet designed for use by drywall installers. The head has a short, partially sharpened blade for trimming drywall, and a round, convex checkered face for driving nails slightly below the surface. The head is typically set at a slight angle to the handle to simplify drywall installation on ceilings.

FRAMING A fairly heavy, wood- or metal-handled hatchet used by framing carpenters. It has a checkered round or square face and a large, sharp blade for rough trimming of wood.

LATHING A light, wood-handled hatchet with a straight, flat-topped blade and a round or square face, used to cut and nail on wood lath prior to plastering.

SHINGLING A wood- or fiberglass-handled hatchet typically having a square face and a narrow blade sharpened on the end and bottom, used for cutting and installing wood shingles and shakes. Some types have adjustable marking pins for measuring and laying out shingle exposure.

hatchway An access opening, such as an attic or crawl space, that is covered by a hatch.

haunch A shoulder cut on a tenon.

hawk A small, square, metal or wooden plate with a vertical handle underneath; used to hold mortar or stucco while it is being applied. Also called a mortar board.

haydite A lightweight aggregate formed by heating certain types of clay until they expand and form pellets; used in concrete blocks and other pre-cast objects.

hc Abbreviation for *heating capacity of air.*

head 1. The top member of a window sash or door frame. See Fig. W-3. 2. The metal portion of a tool such as a hammer or hatchet, milled or forged into a striking surface or other configuration, and then attached to the handle. See Fig. H-1. 3. See s*tatic head.*

header 1. A structural member placed horizontally over a window, door or other framed opening to carry the load over that opening. See Figs. F-4 and F-10. 2. A member placed perpendicular to the joists to which other joists are nailed in framing around a chimney, stairway, or other opening in a floor or ceiling. See Fig. J-2. 3. A brick or stone laid perpendicular to the wall's face, exposing the end. See Fig. B-10. 4. The pipe that runs across the edge of a series of collectors, gathering or distributing the heat-transfer fluid from or to each collector. See Fig. C-6

header block See *building block.*

header hanger See *timber connector.*

header joist See *rim joist.*

header tile See *structural clay tile.*

head joint The vertical layer of mortar between bricks, blocks, or other masonry units. See Fig. B-10.

head lap The distance that the bottom of a shingle or siding board overlaps the top of the one in the row preceding it.

head room 1. The clear area between a floor and a ceiling. 2. The minimum clear area between a flight of stairs and the bottom of the floor above it. See Fig. S-16. 3. The clearance or open space in any area that is accessible, such as an attic or crawl space.

hearth The inner and/or outer floor of a fireplace, commonly made of brick, stone, or tile. See Fig. F-4.

hearth extension A noncombustible prefabricated or site-built pad used to extend an existing hearth to provide additional protection to combustibles.

hearth pad A prefabricated or site-built pad constructed from noncombustible materials, designed for use under a wood stove or pellet stove. Also called a stove mat.

hearth rug A special rug, usually rectangular or half-circular, woven from low-combustion fibers. Designed to protect the floor in front of a fireplace or wood stove from damage from sparks and hot ash.

heartwood See *wood.*

heat-absorbing glass A special type of glass that absorbs certain wavelengths of sunlight. Because it absorbs more solar heat than ordinary glass, summer heat gain in the building is reduced and more winter heat is retained.

heat anticipator See *thermostat anticipator.*

heat exchanger A series of fins, tubes, or baffles, either freestanding, such as a radiator or convector, or located within a furnace, solar storage tank, or other similar heating application. The exchanger presents a large, heat-absorbing surface, which is warmed by the heat flowing through or past it and provides more area to give off that heat to the surrounding air or water. See Fig. C-5.

heat gain The amount of heat that a space gains from warmer outside air through transmission and infiltration.

heating capacity of air The amount of heat required to raise the temperature of 1 cubic foot of air 1°F; abbreviated hc.

heating efficiency A measure of the effectiveness of a heating appliance or system. The three primary efficiency considerations follow.

COMBUSTION EFFICIENCY A ratio of the amount of heat actually available from the combustion of a fuel to the amount of heat potentially available from that fuel under ideal circumstances.

SEASONAL EFFICIENCY A ratio of the amount of heat delivered to a building during an entire heating season to the amount of fuel consumed during that season; lower than steady-state efficiency.

STEADY-STATE EFFICIENCY A ratio of the amount of heat delivered into the distribution system to the amount of heat available in the fuel. This calculation makes allowance for heat loss in the stack and other areas, and is therefore lower than the combustion efficiency.

heating load The amount of energy needed to satisfy a particular building's space-heating or water-heating needs.

heating season That period of the year, usually from early fall to late spring, when a building requires additional space heat for the comfort of its occupants.

heating, ventilating, and air conditioning Of or relating to the entirety of a building's air-handling equipment, including the furnaces, fans, humidifiers, ducts, vents, and other permanent equipment, but generally not portable fans and air conditioners or heaters that are not ducted; abbreviated HVAC.

heat loss The heat inside a building, pipe, duct, etc. that is lost to the colder outside air. The two primary types of heat loss follow.

TRANSMISSION HEAT LOSS The heat loss from heat that is transmitted from a warm area to a cold area, such as from a living room to an unheated basement.

INFILTRATION HEAT LOSS The heat loss from cold air that enters a building and the equal amount of warm air that leaves it; caused by small leaks around windows, doors, etc.

heat of vaporization The amount of heat required to change a liquid into a vapor.

heat pump A heating and cooling appliance that works on the same principles as a refrigerator or air conditioning system. In the heating mode, heat is drawn from the outside air, which even at freezing temperatures still contains considerable heat energy, and is transferred to the inside of the building through a refrigerant. When cooling is desired, the unit reverses itself and draws heat from the inside air, transferring it to the outside. See *cooling system*.

heat resistance The ability of a material to resist the passage of heat through it, the primary consideration when evaluating a material for use as thermal insulation. The terms and formulas associated with heat resistance follow.

K-VALUE The number of Btus that will pass through 1 square foot of a material 1 inch thick in 1 hour at a temperature difference of 1 degree between the surfaces. The lower the K-value, the better the material is as a thermal insulator.

C-VALUE Similar to K-value, except that it is a measure of the conductivity of the actual standard thickness of a material, such as 1/2-inch plasterboard or 3/4-inch plywood; the number of Btus that will pass through 1 square foot of a material of actual standard thickness in 1 hour at a 1-degree temperature difference between the two surfaces. The lower the C-value, the better the material is as a thermal insulator.

R-VALUE The measurement of the resistance to heat flow of a given material. The higher the R-value, the better the material is as a thermal insulator. Because R-value measures how well a material stops heat flow, it is the opposite, or reciprocal, of the C-, K-, and U-values, which measure how well a material *permits* heat flow. Therefore, the following conversion formulas may be used.

$1/K$ = R-value per inch of a uniform material.
$1/C$ = R-value of the standard form and thickness of a material.
$1/U$ = R-value of a combined building component or section.

U-VALUE A measure of the amount of heat that is conducted through the combined materials in a building component or section, such as an entire door or window assembly; the number of Btus passing through 1 square foot of a combined building component or section in 1 hour at a temperature difference of 1 degree between the surfaces. The lower the U-value, the more effective the component is in resisting heat flow. Windows in particular are commonly rated in U-value instead of R-value.

heat sink Any thermal mass that serves to absorb and store solar heat.

heat storage Of or relating to rocks, water, salt, or other materials used to absorb heat and retain it until needed.

heat tape 1. A type of tape containing small electric heating wires that can be wrapped around cold-water pipes to help prevent freezing. 2. A similar type of device attached in a zigzag pattern on top of a roof overhang to prevent the formation of ice dams.

heat transfer The natural process by which heat is exchanged between objects of different temperatures or between an object and the surrounding air. These processes include the following. See Fig. H-4.

CONDUCTION The movement of heat from molecule to molecule through a solid. For example, when the outside of a brick wall is heated, that heat is conducted through the brick to the inside of the wall.

CONVECTION The natural rising and falling of a gas or liquid medium. As the medium (air, for example) absorbs heat, it expands, becomes lighter, and rises; as it cools, it contracts, becomes heavier and more dense, and falls, creating a cycle. This cycle of natural rising and falling is called a convection current or convective loop.

RADIATION The natural flow of heat from a warm surface to a cool surface. The heat is transmitted by electromagnetic wave motions, called rays, and, unlike convection, requires no movement of air. See Fig. S-15.

heat-transfer fluid Air, water, or other medium that is used to transfer and circulate heat in a solar heating system.

heat treatment The use of controlled heating and cooling of a metal or metal alloy for the purpose of improving the metal's strength, hardness, durability, or machining characteristics.

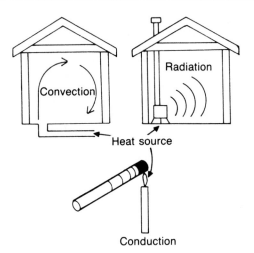

Fig. H-4. Three primary methods of heat transfer

hectare A metric measurement of land area equal to 10,000 square meters, or 2,471 acres.

heel 1. That portion of a rafter that rests on the wall plate. 2. On a crosscut rip saw, that portion of the blade directly beneath the handle. See Fig. H-2.

helix A line forming a perfect spiral, created by extending the line forward around a cylinder at a fixed rate; for example, a screw thread.

helve The handle of an ax or hatchet.

hemicellulose See *wood.*

hemlock See *softwood.*

hemp A sturdy, Asiatic plant fiber used in the making of rope and certain types of cloth.

herringbone A pattern of bricks, tiles, or wood strips in which each row is laid out diagonally and in the opposite direction from the preceding row; a zigzag pattern. See Fig. H-5.

hertz A measurement of electrical frequency equal to one complete cycle per second; abbreviated Hz.

Fig. H-5. Herringbone design

hexagon A polygon having six sides and six angles. In a regular or equilateral hexagon, each angle equals 60 degrees. See Fig. P-8.

hex head A bolt head having six sides, as opposed to a square head.

hex nut A threaded nut having six sides, as opposed to a square nut.

hickey See *conduit bender.*

hidden line A dashed line on a drawing used to represent that which is hidden below the surface of the object being drawn. See Fig. O-4.

hide glue See *adhesive.*

hiding power The ability of a paint or other coating to obscure the surface over which it is applied.

hi-early cement A portland cement or concrete that has a high, early strength.

high-limit switch A device that senses the temperature in electric, oil, and gas-fired furnaces and shuts off the supply of fuel to the furnace when that temperature reaches a certain point.

high-speed steel See *steel.*

highway mesh See *welded wire mesh.*

hinge A movable metal joint consisting of two sections joined by a pin, upon which objects such as doors and lids can open, fold, or swing.

hinge-bound Of or relating to a door that is prevented from closing completely because one leaf of its hinge has been mortised too deeply into the jamb or stile.

hinge connector See *timber connector.*

hip The external angle formed where two adjacent sloping sides of a roof meet. See Fig. R-8.

hip-and-ridge Of or relating to shingles designed specifically for installation as the final covering over a hip or ridge.

hip rafter See *rafter.*

hip roof A roof that rises from all four sides of a building; as opposed to a gable roof. See Fig. R-8.

hog wire See *reinforcing wire.*

holdfast Any of a variety of metal devices used to hold material on a workbench while the piece is being shaped. The typical design is somewhat L-shaped, with one curved leg and one long, tapering leg, designed to be tapped into a predrilled hole in the workbench. Some newer types also use a screw clamp.

hole See *lumber defect.*

holiday A slang term used to describe an area of a painted surface that was missed or only lightly coated; also called a skip, Sunday, or vacation.

hollow-back Of or relating to wide, flat trim that has the center area of the back face hollowed out to allow for a better fit against irregular plaster walls; used mainly for casings and baseboards.

hollow-core door See *door.*

hollow-door anchor See *anchor, hollow wall.*

hollow-ground Of or relating to a cutting tool having a beveled edge that has been ground concave instead of flat.

hollow wall 1. A double masonry wall with an airspace in between the two layers of masonry. 2. Any wall with an air space between the inside and outside faces; also called a cavity wall.

hone See *whet*.

honeycomb 1. Voids left in dried concrete resulting from bad pouring and tamping. 2. A reinforcement of thin sheet material, such as paper, by gluing together the paper so that it forms a series of hollow cells; used primarily between the face sheets in a hollow-core door. See Fig. D-3. 3. See *lumber defect*.

hood A type of ventilator mounted over a cooking surface, usually consisting of an exhaust fan, grease trapping screens, and a light, all in a sheet-metal enclosure. Ducted hoods vent to the outside of the building via a vent pipe; ductless units filter the air that is drawn in, releasing it back into the room. Also called a range hood.

hood mold A type of decorative molding or trim that projects above a window.

hook knife A type of knife, or knife-blade insert for a utility knife, having a relatively broad, flat blade curving into a sharply pointed hook. Used primarily for cutting linoleum, leather, and similar material.

hook nail See *nails, head types*.

hopper A chamber attached to wood- or coal-burning appliances that contains fuel for the appliance and, typically, the means for conveying that fuel to the combustion chamber. Also called a fuel magazine.

hopper window See *window*.

horizontal Parallel to the plane of the horizon; level; extending side to side. The opposite of vertical. See Fig. P-8.

horizontal boring machine A type of stationary power boring tool similar to a drill press except that the bit moves horizontally instead of vertically. Often used for boring dowel holes in cabinets and furniture.

horizontal fixture branch 1. A horizontal length of water pipe that extends from a riser to a fixture or outlet. 2. A horizontal length of waste or soil pipe that connects a fixture to a soil stack.

horse 1. A sawhorse. 2. A braced stack of timbers used to temporarily support a load.

horsepower A unit of electrical energy equal to 746 watts of power.

hose adaptor A fitting used to adapt the female end of a garden hose to the male end of a standard threaded pipe.

hose bibb A water faucet with a male thread to receive a female hose connection. See Fig. A-2.

hose clamp A type of clamp made from a slotted metal band and an attached worm gear that is turned with a screwdriver or nut driver, thus drawing up the band; commonly used to secure a hose over a metal pipe. Also called a worm clamp.

hot-dipped galvanized nail See *nails, finishes*.

hot-melt adhesive See *adhesive*.

hot mop To apply hot molten tar, usually in alternating layers with felt paper, to waterproof a roof, shower pan, or other area.

hot-mopped pan A shower pan formed by applying layers of hot tar and felt over a wood or concrete base.

hot water Water maintained and delivered at a temperature of 120° F or more.

hot-water dispenser A built-in appliance that consists of a small, undercounter storage and heating tank and a faucet, which instantly dispenses hot water at about 200° F for cooking, making beverages, and other uses.

hot-water heating system A method of providing heat to a building heating water in a central unit, passing it through pipes to individual heating units in the rooms, then circulating it back to the central unit for reheating; also referred to as a hydronic heating system.

hot wire Any wire carrying current in an electrical circuit; commonly color-coded black or red, but may be any color except white or green.

hour angle The movement of the sun in an hour, measured in degrees. For each hour from noon, the sun moves 15 degrees, equaling 24 hour angles, or 360 degrees of movement, in one full day.

house drain See *building drain*.

housed stringer A stair stringer in which vertical and horizontal grooves have been cut to receive the ends of the treads and risers.

house tightening Using any of a variety of measures, such as caulking, weatherstripping, storm windows, etc. to "tighten" an existing building and make it more resistant to air infiltration and heat loss.

house trap A large, U-shaped trap, normally with cleanout plugs at the top of both legs of the U, that is inserted in the main sewer line to serve as a trap and main cleanout for the main building drain. See Fig. H-6.

housing tract A large parcel of land broken up into individual building lots and upon which a series of relatively similar homes are constructed, usually by one building company.

hub 1. The bell-shaped end of a piece of cast-iron plumbing pipe that receives the straight end of the next pipe. The joint is then sealed, usually with oakum and lead. 2. A threaded

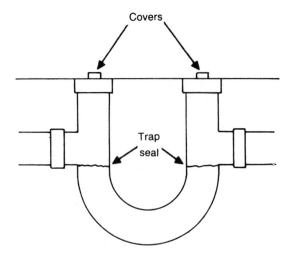

Fig. H-6. Typical house trap

fitting attached to the top of a service panel into which the service entrance conduit is attached. See Figs. S-9 and S-10.

hubless See *no-hub*.

humidifier A device that releases water vapor into a house in controlled amounts to increase humidity; available as small units for individual rooms or larger units which are attached to a central heating system to cover the entire house.

humidistat A device that reacts to changes in the humidity within a building; used to control the operation of a humidifier or air-to-air heat exchanger.

humidity The amount of water vapor in the air. See also *relative humidity*.

HVAC Abbreviation for *heating, ventilating, and air conditioning*.

hybrid solar system A solar heating and/or cooling system that is a blend of both active and passive systems.

hydrate To mix a material with water.

hydrated lime See *lime*.

hydration The chemical process in which a material dries and hardens by taking up water.

hydraulic cement Cement that will set and harden under water.

hydrocarbon A chemical compound consisting of hydrogen and carbon.

hydrometer A tubelike instrument used to measure the specific gravity or density of a liquid.

hydronic Of or relating to an object or system that uses hot water for heat.

hydronic heating system See *hot-water heating system*.

hydrous Containing water; watery.

hypalon A type of rubber-based roof coating that is abrasion resistant and fire resistant, and can be produced in numerous colors for decorative effect. It is applied wet, usually in two applications.

hypotenuse In a right triangle, the side opposite the 90-degree angle.

Hz Abbreviation for *hertz*.

I

I In electrical formulas, the symbol for current.

IAPMO Abbreviation for *International Association of Plumbing and Mechanical Officials*, authors of the Uniform Plumbing Code.

I beam A structural steel member having a cross section that resembles the letter I, commonly used horizontally to carry vertical loads over wide openings, or vertically as a post in some types of steel frame buildings.

ic Abbreviation for *internal cooling*.

ICBO Abbreviation for *International Conference of Building Officials*, authors of the Uniform Building Code.

ice dam A buildup of ice on the roof of a building over the unheated or uninsulated overhangs.

ID Abbreviation for *inside diameter*.

impact noise rating In buildings of two or more stories, an estimate of the impact sound-insulating ability of a floor/ceiling structure; abbreviated INR.

impact sound A sound that is transmitted through a building by vibrations within the structural materials.

impedance The opposition or resistance to electrical flow through certain materials; often placed in a circuit intentionally through the use of certain resistive elements, for example, in order to create heat in a stove.

impeller blade One of the small, pivoting metal arms inside a garbage disposal that act against the food to break it down and force it out into the drain.

imperial gallon A British unit of liquid measure containing 277.42 cubic inches (about 5 U.S. quarts), as compared to the American gallon, which contains 231 cubic inches (4 U.S. quarts).

impost The uppermost part of a column that supports the end of an arch.

inactive door See *door*.

incandescent light A warm, yellow light produced by passing an electrical current through a thin tungsten wire, called a filament, causing it to heat up and glow.

incinerator A device in which waste is destroyed by burning.

inclination In solar system design, the angle of the solar collectors above horizontal.

inclined Angled with respect to horizontal; sloping.

incumbrance A lien, claim, or other right or interest held against another person's real property. Alternate spelling: encumbrance.

indexing Laying out holes, slots, etc. so they are equally spaced around a circular object.

indirect gain system A type of passive solar system in which heat from the sun is stored in one part of the building, and, through natural movement, is used to heat other parts of the building.

indirect lighting   Light from a fixture that is directed onto one or more reflective surfaces, such as a wall or ceiling, which "bounces" it onto the surface being lit.

indirect waste and vent   Of or relating to a plumbing fixture, for example a washing machine, that is not directly hooked into a waste line but drains into and is vented through a floor drain, laundry sink, etc.

individual circuit   See *branch circuit*.

individual-vent   Of or relating to a plumbing fixture or group of fixtures that has its own vent through the roof; also called continuous vent. See Fig. D-14.

indoor air quality   The presence and level of concentration of potentially harmful contaminants inside a building, particularly one with poor ventilation.

inert   Of or relating to a material devoid of active properties; unable or unlikely to form compounds.

infiltration   The intentional or unintentional movement of air into and out of a building through openings in the envelope.

infiltration barrier   See *air barrier*.

infiltration heat loss   See *heat loss*.

infrared   Of or relating to light having a wavelength greater than that of the visible light at the end of the spectrum and lying in the range between visible red light and microwaves.

infrared radiation   Radiation from the sun that has wavelengths slightly longer than visible light and that is felt as heat.

inglenook   A small seat or nook next to a fireplace.

ingress   1. A place of entrance, as into a building. 2. The right of access or entrance, such as a variance granted for an entrance road.

initial set   The first stiffening or setting up of concrete or other wet-applied materials.

inlaid   Of or relating to decorative pieces that are recessed into the surface of an object.

in-line joint   An end-to-end butt joint of two pieces of lumber that is secured by additional pieces of lumber, one nailed or bolted to each side of the joint. See Fig. J-1.

inorganic   Of or relating to a material composed of matter other than animal or vegetable.

inr   Abbreviation for *impact noise rating.*

insert   1. To place or cut one material into another. 2. See *fireplace insert.*

inset   A material or decorative design that is recessed into another material.

inside stop   A narrow piece of trim placed in a groove in one of the side jambs of a double-hung window, between the upper and lower sashes.

insolation   The total amount of solar radiation, including direct, diffuse, and reflected, that strikes a surface exposed to the sky.

insulated glass   A glass unit made up of two or more sheets of glass separated by an air- or gas-filled space.

insulated steel door   See *door.*

insulating asbestos board   See *building board.*

insulating brick   See *brick.*

insulating tape   See *electrical tape.*

insulation, acoustical   Material that is used to absorb the transmission of sound waves, thus reducing noise from a room or building.

insulation baffle   Panels of wood, cardboard, plasterboard, or other materials placed over the rafters at the ceiling level to prevent insulation from clogging the frieze or soffit vents. See Fig. I-1.

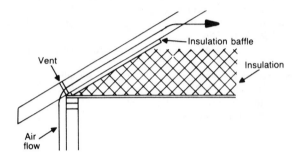

Fig. I-1. Insulation baffle

insulation, electrical   Nonconductive coatings, such as rubber or plastic, that are used to cover the outside of electrical wires, and also in the manufacture of electrical tools, appliances, and devices.

insulation, thermal   Any of a variety of materials that are high in resistance to heat transmission; used in various areas of a building to reduce heat loss and gain. The most common forms and types of thermal insulation follow.

BATTS Flexible strips of insulation in precut lengths, designed to fit between structural members. Batts may be faced with kraft paper or foil and held in place with staples, or may be unfaced and held in place by friction against the members on each side. There are two common batt materials. *Fiberglass* consists of long filaments of spun glass that are loosely woven and cut into various widths and thicknesses. *Rock wool and slag wool* consists of granite rock or furnace slag that is melted at temperatures between 2500° F and 3000° F.

BLANKETS A form of insulation similar to batts, but manufactured in rolls rather than precut lengths.

LOOSE-FILL INSULATION Various thermal insulation materials that are provided in loose, bulk form, usually in bags, to be poured, blown, or placed by hand into walls or attics. There are five common loose-fill materials. *Cellulose is* made of paper products, primarily recycled newsprint or wood fibers, which are shredded and milled into a fluffy, low-density material, then treated with fire-retardant chemicals, usually boric acid. *Fiberglass* is made of the same material as that used in batts, but left loose. *Perlite* consists of a volcanic material, expanded 4 to 20 times by heating to produce a light, cellular material in pellet form. *Rock wool* is made of the same material as that used in batts, but left loose. *Vermiculite* consists of a foliated material, having numerous thin layers, that expands under heat in an accordionlike manner to produce lightweight pellets, which may be used alone as loose-fill insulation, or mixed in plaster or concrete.

RIGID BOARDS Thermal insulation in the form of rigid sheets, often with tongue-and-groove edges, for use on roofs, walls, stem walls, and under slabs. The sheets are usually 2 x 8 feet or 4 x 8 feet, and come in a variety of thicknesses, each with a different R-value. The three most common materials used in rigid boards follow. *Extended polystyrene* consists of polystyrene beads fed into an extruder and melted into a thick fluid, which is then injected with a mixture of gases to foam the fluid into a mass of bubbles, then heat and pressure are applied and the mixture is shaped by the extruder into a board. *Molded or expanded polystyrene* is made of polystyrene beads poured into a mold, then heated, which causes them to swell to fill the mold and fuse together into a solid form. Because it is lower in density and its cells do not contain gases, it has a lower R-value and is more brittle than extruded polystyrene. Commonly referred to as beadboard. *Polyurethane and polyisocyanurate* consists of boards formed from plastic polymers in processes similar to the polystyrenes, but using somewhat different chemicals. The resulting foam sheets contain a number of gas-filled cells and offer a high R-value.

Thermal insulation is also described by the area within the building where it is located, or by its specific application. These areas and applications include the following.

CEILING INSULATION Insulation, usually batts or loose fill, installed above the finished ceiling and designed to prevent heat flow into an unconditioned attic. See Fig. I-1.

DUCT INSULATION Batts or blankets wrapped around heating supply and return ducts to reduce heat loss out of the ducts.

FLOOR INSULATION Insulation, usually batts, installed beneath the building's first conditioned floor area to prevent heat loss into an unconditioned basement or crawl space.

FOAMED-IN-PLACE INSULATION A mixture of urea-formaldehyde resin, a foaming agent/catalyst, and compressed air that is forced through a hose into small holes drilled in walls, where it then fills the wall cavities and cures into a solid mass; used for retrofit insulation in existing residential wall cavities.

PERIMETER INSULATION Rigid sheets or batts placed over a concrete or concrete-block stem wall underneath the building.

PIPE INSULATION Foam jackets or tape used to wrap plumbing pipes in unconditioned areas of a building.

REFLECTIVE INSULATION Thermal insulation with one or both sides covered with a surface having low heat emissivity, such as aluminum foil.

SLAB INSULATION Rigid sheets of insulation installed beneath a concrete slab on grade.

WALL INSULATION Batts, rigid boards, blown loose fill, or foamed-in-place insulation placed into wall cavities to prevent heat flow between conditioned areas and unconditioned areas or the outside.

WATER HEATER JACKET A fiberglass or foam jacket used to cover a water heater tank.

insulator An object of nonconductive material, such as wood or glass, used to support a current-carrying conductor.

intergrown knot See *lumber defects*.

interior chimney That portion of a chimney or flue contained within the building's interior, whether exposed or concealed.

interior finish All that goes into finishing the inside of a building for decoration and comfort, including trim, paint and other wallcoverings, carpet, etc.

interior plywood See *plywood*.

interior trim A wide-ranging term covering all the trim items inside a building, including baseboards, casings, etc.

interlocking weatherstripping A type of door weatherstripping that consists of two metal channels, one mounted on the door and one on the frame. When the door is closed, the two channels interlock to produce a very effective seal.

intermittent duty Of or relating to an electrical system or component that is designed for alternating periods of load and no load; load and rest; or load, no load, and rest.

internal 1. The inner edge, face, or surface of an object. 2. That portion of an object which is protected or concealed from view or exposure to the elements.

internal cooling Of or relating to a type of recessed lighting fixture that does not dissipate its heat upward; the fixture is thus able to be completely covered with insulation without presenting a fire hazard. Abbreviated ic.

internal mass The solar heat storage mass that is contained within a building, including walls, floors, and freestanding mass.

International Association of Plumbing and Mechanical Officials Abbreviated IAPMO. See *Uniform Plumbing Code*.

International Conference of Building Officials Abbreviated ICBO. See *Uniform Building Code*.

intersection In framing, a combination of three studs, or two studs and blocking, used where two walls intersect.

in-the-white Of or relating to natural wood to which no surface finishes, such as paint or stain, have been applied.

intrados In architecture, the inside curve of an arch. See Fig. A-7.

inverted arch A downward-curving arch having the keystone at the lowest point of the curve.

invisible line See *hidden line*.

iron 1. A metallic chemical element, symbol Fe. 2. A widely used metal created from this element. It is malleable, ductile, strongly magnetic, and rusts easily. It is commonly alloyed with other materials, most commonly in the making of steel. 3. The beveled steel cutter used in tools such as planes.

iron pipe See *steel pipe*.

ironwood See *hardwood: lignum vitae*.

irregular coursed Of or relating to odd-sized masonry units that are laid up in rows of different heights.

island A freestanding base cabinet or serving area, located in the middle of a kitchen and not connected to the other cabinet runs.

island kitchen A kitchen design that employs an island in addition to the runs of cabinets.

isolated gain Of or relating to a type of passive solar heating system in which heat is collected in one area, such as a green-house, for use elsewhere in the building.

isometric drawing A method of drafting in which all three dimensions of an object are shown like a picture; a modified version of isometric projection. Vertical lines remain vertical, while horizontal lines are drawn at an angle, commonly 30 degrees. Some adjustment of the object's true dimensions might be necessary in order to make the drawing appear more in proportion.

isometric projection A drafting technique in which the lines of a two-dimensional drawing are "rotated" and projected to form a three-dimensional drawing, with all of the object's dimensions shown in true proportion. See Fig. O-4.

isosceles triangle See *triangle*.

J

jack A mechanical device used for raising a heavy object for a short vertical distance.

jack plane A general-use plane designed for working a board to approximate shape or size.

jack rafter See *rafter*.

jack stud The trimmer that supports a header in a framed wall opening.

jalousie window See *window*.

jamb The wood or metal pieces that form the sides and top of a door or window enclosure. The vertical pieces are referred to as side jambs, or jamb legs, and the horizontal piece as the head, or top jamb. See Figs. D-6 and W-3.

jamb block See *building block*.

jambstone A dressed stone placed vertically at the side of a window or door opening so as to form all or part of the jamb.

J-bend A J-shaped piece of plumbing pipe, used along with a 90-degree elbow in making up a sink trap.

J-bolt See *anchor bolt*.

jig A template or guide used for repetitive cutting or assembly.

jigsaw A portable power tool with a short, narrow blade that operates with an up-and-down motion, perpendicular to the motor; used for cutting curves and other irregular shapes. Also called a saber saw.

joggle A shoulder on a structural member that receives the thrust from a brace or strut.

joiner A term dating back to the late fourteenth century, used to describe a woodworker skilled in making and installing wood finish materials, including moldings, stairs, windows, doors, and similar trim and finish work.

joinery The various joints used in woodworking. See Fig. J-1.

joint The area where two surfaces meet and are attached to one another with nails, screws, mortar, glue, interlocking cuts, or other such means.

joint cement A type of finish plaster, available premixed or as a dry powder for mixing with water; used to embed joint tape, cover nails, and otherwise finish plasterboard.

jointer A stationary power tool having a long, flat, two-piece table and a revolving cutter head mounted horizontally between the table halves; used for planing and squaring lumber edges and faces for more accurate assembly.

jointer plane A hand plane having a long, flat bed; used to true up lumber edges.

joint tape Paper or mesh tape used with joint cement to cover and reinforce the seams in plasterboard panels.

joint venture A construction project whereby two or more contractors jointly act in the capacity of prime contractor.

joist One of a parallel series of structural members supported by beams, girders, or bearing walls and used to support floor and ceiling loads. See Figs. B-2 and J-2.

joist bridging See *bridging*.

joist cleat A strip of wood secured to a girder to help support the ends of floor joists.

joist hanger See *timber connector*.

joist hanger nail See *nails, common types*.

J-trap A fixture trap shaped like the letter J. See Fig. B-5.

jumbo brick See *brick*.

jumper wire A short length of heavy-gauge wire clamped to the cold-water pipes on either side of a water meter, water filter, or dielectric union to ensure the use of the entire run of pipes as a ground, without interruption, when the water pipes are used as a ground for the building's electrical system.

junction box See *electrical box*.

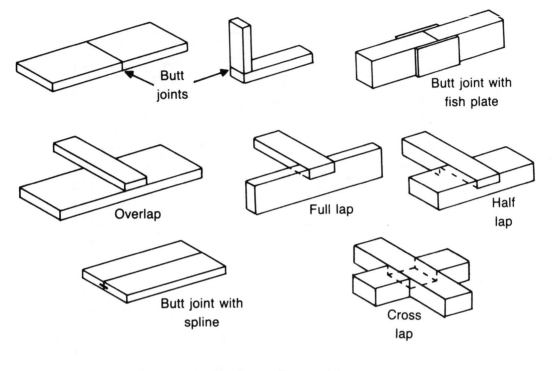

Mortise-and-tenon joints

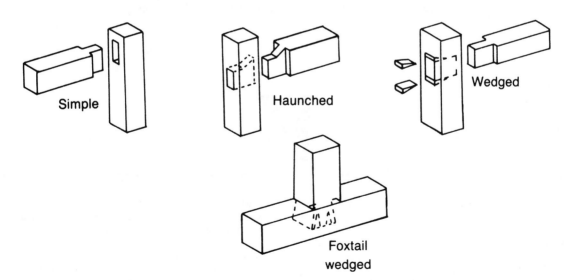

Fig. J-1. Common examples of wood joinery

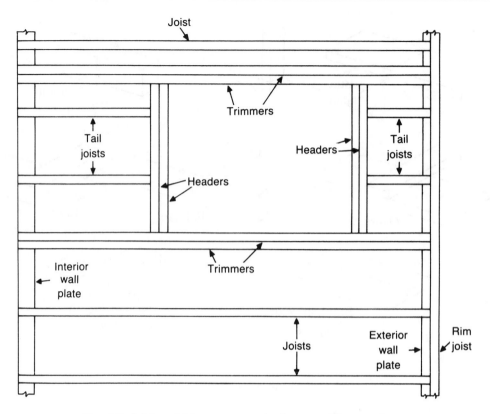

Fig. J-2. Joists and trimmers around a floor or ceiling opening

KD 1. Abbreviation for *knocked down*. 2. Abbreviation for *kiln dried*.

Keene's cement plaster See *plaster*.

kerf The groove or space created by the cutting action of a saw blade. See Fig. K-1.

kerfing 1. Saw cuts made lengthwise along a piece of wood to relieve stress and prevent warping, or for decorative effect. 2. Saw cuts made widthwise across the back of a piece of wood to allow it to bend.

key 1. That part of the scratch coat of plaster or stucco which is forced through the lath, locking the coating against the surface. 2. The grip of one coat of plaster or stucco to the coat that preceded it. 3. A small length of square metal rod fit into a slot on a shaft and a corresponding slot on a pulley or wheel; used to prevent the pulley from rotating on the shaft. 4. A formed or cast projection on a beam, foundation wall, or other structural member designed to fit into the corresponding keyway of another member for a more secure connection. 5. A groove or channel formed in a concrete footing to interlock with and support a subsequently poured concrete wall. See Fig. D-10.

keyhole saw See *handsaw*.

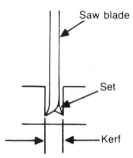

Fig. K-1. Saw kerf

keystone The central, wedge-shaped stone or brick in an arch that locks the others together. See Fig. A-7.

keyway A slot that is cut or formed into a shaft, wheel, pulley, foundation, or other object to receive a key. See Fig. D-10.

kickback The sudden and violent movement of a piece of wood, caused when it becomes trapped between the guide fence and a saw blade or other cutting surface and is thrown back by the rotation of the cutter.

kicker 1. A structural member that is used to support a roof purlin. See Fig. R-5. 2. Any relatively short supporting member.

kicker stake A concrete form stake placed so as to angle down from the top of the form board; used to provide extra support to the form and to prevent the top of the form from spreading out when filled.

kick plate A large, decorative, metal plate placed on the lower part of a swinging door to protect the door from damage when it is pushed open with a foot, cart, etc.

kiln Any of various types of ovens used to dry wood, bake bricks or pottery, etc.

kiln dried Of or relating to wood that has been dried to a constant weight in controlled artificial heat, usually 212°F to 221°F, resulting in a more uniform moisture content and less warpage than air-dried lumber. Abbreviated KD.

kilowatt A unit of measure equal to 1,000 watts. Abbreviated kW.

kilowatt-hour The unit by which electricity is metered and sold by a power company, equal to 1,000 watts of electricity consumed in 1 hour. Abbreviated kWh.

kinetic energy See *energy.*

king post The main vertical member in certain types of trusses. See Fig. T-7.

king stud The stud next to a door or window opening to which the trimmer and header are nailed. See Figs. D-6, F-10, and W-3.

kitchen layout The arrangement of residential kitchen cabinets and appliances typically in one of four standard configurations. See Fig. K-2.

knee brace 1. An angular brace that supports a horizontal member and helps to prevent racking. 2. See *timber connector.*

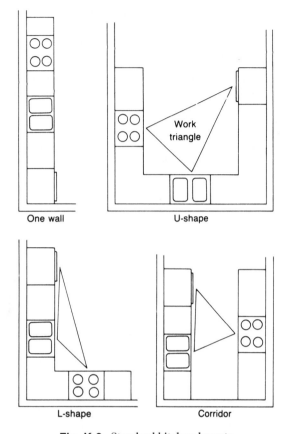

Fig. K-2. Standard kitchen layouts

knee wall A short wall used in an attic, either to provide support for the rafters or to provide vertical wall space if the area is to be occupied. See Fig. C-15.

knife grade The consistency of a material that is required to allow it to be applied with a putty knife.

knob and tube Of or relating to an older method of electrical wiring, no longer permitted for use in today's construction except to repair existing systems, that used individual conductors for hot and neutral, which were secured to the studs and joists with round porcelain insulators, called knobs, and which were protected where they passed through the wood by being run through hollow porcelain enclosures, called tubes. The conductors were either jacketed with a woven cloth insulation or left bare, and splices were made as needed and wrapped with electrical tape, but were seldom enclosed in a box. No ground conductor was used with this system.

knocked down Of or relating to an object that is prefabricated but shipped unassembled, requiring assembly according to accompanying plans or instructions. Abbreviated KD.

knockout An area on a service panel or electrical box that can be removed to accommodate wiring. Metal boxes have circular rings that are partially cut through, while plastic boxes have thin areas that can be easily broken out. See Fig. E-2.

knockout seal A plastic or metal cover that snaps or clamps into place to cover an open, unneeded knockout hole in an electrical panel or box.

knot See *lumber defect.*

knotty pine See *softwood.*

knuckles One of the loops of metal on the edge of a hinge leaf that receive the pin. See Fig. B-12.

knurl The roughed, cross-hatched area on a tool such as a nail-set or on a thumbscrew or nut designed to help keep the user's fingers from slipping. See Fig. N-1.

knurled nails See *nails, shank types.*

kraft paper A paper-making process that uses caustic soda and other chemicals to remove the lignin from the wood pulp, resulting in a very strong paper, used extensively in construction. The name is derived from the German word *Kraft*, meaning strength.

K-value See *heat resistance.*

kW See *kilowatt.*

kWh See *kilowatt-hour.*

L

label A drip cap that extends horizontally over an opening, usually with ends that project past the opening and turn downward.

lacquer A fast drying wood finish derived from tree sap mixed with resin, solvent, and other ingredients. It is usually clear but can be tinted to a variety of colors. It is usually applied by spray over a sanding or lacquer sealer.

ladder hook A large, L-shaped metal bracket that clips to the top rung of a straight ladder, allowing the ladder to be hooked over the ridge when needed for working on a roof.

ladder jack An L-shaped, diagonally braced metal bracket that hooks over the rungs of a straight ladder, used in pairs with two ladders to support a light scaffolding.

ladder rating A rating for stepladders and extension ladders based on the amount of weight they will hold. Type I (industrial) is rated for 300 pounds. Type II (commercial) is rated for 250 pounds. Type III (household) is rated for 200 pounds.

lag bolt A large, pointed, coarse-threaded screw with a square or hex head; used for heavy-duty fastening in wood or with a lag shield in masonry.

lag shield See *anchor, masonry.*

laitance The soft layer of cement and water that rises to the top of freshly poured concrete; sometimes called cream.

lally column A round, hollow steel member, sometimes filled with concrete, used to support a beam or girder.

laminate To glue two or more surfaces together, such as the veneers in a sheet of plywood.

laminated glass A type of safety glass consisting of two sheets of glass bonded to a clear inner sheet of thin plastic, offering a high resistance to impact.

lamp A device manufactured to create a source of light. Commonly called a *bulb* when used in incandescent fixtures, and a *tube* when used in fluorescent fixtures.

lampblack A fine, black soot produced by the incomplete combustion of carbon-based materials, primarily lamp oil. Sometimes used as a pigment in ink and other material.

lamp holder A porcelain fixture designed to mount directly to a variety of outlet boxes and hold a single, unshaded lightbulb. Some types contain a switch operated by a pull chain.

landing A platform between flights of stairs or at the termination of a flight of stairs. See Fig. S-17.

landing tread A stair tread used at the start of a landing, or along a level balcony.

landmark See *bench mark.*

lap The amount or distance that the edge of one object covers the edge of another, as with shingles or siding.

lap cement A liquid asphalt cement used to glue and waterproof the overlaps in roll roofing or asphalt shingles.

lap joint A connection formed between two members by overlapping them and then securing them together. See Fig. J-1.

lap siding See *bevel siding.*

latch bolt The part of a lockset that engages the strike plate to hold the door closed, and retracts when the knob is turned to allow the door to open.

latch unit A mechanism consisting of the latch bolt, guard bolt, and front plate that is inserted through a hole in the edge of a door to engage the lock mechanism.

latent defect A defect in materials or workmanship that is not immediately apparent to normal inspection, but which may appear later due to age, use, exposure, or other action.

latent heat A change in heat content that takes place without a corresponding change in temperature.

lateral force A force acting against the side of an object, such as the wind against a building.

lateral support Bracing, blocking, or other reinforcement used to counteract lateral forces.

latewood See *wood: summerwood.*

latex paint Paint blended from latex resins in a water base. See *paint resin: latex resin.*

latex resin See *paint resin.*

lath One of various strips of wood, sheets of metal or wire net, or perforated gypsum panels that are attached to a building's framework to serve as a base for plaster or stucco.

lath and plaster Of or relating to a building having interior walls finished with plaster over a wood lath base.

lathe A stationary power tool that holds a length of wood or metal at the center of each end and rotates it horizontally, allowing it to be cut into a variety of cylindrical shapes.

lathing hammer See *hammer.*

lath nail See *nails, common types.*

latitude The angular distance of a given point north or south of the earth's equator, measured in degrees along a meridian.

latitude lines Lines representing latitudes and running parallel to the equator.

lattice 1. A structure consisting of wood or metal strips that cross or are interwoven and form regularly spaced openings. 2. A thin, flat molding, rectangular in section, used to build a lattice or to conceal butt joints.

lavatory Any of a variety of sinks designed for use in a bathroom.

layout line 1. A chalk or pencil line drawn out on a board or other material for cutting, drilling, etc. 2. The chalk line on a building's floor to indicate placement of a new wall.

lazy susan 1. A corner base or wall cabinet having one or more revolving shelves for easy access to the cabinet's contents. 2. A revolving dish or tray.

leach Any process, such as that of a septic system, in which liquid is filtered through another medium such as sand or gravel and separated from solids.

leach field See *disposal field.*

lead caulking anchor See *anchor, masonry.*

leader See *downspout.*

lead-lined door See *door.*

leaf guard Strips of screen installed over the top of a gutter to keep out leaves.

leaf strainer A rounded wire device inserted into the hole in a gutter where it empties into the downspout; designed to prevent leaves and other debris from entering the downspout.

lean-to roof See *shed roof.*

ledger A strip of wood secured to vertical wall framing to support horizontal members such as joists or lookouts; also called a ribbon. See Figs. C-4 and P-3.

leeward On the side opposite from where the wind is blowing; protected from the wind. Opposite of windward.

left-hand door See *door.*

left-hand reverse door See *door.*

left-hand thread A screw or bolt that is threaded in a counterclockwise direction; commonly found on power tool arbors so that the tool's rotation tends to tighten, rather than loosen, the cutter attached to it.

lessee In a lease agreement, the lessee is the party who rents space from someone else, the *lessor.*

lessor In a lease agreement, the lessor is the party who owns and rents out space to someone else, the *lessee.*

let in To cut out an area in one member to allow for the insertion of another member. See Figs. B-2 and P-3.

let-in brace See *corner brace.*

level 1. An object or line that is exactly horizontal, having no slope. See Fig. P-9. 2. A spirit level. 3. Having a flat, even surface with no irregularities.

level cut Any horizontal cut; a horizontal cut on an angled member, such as a rafter. See Fig. R-5.

leveling board A long, straight board on which a level is placed to allow for leveling over a greater length than the level alone can cover.

leveling rod A flat bar marked in feet, inches, and fractions of an inch, usually in two interlocking pieces that can extend to 12 feet in height, and having an adjustable, locking target; used with a builder's level or transit as a target for measuring and checking heights.

leveling stake A stake or peg driven into the ground at a fixed height and used to gauge the depth of fill dirt, poured concrete, etc.

level transit A leveling instrument similar to a builder's level, except that the telescope can pivot vertically as well as horizontally, allowing for accurate sighting of plumb as well as level lines.

lewis A metal device that fits and locks into a prepared hole in a large stone and is then attached to a crane for lifting the stone into position.

lien A claim or incumbrance placed against another person's property as security for the payment of a debt.

lien release A document in which a person, usually the contractor, acknowledges payment and releases his or her right to lien.

lien waiver A document in which a person gives up his or her right to place a lien.

life-cycle cost A method of analyzing the relative costs of different appliances, heating systems, or other items by predicting the initial price, fuel use, and maintenance and operating costs for the life of each object.

light 1. A pane of glass in a window or door. 2. The space in a sash that receives a pane of glass. 3. Radiant energy falling within the range of wavelengths that stimulate the organs of sight.

light box　A decorative wall- or ceiling-mounted enclosure containing one or more light fixtures concealed behind translucent diffuser panels.

light construction　See *light framing*.

light fixture　A device that houses and operates a lamp in order to produce light. The common types of light fixtures are: recessed, wall or ceiling mounted, swag (hanging), and portable.

light framing　Conventional wood framing used in residential and small commercial buildings, as opposed to larger commercial and industrial work involving greater loads and the use of structural steel; also called light construction.

light scoop　A smaller version of the shed dormer; used to provide light and ventilation to an attic or upstairs room without substantially increasing the headroom. See Fig. R-8.

light tight　Of or relating to an enclosure that is sealed in such a manner as to prohibit the entrance or escape of light.

light well　An open shaft used to bring daylight and ventilation to interior portions of a large building that are not otherwise exposed to the outside.

lignin　See *wood*.

lignum vitae　See *hardwood*.

lime　Calcium oxide, prepared by calcining limestone or calcium carbonate. In its dry form it is known as quicklime, common lime, or anhydrous lime, and becomes slaked or hydrated lime when mixed with water. Used extensively in making cement, plaster, mortar, and many other common building materials.

lime burning　See *calcine*.

limewash　An early type of paint, made from a combination of slaked lime and water.

liming　Finishing wood, especially oak, by applying a thin coat of white sealer or thinned white paint, wiping it off while still wet, and finishing with a clear sealer. The wood is left with a faint white glow that does not obscure the grain.

linden　See *hardwood: basswood*.

linear foot　One foot of length, regardless of an object's width or thickness, as opposed to a square or cubic foot; also called a running foot.

linear measurement　A one-dimensional measurement taken along a lengthwise line without regard to width or depth.

line blocks　Small, L-shaped plastic blocks used in masonry construction that are hooked over the corner bricks. A line is stretched between them to serve as a guide for the entire course.

line level　A type of small spirit level with a hook on each end; designed to be hung on a string for establishing a level line over a long distance.

lineman's pliers　Heavy-duty pliers that are designed primarily for electrical wiring, having wide, serrated jaws, a wire cutter, and insulated handles.

line wire　A series of parallel, horizontal wires, usually 6 inches on center, that are attached to the framing of an exterior wall as supports for stucco.

linoleum　A resilient floor covering made of oxidized linseed oil and resin with pigment, applied to a felt or burlap back; usually manufactured in 12-foot-wide rolls, but also available in 6- and 9-foot rolls and as square tiles.

linseed oil　A yellowish oil produced from flaxseed and used in the making of paint, varnish, linoleum, and other products, or by itself as a wood finish. Common types follow.

BODIED LINSEED OIL　Linseed oil that has been thickened by the addition of heat or chemicals, and that ranges in viscosity from slightly thicker than oil to a near jelly.

BOILED LINSEED OIL　Linseed oil to which lead, cobalt or manganese salt has been added to make the oil harden more rapidly when spread in thin coatings.

lintel　1. A structural unit, such as a steel angle iron or precast masonry unit, used to carry the weight of masonry construction over an opening. See Fig. F-4.　2. A header.

lintel block　See *building block*.

lip　A rabbeted edge that overlaps another member, such as on a cabinet door. See Fig. C-1.

lip molding　See *wainscot cap*.

liquefied petroleum gas　A type of super-cooled, heavier-than-air propane gas; used in some types of heating systems. Abbreviated LPG.

liquid absorber plate　An absorber plate mounted in a collector for use with a liquid solar heating system, and usually consisting of tubes soldered between two metal plates or two sheets of metal bonded together to form channels.

liquid system　A solar heating system in which a liquid, usually plain or chemically treated water, is pumped through a collector to a storage tank and back in an open or closed loop. The pipes carrying the hot liquid pass through convector heaters or past a forced-air blower in order to transfer heat from the liquid into the building.

listed　Equipment that has been tested and approved by an independent testing agency and shown to meet minimum national standards.

lite　Alternate spelling of light. See *light*.

liter　In the metric system, the basic measuring unit for capacity, equal to 1.05 U.S. quarts.

live load　Any variable, movable load to which a structure may be subjected, including furniture, people, supplies, etc.

live wire　See *hot wire*.

load bearing　See *bearing*.

load-bearing tile　See *structural clay tile*.

location　The area where an electrical device is installed for operation. Locations are identified as the following.

DAMP LOCATION　A location that is partially protected, such as under a canopy or porch overhang, but still subjected to some moisture.

DRY LOCATION　A location that is not normally subjected to wetness.

WET LOCATION　A location that is underground, embedded in concrete or masonry, in direct contact with the earth, subject to saturation with water or other liquid, or completely exposed to the weather.

lock   The mechanism used to latch and secure a door. There are four basic types.

CYLINDRICAL LOCK   The most common lock type in use today, having a knob-operated cylinder that moves the latch. It may or may not be keyed.

MORTISE LOCK   An older style of lock consisting of a large, square enclosure that contains a knob-activated, spring-and-lever mechanism that moves the latch, and a separate deadbolt. The entire enclosure must be mortised into the edge of the door. Seldom used today because of cost and installation time.

RIM LOCK   A lock having a door-mounted lock case that is lever operated from the inside and key operated from the outside. The latch engages a strike plate that is mounted on, rather than in, the door frame. Also called a night latch.

TUBULAR LOCK   An older style of knob-operated cylinder lock having a square shaft that operates the wedge-shaped latch. It cannot be equipped with a key lock.

lock block   A block of wood laminated to the inside of both stiles of a flush, hollow-cored door to which the lockset is installed. See Fig. D-3.

locking stile   The stile of a paneled door to which the lock is attached; also called a lock rail.

locknut   1. In plumbing, a nut placed over one piece of a pipe or fitting that engages the threads of another pipe or fitting to lock the two together. See Fig. B-5. 2. In electrical wiring, a metal ring having internal threads and a series of projections around the outside; used to secure conduit or conduit fittings to a box or another fitting. See Fig. S-10. 3. A nut having a fiber insert that seals onto the male threads of a bolt to prevent loosening. 4. A second nut on a bolt that is tightened down on top of the first nut to prevent both from loosening.

locknut, grounded   A threaded locknut having an attached ground screw for connecting to a ground wire; used with metal conduit in place of a regular locknut in order to provide a means for grounding the conduit.

lock rail   See *locking stile*.

lockset   The complete lock mechanism for a door, including knobs, latch unit, cylinder, etc. There are three basic types of locksets.

ENTRANCE LOCKSET   A lockset that is key operated from the outside, with a button or latch inside; primarily used for exterior doors.

PASSAGE LOCKSET   A lockset that has no locking mechanism; used primarily for interior doors such as closets.

PRIVACY LOCKSET   A lockset that can be activated by a button from one side only. A release, requiring a special tool, is provided on the other side for emergencies. Commonly used on bathroom and bedroom doors.

longitude   The distance in degrees of a given position east or west of the prime meridian, which passes through Greenwich, England.

longitude lines   Lines representing longitudes and running perpendicular to the equator; also called meridians.

longitudinal   Parallel to the long axis of an object; lengthwise.

longitudinal beam   A beam that runs parallel to the long axis of a building. See Fig. O-1.

longitudinal section   In drafting, a drawing made of a section cut through the long axis of an object.

long oil resin   See *oil resin*.

long-sweep fitting   A waste-line fitting, such as an elbow, having a long radius of turn to allow a smoother flow of solid waste.

longwave radiation   See *solar radiation and greenhouse effect*.

lookout   1. A structural member running from the outside wall to the lower end of a rafter; used to support the plancier or soffit. See Fig. C-4. 2. In a gable roof, a horizontal member attached to the last two rafters at the gable end and extending out to support the barge rafter. See Fig. R-3. 3. A short wooden bracket that supports the overhanging portion of a second story.

loop vent   See *back vent*.

loose-fill insulation   See *insulation, thermal*.

loose knot   See *lumber defect*.

lot   A parcel of land upon which a structure is built.

lot line   See *property line*.

louver   One of a series of movable or fixed, angled slats set horizontally in a frame to provide ventilation while obscuring vision or blocking the weather.

louver window   A small, slatted window placed high in a gable end or at the top of a tower, used primarily for ventilation. See Fig. F-5.

louvered door   See *door*.

low-flow shower head   A type of shower head designed to limit or restrict the flow of water through it while still maintaining an adequate amount of pressure; used to help conserve water.

low steel   See *steel: mild steel*.

low-water cut off   Of or relating to a type of valve used with steam or hydronic heating systems that is designed to shut the system down if the water level drops below a certain point.

LPG   See *liquefied petroleum gas*.

L-shaped   Anything having two distinct arms, wings, or other sections at right angles to each other, forming a shape which resembles the letter L.

L-shaped kitchen   A kitchen having all of the cabinets and appliances arranged along two adjacent walls in an L-shape. See Fig. K-2.

L-type stairs   See *stairs: platform stairs*.

lug   A projection for holding, gripping, or supporting something, usually provided on various threaded objects to aid in tightening or loosening.

lug sill   A windowsill that projects past the actual width of the window on both sides.

lumber   The end product of a saw and planing mill, in which a tree has been sawn to width, length, and thickness, then dried. Lumber is commonly available at lumberyards in a variety of standard sizes and in several different forms, which follow. When lumber is ordered, it is commonly referred to first by the thickness in inches, then the width in inches, then the length in feet; for example 2 x 4 (inches) x 10 (feet).

ACTUAL SIZE The actual finished dimensions (thickness & width) of a piece of lumber, as opposed to its nominal size. For example: A 2 x 4 (nominal size) is 1-1/2 x 3-1/2 (actual size).

BOARD LUMBER Lumber of any length that has a thickness of less than 2 inches and is 2 inches or more in width.

DIMENSION LUMBER Lumber of any length that has a thickness of 2 inches up to, but not including, 5 inches and is 2 inches or more in width.

FACTORY AND-SHOP LUMBER Lumber that will be cut up for use in further manufacturing; graded by the percentage of area which will yield pieces of a particular size and/or quality.

GREEN-PLATE LUMBER See *lumber: pressure-treated lumber.*

MATCHED LUMBER Lumber that has been milled to form a rabbet, tongue and groove, or other edge configuration, which allows it to mate with the edge of a similarly milled board.

NOMINAL SIZE The standard commercial size by which lumber is known and sold, as opposed to its actual size.

POST LUMBER Lumber that is 4 x 4 inches or larger in width and thickness and square in cross-section.

PRECUT STUD A piece of lumber, usually a 2 x 4, that has been precut at the mill to a length of 92-1/4 inches. When installed in a wall with a 2 x 4 sole plate and two 2 x 4 top plates, it yields a rough wall height of 96-3/4 inches, which is designed for the use of standard 4 x 8-foot sheets of wallboard, as it allows for a thickness of 1/2 or 5/8 inch on the ceiling and 96 inches on the wall, as well as 1/4 or 1/8 inch of slack.

PRESSURE-TREATED LUMBER Lumber that has been chemically treated with a wood preservative to prevent damage from insects, fungi, and other forms of biological decay. The chemical is forced into the wood under high pressure to ensure complete saturation, making this type of lumber suitable for burial, contact with concrete, and all outdoor uses. Often called green plate due to its greenish tint and its common use as a sill plate.

ROUGH LUMBER Lumber that has been cut at the mill, but has not been planed smooth; commonly used for a rustic look in certain outdoor applications, such as landscaping. Actual sizes vary; i.e., a rough 2 x 4 will average 1-3/4 to 2 inches in thickness and 3-3/4 to 4 inches in width.

STRIP LUMBER Lumber less than 2 inches in thickness and less than 3 inches in width.

STRUCTURAL LUMBER Lumber that is intended for heavy construction use, such as posts, columns, and timbers; lumber over 2 inches thick.

SURFACED LUMBER Lumber that has been cut at the mill, then planed smooth on all four sides. Lumber is also occasionally planed on only one, two, or three sides.

TIMBER Lumber of any length that is 5 or more inches across at its smallest dimension.

YARD LUMBER Lumber that is intended for light framing and finish work, including studs and joists, siding, flooring, etc.

lumber defect One of various irregularities and abnormalities in finished lumber that in some way affect its strength, durability, or appearance. These defects follow. See Fig. L-1.

BARK POCKET A cavity within the wood containing bark.

BLUE STAIN A blue or gray discoloration of the sapwood caused by a fungus growth in unseasoned lumber. It is especially common in pine and affects the appearance but not the strength of the lumber.

CHECK A shallow or surface separation of a piece of wood along the grain, caused when the surface of the wood dries faster than the interior. It is shorter and less damaging than a split and can be of one of two types. An *end check* is a check beginning at the end of a piece of wood. A *through check* is a check extending from one face of a piece of wood through to the opposite face.

COMPRESSION WOOD Lumber obtained from the lower side of a leaning or bent tree; usually darker than normal, it has

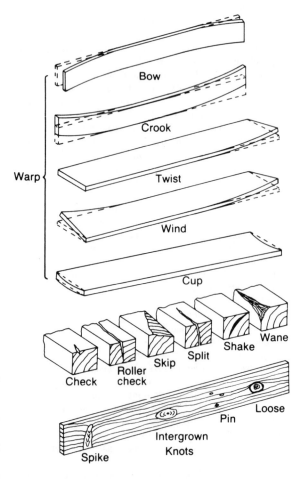

Fig. L-1. Lumber defects

greater shrinkage along its length and erratic, undependable strength.

DECAY The disintegration of wood or other material through the action of fungi.

END CHECK See *lumber defect: check.*

HOLE A defect in lumber caused by wood-boring insects or worms or by rough handling by equipment during logging or milling. Holes are rated like knots as being pin, small, medium, or large in size, and will lower the grade of the finished lumber.

HONEYCOMB Separation of the wood's fibers along the grain in the interior of the tree.

KNOT The portion of a tree branch or limb that appears in the face or edge of a finished piece of lumber; considered a defect in structural lumber, although it has decorative value for paneling and certain other uses. The wood surrounding the knot will be dense and hard. Knots fall into three basic categories. An *encased or loose knot* is produced by a branch that was dead. This type of knot will usually loosen and fall out, leaving a hole in the wood. An *intergrown or tight knot* is a sound, relatively circular knot resulting from a live branch. A *spike* is a long, usually pointed knot resulting from a branch that was sawn along its axis instead of across it. Spike knots can be either encased or intergrown. Knots are also described by size, which is the average of its length and width. A pin knot is less than 1/2 inch; a small knot is less than 3/4 inch; a medium knot is less than 1-1/2 inches, and a large knot is greater than 1-1/2 inches.

PECK An area of missing or disintegrated wood, resulting from advanced decay in a localized area of a living tree, that stops spreading after the tree is cut. Most common in cedar and cypress, pecks seriously impair a board's structural value, and "pecky" lumber is primarily used only as a decorative wall or ceiling covering, or for fence boards.

PITCH POCKET An opening between the tree's annual rings containing liquid or solid pitch.

ROLLER CHECK A crack in the wood caused when a cupped piece of lumber is flattened as a result of passing through machine rollers during the manufacturing process.

SHAKE A crack that occurs along the grain and between the annual growth rings before the tree is cut, usually as the result of an abrupt change between spring and summer growth.

SKIP A low or depressed area left unsmoothed by the planer knives.

SPLIT A separation along the grain of a piece of wood caused when the wood's surface dries faster than the interior, usually starting from the end. It is longer and deeper than a check.

TENSION WOOD Lumber produced from the upper side of a leaning or bent tree characterized by greater shrinkage along its length, erratic strength, difficult nailing, and a fuzzy cut surface.

THROUGH CHECK See *lumber defect: check.*

WANE An area of bark along a board's edge that reduces that board's usable width.

WARP Any deviation in a board from a perfectly true surface. Depending on how and where the deviation occurs, it is referred to as one or a combination of the following. A *bow* is a deviation flatwise in a piece of lumber. If an imaginary line is drawn from one end of a board to another and the board's face deviates from that line, the board is bowed. A *crook* is a deviation edgewise in a piece of lumber. If an imaginary line is drawn from one end of a board to another and the edge deviates from that line, the board is crooked. A *cup* is a deviation across the face of a piece of lumber. If an imaginary line is drawn from one edge of a board to another and the face deviates from that line, the board is cupped. A *twist* is a deviation in a piece of lumber, either flatwise or both edgewise and flatwise. If the board is laid on a flat surface, three corners will touch, while the fourth will be elevated. A *wind* is a type of warp in which the board tends to curl up. If the board is placed on a flat surface, it will rest on two diagonally opposite corners, the other two corners being elevated.

WIND SHAKE A separation of the annual rings that occurs while a tree is growing and usually appears as a check or wane in milled lumber.

lumber grading Rating lumber based on appearance and strength at the time it is milled, and stamping the lumber accordingly. The terminology and rules for grading can vary widely, depending on the species of wood, its intended use, and the association of lumber producers that governs the region where the lumber is milled.

lumen A measurement of light quantity equal to the amount of light falling on 1 square foot of a surface 1 foot from a standard candle. If measured directly from a light source, expressed in footlamberts; if measured at the point where the light is used, expressed in footcandles.

luminaire A complete lighting unit, including the lamp, socket, reflector, globe, wiring, and all other parts.

luminous ceiling panel A large enclosure recessed into or surface mounted on a ceiling, usually containing several fluorescent lights behind plastic diffusers, creating a broad, well-diffused glow of light.

luminous flux 1. That portion of the radiant energy visible to the human eye, producing the sensation of light. 2. The time rate of the flow of visible light, expressed in lumens.

M

M   The Roman numeral for 1000, often used when describing materials or other things in units of 1000.

macadam   A road-surfacing material comprised of small pieces of broken rock in compressed layers, often mixed with tar or asphalt.

machine burn   A scorched or burned area on a piece of wood, usually caused by the friction of a dull saw blade or other worn cutter.

machine thread   A screw thread designed for use with a nut or similar fastener, or for use with a compatibly threaded object.

macroclimate   The larger, overall climate for a particular region, as opposed to a microclimate; also called regional climate.

made ground   Any area of land created by filling in natural or excavated depressions.

magnesium rod   See *anticorrosion anode.*

magnetic catch   A small metal plate attached to a cabinet door that contacts a magnet on the cabinet frame; used to keep the door closed.

magnetic direction   A geographical direction as indicated by a magnetic compass.

mahogany   See *hardwood.*

main   1. The primary water or sewer line in a building from which the branch lines are taken. 2. The primary water or sewer line in a community from which branch lines for individual buildings are taken.

main cleanout   An opening provided in the main building drain line near the area where the line leaves the building; used to gain access to the line for clearing obstructions. See Fig. D-14.

main disconnect switch   A heavy-duty switch located in the service panel used to shut off all electricity to the building at one time. Depending on the age and type of service panel, it might be a large, external handle; a nonmetallic pull-out block with cartridge fuses; or a circuit breaker rated for the amperage of the entire system. See Fig. S-10.

main drain   See *building drain.*

main shutoff   A water valve located on the main water line, usually where it enters the building; used to shut off all water coming into the building.

main stairs   See *stairs.*

male/female   Of or relating to a method of designating parts that slide or thread into each other. The male part or thread is external, designed to be inserted into a corresponding slot or internally threaded hole, which is female.

malleable   Of or relating to metal or other materials that are capable of being easily machined or otherwise worked with.

mallet   See *hammer.*

mansard roof   A modified type of hip roof in which the slope to all four sides is broken into two slopes, the upper slope being nearly or completely flat and the lower slope pitching sharply. Primarily used on two-story homes to provide greater headroom on the second floor. See Fig. R-8.

mantel   1. The shelf above a fireplace opening. See Fig. F-4. 2. The trim that surrounds a fireplace opening.

mantelpiece   The complete framework of mantel and moldings that surrounds a fireplace opening. Also called a chimney-piece.

manufactured home   Any chassis-mounted dwelling intended for transportation in one or more sections, to be erected on blocks or a fixed foundation, and having a minimum width of 8 feet, a minimum length of 40 feet, and a minimum floor area of 320 square feet. Also referred to as a mobile home.

maple   See *hardwood.*

marking gauge   A type of measuring and marking device consisting of a bar with a pencil or steel point in one end and a movable guide block, which is guided along the edge of the piece being marked while the point enscribes a line parallel to the board's edge.

marquetry   Decorative work formed from small, hand-shaped pieces of wood, often inlaid with stones and other materials.

masking tape   Adhesive-coated paper tape, available in a variety of widths; primarily used for masking off areas not to be finished when painting.

Masonite   A brand name of hardboard. See *hardboard.*

masonry   Stone, brick, concrete, concrete block, or other similar rock- and earth-based building units or materials.

masonry bit   A carbide-tipped drill bit designed specifically for drilling into brick, concrete, and other masonry.

masonry cement   Portland cement made of high calcium limestone, gypsum, and an agent to help entrap air, providing a mortar that is more pliable and shrinks less than one made with standard portland cement.

masonry unit   An individual brick, tile, stone, block, etc. used in building.

mass   1. The density of a material in relation to its volume. The heavier a material is, the greater its mass. 2. See *thermal mass.*

mastic   Any of a variety of thick, semiresilient materials used for bonding certain wall and floor covering products, or as a weatherproofing sealant. Most provide a flexible seal and remain pliable with age.

matched lumber   See *lumber.*

material schedule   See *schedule of materials.*

materials list   A list of the type and quantity of materials required for a specific project; also called a bill of materials.

materialman An individual or firm who furnishes materials or supplies for a construction project, but performs no actual labor on the job.

mBh A quantity of heat equal to 1,000 Btus per hour.

mBtu See *Btu*.

mcm Abbreviation for *1,000 circular mills*, a designation given to electrical conductors larger than 4/0 AWG to indicate the conductor's cross-sectional diameter. One circular mill is equal to a circle one mill in diameter; so 300,000 circular mills would be stated as 300 mcm.

mdo For plywood, the abbreviation for *medium density overlaid*. See *plywood siding*.

mean The average or intermediate point; the point halfway between two extremes.

mean daily temperature The average of the maximum and minimum daily temperatures; used in determining degree days.

measuring tape A strip of cloth or steel contained in a case and marked off in graduations of feet, inches, and parts of an inch; used primarily for linear measurements.

mechanic Any skilled person who uses tools in his work.

mechanical equipment Generally, all the heating, air conditioning, ducting, and ventilating equipment and materials in a building.

mechanical fastener Any of a wide variety of fastening devices, including screws, nails, pins, cleats, bolts, etc., that make a physical connection between two objects by extending into, onto, or through both pieces being joined; does not include glue, cement, or similar materials.

mechanical plated nails See *nails, finishes*.

mechanic's lien A lien placed against the owner of a piece of property by a contractor, subcontractor, or material supplier because payment has not been made for labor or materials supplied for work done on the property.

medallion A raised ornamental piece; used as a decoration on doors, walls, etc.

medium-density overlaid plywood See *plywood siding*.

medium oil resin See *oil resin*.

medullary ray See *wood*.

meeting rail 1. In a double-hung window, the top rail of the bottom sash and the bottom rail of the top sash, which meet in the middle of the window; also called a check rail. See Fig. W-3. 2. Any meeting point of two sashes.

melting point The temperature at which a specific solid changes to a liquid.

member Any part or element of a building, particularly one that carries a structural load.

memory The ability of a material to return to its former shape after having been compressed, expanded, or otherwise deformed.

mending plate 1. A drilled metal plate in a straight, L, T, angle, or other configuration used to repair and reinforce various wood and metal joints. See Fig. M-1. 2. A straight wood plate used to reinforce stile and rail joints in large doors.

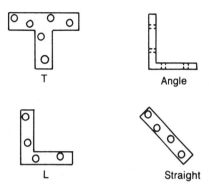

Fig. M-1. Common mending plates

mercury switch A noiseless electrical switch that uses a small, sealed tube of mercury to complete the circuit when the switch is operated.

mercury-vapor light See *gaseous-discharge light*.

mesh 1. The number of openings per linear inch of a screen or sieve. 2. An arrangement of intertwining or interlocking material making up a screen or sieve.

Metalbestos The brand name of a type of interlocking, doublewalled pipe having a heat-resistant layer of asbestos between the walls; commonly used for flue and chimney pipes.

metal corner L-shaped, tapered sheet-metal pieces, usually with a bottom tab, used to finish off the exterior corners of horizontal siding or siding shingles.

metal lath Metal in sheets which are slit and stretched to form a regular series of openings; used as a backing for plaster and stucco, particularly overhead. Sometimes called expanded metal.

metal stud A wall stud formed from sheet metal, U-shaped in section, with preformed holes for electrical conduit and pipes. They are locked into top and bottom channels, which act as the wall plates.

metal tie A steel rod formed into one of various shapes and coated with cement; used to connect two separate masonry walls in cavity wall construction.

meter 1. In the metric system, the basic measuring unit for length, area, and volume; equal to 39.37 inches. 2. An instrument used to measure the flow of gas, electricity, liquid, etc. See Fig. S-9. 3. An instrument for measuring time, distance, speed, height, etc.

meter socket A device, used with an electrical service panel, consisting of four copper clips that secure and activate an electric meter, along with screw terminals that receive the wires coming in from the power company and going out to the service panel. See Fig. S-10.

metric system A decimal system of weights and measures that uses as its basic units the gram for weight and mass; the meter for length, area, and volume; and the liter for capacity.

microclimate   Weather conditions, including sun and wind patterns, that are unique to one particular building or parcel of land. Influenced by the topography of the land, trees, fences, walls and other wind and sun screens, neighboring buildings, nearby water, etc.

microwave   1. An electromagnetic wave in the frequency range of approximately 1,000 to 30,000 megacycles. 2. The common name for a microwave oven.

microwave oven   An appliance that utilizes microwaves to excite the molecules in food and cause them to rotate extremely rapidly, creating heat-producing friction within the food and causing accelerated cooking.

mil   A unit of measurement equal to 1/1000 (.001) inch.

mild steel   See *steel*.

mill   A factory that is engaged in either the formation of lumber from trees or the manufacture of doors, window frames, trim, molding, and similar building materials from lumber.

milli-   Thousandth.

milliammeter   A special type of ammeter designed to measure electrical current in milliamperes.

milliampere   One one-thousandth of an ampere.

millimeter   A unit of length equal to 1/1000 (.001) meter.

millwork   Generally, any building material that is the end product of a planing mill or woodworking shop, including doors, window and door frames, mantels, panels, molding, trim, cabinets, etc.; usually not applied to flooring or siding.

mineral aggregate   Fine, naturally or artificially colored aggregate derived from stone, gravel, slag, or other sources; commonly used for coating roofs and roofing materials, decorating concrete, and other applications.

mineral fiberboard   See *building board*.

mineral-fiber shingle   See *asbestos cement shingle*.

mineral spirits   A liquid petroleum product having similar properties to gum turpentine; used as a thinner and cleaner for certain finishes.

mineral wool   A wool-like material obtained primarily from molten blast-furnace slag that has been processed into a fiber; used as thermal or acoustical insulation and available as batts or loose fill.

miscible   Of or relating to two or more materials that are capable of being mixed.

miter   To join two pieces at an angle that bisects the finished angle. For example, for two pieces to form a 90-degree angle, the miter cuts on the two pieces would each be 45 degrees. See Fig. M-2.

miter box   Any of a variety of power- or hand-operated tools used to accurately hold and cut a miter on a piece of wood. See Fig. M-3.

miter clamp   An L-shaped clamp used to hold two 45-degree mitered pieces at a right angle for gluing or nailing; used in making picture frames, etc.

miter square   A metal square in which the two legs are fixed at a permanent 45-degree angle.

mixing valve   A faucet body used for a bathtub or shower to

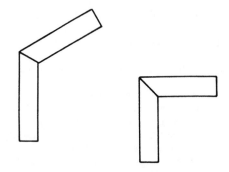

Fig. M-2. Miter joints

provide for hot and cold water to enter the valve separately, where they combine in adjustable quantities before flowing out through a single pipe to the tub spout or shower head.

modular cabinets   Individual prefabricated cabinets available in a wide selection of standard sizes and types that are joined together to form a run.

modular coordination   The designing of a building based on a common unit of measurement, called a module, which accommodates the use of building materials based on the same module for faster assembly and less waste. Examples of common modules include 16, 24, 48, and 96 inches.

module   See *modular coordination*.

modulus of elasticity   The ratio of stress to strain, being an indicator of a material's resistance to a load; also called E-Value.

modulus of rupture   The point at which a brittle material fails from tensile stress while being bent.

Mohs' scale   A scale for rating the hardness of rock and stone using ten minerals as examples, each assigned a number from 1 (talc) to 10 (diamond). Each numbered grade will scratch any lower-numbered grade.

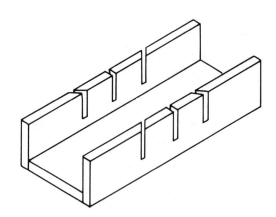

Fig. M-3. Typical miter box

1 - Astragal
2 - Backband
3 - Base
4 - Base
5 - Reversible base
6 - Base molding
7 - Base shoe
8 - Corner guard
9 - Bed molding
10 - Casing
11 - Casing
12 - Mullion casing
13 - Cove
14 - Crown

15 - Drip cap
16 - Fillet
17 - Lattice
18 - Picture molding
19 - Quarter round
20 - Half round
21 - Round (closet pole)
22 - Screen bead
23 - Shingle mold
24 - Stool
25 - Stool
26 - Stop
27 - Wainscot cap
28 - Water table

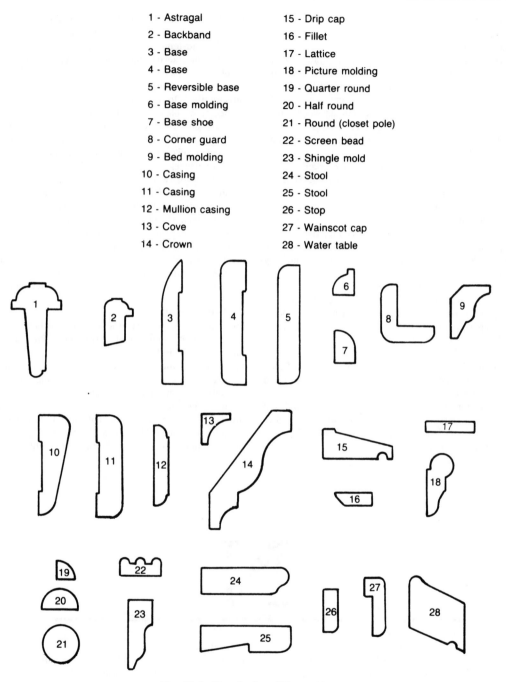

Fig. M-4. Standard molding patterns

moisture   Water or other liquid that is diffused in relatively small quantities.

moisture barrier   See *vapor barrier.*

moisture content   The amount of moisture in a piece of wood, expressed as a percentage of its oven-dry weight. A sample piece is weighed, dried at 212° F until weight loss ceases, then weighed again. Its moisture content is derived by subtracting the dry weight from the initial weight, then dividing the result by the dry weight. Framing lumber has a moisture content of about 15 percent; cabinet and furniture lumber is about 7 to 10 percent.

molded polystyrene   See *insulation, thermal: rigid boards.*

molder   A machine with a variety of interchangeable cutters, used to produce moldings.

molding   A strip of wood or other material having a rounded or otherwise decorative surface; used to conceal joints or to accent and highlight other surfaces. See Fig. M-4.

molding plane   A woodworking plane with interchangeable cutters in various shapes, used for producing moldings by hand.

Molly bolt   See *anchor, hollow wall.*

monolithic   Of or relating to a concrete structure that has been poured in one continuous operation and has no joints.

mop   To apply hot liquid tar or asphalt with a mop or large brush, as when coating a roof.

mopboard   See *baseboard.*

Morse taper   A standardized taper of approximately 5/8 inch per foot; commonly used on drill shanks, lathe centers, and other similar objects.

mortar   A mixture of portland or masonry cement, sand, and water, often with the addition of lime; used to bond together masonry units such as bricks, stones, or concrete blocks.

mortarboard   See *hawk.*

mortise   A slot or other recess cut into a piece of wood to receive a tenon, hinge, or other object. See Fig. J-1.

mortise-and-tenon   Of or relating to a woodworking joint in which one piece is mortised to receive the matching tenon of another piece. See Fig. J-1.

mortise lock   See *lock.*

mosaic tile   One of a series of small ceramic tiles arranged in decorative patterns and attached to a backing in regular-sized sheets, which are applied and grouted like regular tile.

motor controller   Any of a variety of electrical devices used to provide and control the incoming current necessary to start and run an electric motor.

mouse   A small weight attached to a cord; used to aid in pulling an object through an inaccessible opening.

movable insulation   Insulating panels, blankets, shutters, and other devices that are used to cover doors, windows, and skylights to stop heat loss, but that can be easily moved or taken down to permit heat gain.

mud   1. A nickname for the joint cement used for plasterboard. 2. A nickname for wet concrete, mortar, stucco, or similar materials. 3. Dirt and water combined in a semiliquid, semi-solid state.

mud sill   See *sill.*

mullion   A thin, nonstructural bar or divider between window or door units, frames, or other similar openings. See Fig. F-2.

multiple-use window   See *window.*

multipurpose tool   A type of pliers for electrical work, available in a variety of styles and usually containing five or more tools in one, including wire cutters and strippers, crimper, and screw cutter.

muntin   The vertical and horizontal members that divide the individual panes of glass in a window. See Fig. W-3.

muriatic acid   A commercial form of hydrochloric acid.

N

nail bags   Any of a variety of bags or pouches made of leather, canvas, or nylon, worn on a wide belt. used to hold nails and small tools while the carpenter is working.

nailer   A piece of wood secured to a surface, such as a wall or ceiling, and used as a fastening surface for other members.

nail hammer   See *hammer: claw hammer.*

nailing concrete   A soft type of concrete containing no large aggregate, into which nails can be driven with relative ease.

nail punch   See *nail set.*

nails   One of various mechanical fasteners in a wide selection of sizes, shapes, materials, and finishes. All nail types share three common features: a point, a shank, and a head, manufactured in different combinations to suit particular applications.

nails, common types

BOX NAIL A nail that is lighter and has a smaller diameter than a common nail, with less tendency to split the wood; used in framing applications where shear strength is not a consideration, as for nailing subfloors and in many types of assembly work.

CASING NAIL A nail that is similar to a finishing nail in style and application, but with a tapered head and a heavier shank.

COMMON NAIL The most popular type of nail for general carpentry and framing, having a large-diameter shank and good shear strength, and available in a wide selection of sizes, styles, and finishes.

CONCRETE NAIL A nail made of tempered steel in round, square, and fluted shank styles; used primarily for fastening wood members to concrete and masonry. Also called a masonry nail.

CUT-FLOORING NAIL A flat nail that tapers up from a dull, squared-off point to a roughly square head; used for blind-nailing through the tongue of a piece of flooring to minimize splitting.

DRYWALL NAIL A ring-shank or cement-coated nail, usually with a cupped head, designed specifically for installing drywall sheets.

FINISHING NAIL A nail having a small, slightly rounded head and designed to be driven below the surface so the resulting hole can be filled and finished; used for finish and trim applications.

FLOORING NAIL A hardened, spiral or ring-shanked nail used for installing hardwood flooring.

FURRING NAIL An aluminum or galvanized steel nail with a broad head and a thick fiber washer attached; used to install materials that must remain spaced away from their supporting members, such as stucco netting.

JOIST HANGER NAIL A short, hardened nail having a large diameter, barbed shank, and very high shear strength, designed for fastening wooden members into timber connectors.

LATH NAIL A small, blued nail used for installing wood lath; also called a plasterbase nail.

MASONRY NAIL See concrete nail.

PLASTERBASE NAIL See lath nail.

ROOFING NAIL A galvanized nail with a very broad head and a barbed shank, designed for installing asphalt and composition shingles, rigid insulation, and other materials with relatively soft surfaces.

SHINGLE NAIL A galvanized nail with a slender shank and a relatively small head; used to install wood shingles and shakes.

SIDING NAIL A galvanized nail, usually with a barbed or ring shank, used to install wood siding.

STAPLE A U-shaped wire fastener with two points, primarily used for installing fencing and certain types of wiring and netting.

UNDERLAY NAIL A ring-shank nail designed for installing plywood or particle board underlayment over a wood subfloor.

WASHER NAIL Any of a variety of nails having a washer, commonly of rubber, plastic, vinyl, or lead, located beneath the head. The washer flattens out under the head as the nail is driven, providing a weathertight seal over the nail hole.

Commonly used with plastic and metal roofing, and in similar applications.

nails, finishes

ALUMINUM NAIL A nail manufactured from aluminum having the advantage of being rustproof, but needing a thicker shank than steel nails to prevent bending; used for installing a variety of wood and nonwood materials in exterior applications.

BLUED NAIL A nail that is sterilized by heat until an oxidation coating is formed, offering good temporary rust protection for interior applications.

BRIGHT-FINISHED NAIL A nail having a bright, uncoated steel finish; used where corrosion resistance is not required.

CEMENT-COATED NAIL A nail having a semipermanent resin coating designed to give the nail greater holding power.

COLORED NAIL A small-diameter finish nail with a dyed, painted, or anodized finish, available in a variety of colors; mainly used with prefinished paneling and moldings to eliminate the need for countersinking and finishing.

DOUBLE-DIPPED GALVANIZED NAIL See nails, finishes: hot-dipped galvanized nail.

ELECTROPLATED NAIL A nail having a corrosion-resistant coating applied by immersing the nail in an electrolytic solution, then depositing a shiny film of zinc on the nail by an electrical current from zinc anodes. The resulting coating is relatively thin and best suited for indoor applications.

GALVANIZED NAIL A nail having a thick, somewhat rough coating of zinc to increase its corrosion resistance. Zinc chips are sprinkled over steel nails in a drum, then the drum is rotated in a furnace to melt the zinc over the nails. Also called hot-galvanized. In common usage, galvanized often refers to a nail that has been zinc-coated using any of a variety of methods.

GOLDEN-GALVANIZED NAIL See nails, finishes: mechanical-plated nail.

HOT-DIPPED GALVANIZED NAIL A nail that has been dipped into a vat of molten zinc for corrosion resistance. The resulting coating is thick and uniform, and is considered by many to be the most effective galvanizing process. The nails can be dipped a second time for additional protection, and are then commonly referred to as double dipped.

HOT-GALVANIZED NAIL See nails, finishes: galvanized nail.

MECHANICAL PLATED NAIL One of various cold-plated nails that are rotated in a barrel with zinc dust and small glass pellets, which hammer the zinc into the nails, then are immersed in a chromate rinse, giving them their characteristic gold or greenish color. Also called peen-plated or golden-galvanized.

QUENCH-HARDENED NAIL A nail that has been heated, quenched, and tempered to increase its resistance to bending; often used with hardwoods and concrete.

PEEN-PLATED NAIL See nails, finishes: mechanical-plated nail.

VINYL-COATED NAIL A nail that has a thin coating of vinyl resin, usually green in color, to provide for easier driving and some resistance to pulling out; used primarily for framing applications.

nails, head types

CHECKERED HEAD A head etched with thin, criss-crossing lines; designed to prevent the hammer from sliding off the nail head, although its value is somewhat questionable.

CUPPED HEAD A slightly concave head used on drywall nails to permit the easier application of joint compound over the head.

DOWEL HEAD See *nails, head types: headless nail.*

DUPLEX NAIL A nail designed for temporary work such as forms and braces, having two heads spaced about 1/4 inch apart. The nails are driven as far as the first head, leaving the second head exposed for easy removal.

FLAT COUNTERSUNK HEAD A head that is similar to a flat head, but having a slightly tapered underside, allowing it to be driven in flush with the surrounding surface; commonly called a sinker.

FLAT HEAD The common, general-purpose head found on most nails, suitable for a variety of applications.

HEADLESS NAIL A nail having no head at all and little holding power; sometimes used as a finishing nail. Also called a dowel or dowel head.

HOOK HEAD An L-shaped head commonly found on nails used for barrels and kegs.

OVAL HEAD A half-rounded head that remains above the surrounding surface after driving, providing a surface over which other objects can slide.

SET HEAD A small, rounded or tapered head, commonly used on casing and finishing nails.

SINKER HEAD See *nails, head types: flat countersunk head.*

TIE-DATED HEAD A head that has two numbers either stamped or embossed on it, indicating the year in which it was driven. The name is derived from their common use as an indicator of the age of a railroad tie.

UMBRELLA HEAD A broad, flat head with a raised area in the center, designed to seal over the nail hole when installing metal roofing.

nails, point types

CHISEL POINT A tapered, flattened point; used primarily on large spikes, making it easier to drive them into heavy timber.

DIAMOND POINT A common, general-purpose point featuring easy starting for most uses in wood. There are two variations. The *blunt diamond point* is a shorter, less sharp version of the standard diamond; good for use with softwoods to help prevent splitting. The *long diamond point* is a longer, sharper version of the standard diamond for faster starting.

NEEDLE POINT The sharpest and easiest to start of all the point types.

POINTLESS NAIL A nail having the most blunt of all the points, offering the greatest protection against splitting because of the point's tendency to cut through the wood's fibers rather than follow the grain.

SIDE POINT A point that extends down one side of the shank rather than being centered; good for use when the nail will protrude through the wood and be cinched over.

nails, shank types

ANNULAR-THREADED SHANK See *nails, shank types: ring shank.*

BARBED SHANK A shank having small notches along its entire length that angle up toward the head, offering greater holding power than the common smooth shank.

DRIVE SHANK See *nails, shank types: spiral-threaded shank.*

FLUTED SHANK A shank similar to a spiral shank but with shallower grooves and less holding power. The grooves on a spiral-threaded nail stop somewhat below the head, while those on a fluted-shanked nail usually extend all the way up.

KNURLED SHANK A shank used primarily for soft masonry, having slightly spiraling grooves that run roughly parallel with the line of the shank, tending to cut through the masonry as they enter, minimizing the risk of cracking.

RING SHANK A type of shank having a series of small rings that taper slightly up toward the head, tending to spread the wood fibers and then lock into them when the nail is driven, offering excellent resistance to pulling out.

SCREW SHANK See *nails, shank types: spiral-threaded shank.*

SMOOTH SHANK A shank having no marks at all, except for the small notches up near the head, thus giving the least amount of holding power of all the shanks.

SPIRAL-THREADED SHANK This type of shank has a series of ridges that spiral up toward the head, causing the nail to turn like a screw when driven, providing very good holding power; also called a screw nail or drive nail.

nail set A small hand tool resembling a punch that is used to set the head of a finishing nail below the wood's surface; available in various tip sizes for use with different sizes of nail heads. See Fig. N-1.

Knurled

Fig. N-1. Nail set

nail spinner A tool that is mounted in the chuck of an electric drill to hold small brads and finish nails and spin them into the wood instead of being hammered in, thus lessening the chances of splitting.

naphtha A volatile, colorless liquid distilled from petroleum; used as a solvent, cleaning fluid, etc.

National Coarse Of or relating to a standardized series of coarse machine screw threads. Abbreviated NC.

National Electrical Code A publication of the National Fire Protection Association that establishes the standard methods and materials for electrical installations, and that works toward the practical safeguarding of people and property from the potential hazards of electricity. It forms the basis of most of the electrical codes in use today. Abbreviated NEC.

National Fine Of or relating to a standardized series of fine machine screw threads. Abbreviated NF.

National Fire Protection Association An association of engineers, electricians, and other qualified personnel acting under the auspices of the American National Standards Institute who meet annually to establish standards for electrical safety, which are published annually as the National Electrical Code. Abbreviated NFPA.

National Pipe Thread The designation of a series of standardized threads for pipe and pipe fittings. Abbreviated NPT.

natural Of or relating to a wood surface that has not been stained, painted, or otherwise finished in any way which alters the wood's natural color or grain.

natural finish A clear finish such as varnish, lacquer, oil, etc., that is used to protect a wood's surface without changing its color or grain to any great degree.

natural gas A lighter-than-air gas, primarily referring to methane; used in many types of heating and cooking appliances.

NC See *National Coarse.*

neat Containing no additives.

NEC See *National Electrical Code.*

needle-point nail See *nails, point types.*

negative pressure Air pressure within a building or device that is lower than the pressure outside.

neon light See *gaseous discharge light.*

neoprene sealant See *sealant.*

net floor area A building's actual, usable floor space, arrived at by subtracting the area used by walls, stairs, columns, etc., from the gross floor area.

neutral wire A grounded conductor used in an electrical circuit to provide a path along which electricity can return to its source, thereby completing its circuit. Except in certain special switching applications, it is never connected to a switch, fuse, or circuit breaker, and is always color-coded white.

newel The main post to which the end of a stair railing, or any other railing, is fastened. See Fig. S-18.

newel post See *newel.*

NF See *National Fine.*

NFPA See *National Fire Protection Association.*

n.g.r. stain Non-grain-raising stain, made with fast-drying solvents to eliminate swelling or raising the wood's grain when applied.

nib 1. One of the cutting points of an auger bit. 2. A small, cone-shaped projection in a finished surface, caused by dust or fiber particles that become embedded in wet paint, varnish, or other finishing materials.

night latch See *lock.*

night setback thermostat See *clock thermostat.*

nipple 1. A piece of plumbing pipe 12 inches or less in length and threaded at both ends; used to join fittings. 2. A hollow, male-threaded piece of brass, steel, or plastic tubing; used to join parts in lamps, faucets, and other objects.

NM cable See *nonmetallic sheathed cable.*

nocturnal cooling Of or relating to a passive cooling system in which heat from warm surfaces radiates out to a clear night sky.

nogging Filling the spaces between structural members with brick, stone, or other masonry.

no-hub Of or relating to cast-iron plumbing pipe that is straight on both ends instead of having a hub. No-hub pipes are joined with neoprene gaskets and metal clamps instead of molten lead to simplify the assembly. Also called hubless. See Fig. B-3.

no-hub clamp See *band clamp.*

noise reduction coefficient The sound absorption ability of an acoustical material at four different frequencies—250, 500, 1000, and 2000 cycles per second—representative of most common household noises. Abbreviated NRC.

nominal size The standardized dimensions by which certain building materials are specified, differing from their actual dimensions. For lumber, the rough size before milling; for example, a 2- x 4-inch piece of lumber (nominal size) is 1-1/2 x 3-1/2 inches (actual size). For masonry, the installed unit including mortar joints; for example, a 4- x 8- x 16-inch concrete block (nominal size) is 3-5/8 x 7-5/8 x 15-5/8 inches (actual size).

non-bearing Of or relating to any structural member that supports no load other than its own weight.

nonconductor A material that has a high resistance to the flow of electricity or heat.

nonconformance Any structure or use of a structure that does not conform with the zoning for the area where the structure is located.

nonferrous Any metal not made of or containing iron.

nonmetallic sheathed cable A widely used type of electrical cable consisting of two or more individually insulated copper conductors and usually a bare ground conductor, grouped together and wrapped with a flexible plastic outer jacket. Abbreviated as NM cable, and commonly referred to by the trade name Romex.

nonrenewable energy source Any energy source having a limited, finite supply, such as oil, gas, coal, and nuclear fuels such as uranium.

nonvolatile Of or relating to a product or portion of a product that does not evaporate at ordinary temperatures.

Norman brick See *brick.*

no-shock receptacle A special type of electrical receptacle that is equipped with spring-operated covers, which rotate 1/4 turn when the plug is removed to seal off the receptacle's opening.

nosing 1. The portion of a stair tread, usually rounded, that projects over the riser. See Fig. S-18. 2. Any similar projecting edge of a molding or drip.

nosing strip A bullnosed piece of molding placed on the edge of a stair tread or other board as a finish piece.

notched trowel A trowel having notches or unsharpened teeth along one or more of its edges; used for spreading various types of adhesives. The size and shape of the notch should be matched to the type of adhesive being used.

NPT See *National Pipe Thread.*

NRC See *noise reduction coefficient.*

nut driver A hand tool resembling a screwdriver, used to drive and remove nuts and hex head screws; available in various head sizes.

O

O The letter designation for the fixed (inoperable) sash in a sliding window.

oak See *hardwood.*

oakum A loose, hemp fiber material obtained from old hemp rope that is untwisted and pulled apart; used as a caulking material for large joints and cracks, or as a base for the lead when joining hubbed cast-iron pipe.

oblique angle Any angle other than a right angle, being either acute or obtuse. See Fig. P-8.

obscure glass Glass that has been treated by sandblasting, chemical etching, or other means to render it translucent; commonly used in bathroom windows.

obtuse angle An angle greater than 90 degrees. See Fig. P-8.

OC See *on center.*

Occupational Safety and Health Administration An agency of the United States Department of Labor charged with the establishment, implementation, and enforcement of mandatory job-safety and health standards. Abbreviated OSHA.

octagon A polygon having eight sides and eight angles. In a regular or equilateral octagon, each angle equals 45 degrees. See Fig. P-8.

OD See *outside diameter.*

offset Of or relating to an indentation, jog, or other deviation from a straight line.

offset ratchet screwdriver A type of screwdriver for hard-to-reach areas, having an interchangeable screwdriver tip at right angles to a reversible, ratcheting handle.

offset screwdriver A type of screwdriver having the tip bent at right angles to the shaft, allowing greater turning leverage and easier access into restricted areas.

ogee A molding or shaped edge having a profile in the form of the letter S or a series of reversed curves.

ogee gutter A metal rain gutter having an ogee shape to its face.

ogive In architecture, an arch that comes to a point at the top.

ohm The basic unit of the resistance to electrical flow. A resistance of 1 ohm passes a current of 1 ampere in response to an applied potential of 1 volt. Represented by the letter R or by the Greek letter Ω.

ohmmeter An electrical testing device used to determine a circuit's resistance in ohms.

oil-based caulking compound See *caulking compound.*

oil-based paint Any of a variety of paints that use oil, especially linseed oil, as a vehicle for spreading the pigment and require a compatible thinner for removal and cleanup.

oil-based stain See *stain.*

oil burner filter A filtering device installed on a fuel-oil supply line, designed to remove dirt, sediment, and other contaminants from the oil before it enters the furnace.

oil burner pump A pump that is used to force fuel oil through a nozzle, atomizing the oil and spraying it into the air tube to mix with air before being burned.

oil-fired heating system A heating system that disperses fuel oil and mixes it with air, then injects the air/fuel mixture into the furnace's combustion chamber, where it is burned.

oil length See *oil resin.*

oil resin Any resin that is mixed with linseed or similar oil for making paint, varnish, and other materials. Described by the relative amount of oil used in the formula.

SHORT OIL RESIN An oil resin having up to 20 gallons of oil per 100 pounds of resin.

LONG OIL RESIN An oil resin having more than 30 gallons of oil per 100 pounds of resin.

Longer oil resins dry relatively soft and flexible and are preferred for construction Shorter oil resins dry hard and brittle, and are mostly used for furniture and appliances.

oil soluble Of or relating to a material that is capable of being dissolved in oil.

oilstone A fine, hard stone that is coated with a thin oil for sharpening and finishing cutting tools such as chisels. It can be natural, such as the Arkansas stone, or man-made of silcon carbide or aluminum oxide. Also called a whetstone.

on center Of or relating to the spacing of framing members, based on the distance from the center of one member to the center of the next. Abbreviated OC.

one-hour wall A fire-resistant wall, usually made up of 5/8-inch, type X plasterboard over conventional framing, designed to withstand direct flame for one hour. Commonly found in residential construction underneath stairways and between the living space of a house and the attached garage.

one-wall kitchen A kitchen in which all of the cabinets and appliances are arranged on one wall. This layout is typical for an apartment or small house. See Fig. K-2.

opaque Incapable of allowing the passage of light; not transparent or translucent.

open beam 1. Of or relating to a ceiling characterized by heavy, solid beams used as rafters, with the underside of the roof boards exposed to the inside and stained or varnished. See Fig. O-1. 2. Of or relating to a nonstructural method of decorating a ceiling using solid, boxed, or false, synthetic beams.

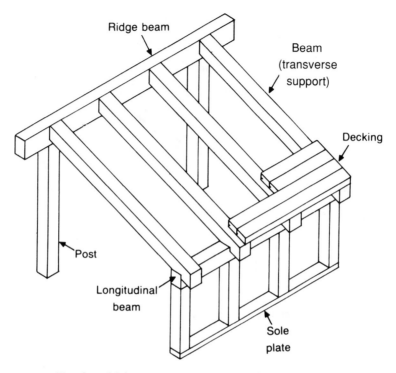

Fig. O-1. Main components of an open beam ceiling

open cell Of or relating to any material containing a number of open and interconnected pockets or cells; the opposite of closed cell.

open circuit In electricity, a circuit that has a switch in the open, or off, position, preventing the continuous flow of electricity.

open cornice A cornice in which the rafter tails and the roof sheathing are left exposed from the underside. See Fig. O-2.

open grain See *grain*.

opening In construction and energy conservation, a hole in a building's envelope, whether intentional, such as a window or door, or unintentional, such as a crack or hole.

open loop Of or relating to a liquid solar heating system used in conjunction with a drain-down system, in which both ends of the system of pipes open directly into a storage tank. See Fig. O-3.

open sheathing See *skip sheathing*.

open stair See *stairs*.

open time The amount of time that a material such as glue, paint, etc., can be applied and worked with before it sets up.

open valley A gutter created by the intersection of two sloping roofs.

Orangeburg See *fiber pipe*.

orange peel A defect in spray painting in which the paint does not level out, leaving a slightly rough, pitted surface, similar to the skin of an orange. It is usually caused by applying the paint too heavily.

orbital sander A portable power tool having a cushioned pad to which sandpaper is attached and that revolves in a tight, orbital motion, resulting in less cross-grain damage while sanding.

order The structure and appearance of classic architectural columns. See *column*.

organic 1. Pertaining to a substance composed of or related to animal or plant matter. 2. Pertaining to a compound containing carbon.

oriel window A window that is corbeled or cantilevered out from the face of a building.

orientation 1. The direction in which a house faces, a prime consideration in solar design. 2. The direction a solar collector faces.

oriented strand board See *building board*.

orifice A small opening of specific size, such as at the end of a pipe or nozzle.

original contractor The person or firm who first contracts with a property owner for a construction project. He deals directly with the owner on all matters related to the project.

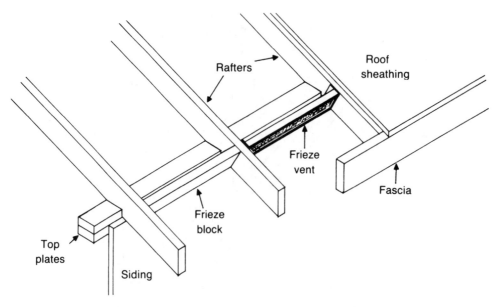

Fig. O-2. Components of a typical open cornice

O-ring   A narrow, round, rubber or nylon ring that fits into a recessed groove; used as a seal between parts to make the joint leakproof.

ornamental connector   See *timber connector.*

orthographic projection   The basic method of two dimensional drafting whereby an object is drawn by extending parallel and perpendicular lines out from two or more sides of the object, forming drawings which depict the object completely. Top, front, and end views are usually enough, but all six sides can be drawn if necessary for detail. See Fig. O-4.

OSB   See *building board: oriented strand board.*

OSHA   See *Occupational Safety and Health Administration.*

outbuilding   A structure located near a house but not attached to it, such as a shop, garage, barn, etc.

outlet   1. A box in which conductors within a circuit end, providing a point where fixtures, receptacles, and other electrical equipment can be attached to obtain current. 2. Commonly, a receptacle. 3. Any exit, port, or other opening at the end of a pipe, duct, or other passageway.

out of true   Not level, plumb, or square, otherwise deviating from a true plane.

outrigger   An extension of a rafter beyond the wall line; usually a piece of lumber added to the end of a common rafter to form a cornice or roof overhang.

outside diameter   The distance around a circular object such as a pipe, measured from the outside. Abbreviated O.D.

oval nail   See *nails, head types.*

oven cabinet   A tall cabinet designed to receive a built-in wall oven, microwave oven, or both.

oven dried   See *kiln dried.*

overcurrent   Any electrical current that exceeds the rated capacity of a conductor or piece of equipment.

overcurrent protection device   A fuse, circuit breaker, or other device that shuts off the electrical current to a circuit when the power flowing through the device exceeds its preset limit. See Fig. S-10.

overhang   1. That part of the roof which extends out over the side wall. See Fig. R-3. 2. The horizontal distance of a roof overhang, measured from the side wall to the fascia. See Fig. R-6. 3. Any object or structure that extends horizontally or diagonally past the vertical surface to which it is attached.

overhead door   A large door that opens upward on a pivot or curved track, so as to be horizontal overhead when fully opened.

overhead service   Electrical service that is provided to a building via overhead wires extending down from the utility poles to the service panel. See Figs. S-9 and S-10.

overload   The operation of electrical equipment or conductors in excess of full-load capacity, which, if continued for a sufficient length of time, would result in damage or dangerous overheating.

oxidize   To combine with oxygen.

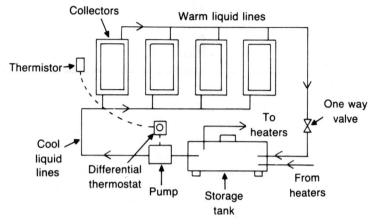

Fig. O-3. Open-loop solar heating system

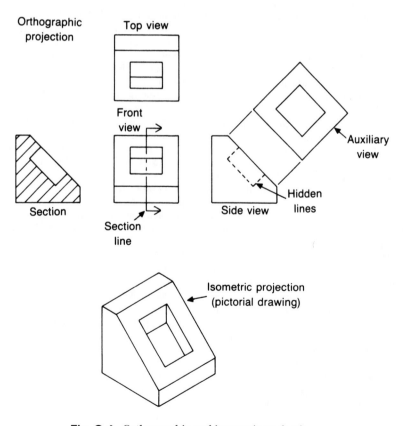

Fig. O-4. Orthographic and isometric projection

P

packing 1. A flexible, wirelike material that is wrapped around a faucet stem to prevent leaking, located beneath the packing nut. 2. Filling in the cavities in a masonry wall.

packing nut A nut that is screwed down over the stem of a valve to hold the packing in place.

padded nail A curved, J-shaped nail with a thick vinyl coating; used for supporting and securing copper pipes.

paddle fan A type of electrically driven, ceiling-mounted fan having large, slightly angled blades; used to circulate the air within a room.

paint A colored, adhesive coating applied over a wide variety of surfaces for protection and decoration. The main ingredients in a paint mixture follow.

DRIER Chemicals added to speed the formation of the paint film.

DYE A coloring agent used in place of pigment when the finished surface coating must be transparent but colored.

EXTENDER One of various additives that increase the thickness of the paint, improve abrasion resistance, and control the degree of gloss.

PIGMENT A coloring agent that is added to the mixture to give the paint its actual color.

VEHICLE The liquid portion of the paint, which allows the paint's pigments to be spread onto the surface; also called the medium.

paintable silicone sealant See *sealant: silicone sealant.*

painter's caulk See *caulking compound.*

paint gauge An instrument used to measure the thickness of a paint coating.

paint hook An S-shaped metal hook, one side of which hooks over a ladder rung, the other side of which holds a can of paint.

paint pan A slanted, formed sheet-metal or plastic pan with a reservoir at one end; used to hold paint and provide a surface for spreading the paint onto a paint roller.

paint remover A mixture of active solvents and other materials used to lift and remove paint, varnish, and other finishes.

paint resin One of various synthetic organic compounds that have replaced many of the natural oils formerly used in paint formulas. Some of the most common paint resins in use today follow.

ACRYLIC RESIN A durable, ultraviolet-resistant resin; often used as an additive in latex paints and factory-baked enamel.

ALKYD RESIN An organic resin having an alcohol and acid base; commonly used for exterior paint and baked enamel paint because of its excellent resistance to ultraviolet light, abrasion, and moisture. Often blended with other resins to further improve the qualities.

EPOXY RESIN A resin having superior hardness, durability, and adhesion and an excellent resistance to weather, abrasion, and chemicals; commonly used in masonry paints because of its resistance to the alkali present in concrete and mortar.

LATEX RESIN An elastic resin such as styrene-butadiene rubber (SBR), often blended with acrylic resins in a water vehicle, and offering good resistance to fading and peeling, easy soap and water cleanup, and good properties of expansion and contraction.

URETHANE RESIN A resin having qualities and characteristics similar to epoxy; commonly used in factory enamels and masonry paint.

paint roller A lightweight roller mechanism with a plastic or wood handle and a removable, material-wrapped cover; used to rapidly apply paint to flat surfaces such as walls or ceilings.

pale See *picket.*

pancake box A 1/2-inch-deep, round, electrical fixture box, designed to be recessed into plasterboard or plaster directly on a stud or other backing.

pane An individual section or piece of glass in a window or door. See Fig. W-3.

panel 1. A large, thin sheet of wood, plywood, hardboard, plastic, or other material. 2. A thin piece of material, usually wood or plywood, that is inserted into grooves in the surrounding stiles and rails, as in a door. See Fig. D-3. 3. A flat area that has been recessed into the surrounding area, as on a wall or ceiling, usually highlighted by moldings or other decoration. 4. To build or cover a surface with panels of material.

panel adhesive See *adhesive.*

panel door See *door.*

paneled framing One or more panels held in place by surrounding stiles and rails; used in door, cabinet, and furniture construction.

paneling Generally, relatively thin, 4- x 8-foot sheets of plywood or hardboard that have been factory stained or painted or have been covered to simulate wood grain; used as a decorative wall or ceiling covering. Also referred to as prefinished paneling.

paneling nail One of various slender finishing nails, usually 1 to 2 inches in length and having a deformed shank, available in a selection of colors; used to fasten prefinished paneling, moldings, etc.

panelized construction Large, prefabricated panels, usually of plywood over a lumber frame, that are installed as a unit for faster construction; most commonly used for flat roof construction on large buildings.

panelized roof hanger   See *timber connector*.

pan form   A metal or fiberglass form resembling an inverted square pan; used with steel rebar as a base for a concrete slab, particularly between floors of a building.

panic hardware   A horizontal push bar extending the entire width of a door, allowing the door to be rapidly unlatched and opened.

pantry cabinet   A tall cabinet used for food storage, normally containing a series of adjustable fixed or rotating shelves, often with racks mounted on the doors.

paperboard   See *building board*.

parabolic reflector   See *concentrating collector*.

parallel   1. Of or relating to lines or objects that extend out in the same direction and always remain an equal distance apart. See Fig. P-8. 2. Of or relating to a method of plumbing a solar heating system that directs the heat-transfer fluid from the storage area through only one collector and back to storage.

parallel connection   In electrical wiring, a set of connections that provides the electricity more than one path to follow. It is the common method of house wiring, whereby hot and neutral wires extend from outlet to outlet, branching off to individual fixtures and receptacles as needed.

parallelogram   A closed figure having four straight sides, with opposite sides being parallel. See Fig. P-8.

parapet   A low railing or wall along the edge of a roof, terrace, bridge, etc.; used for protection.

pargeting   See *parging*.

parging   Applying a thin coat of plaster or cement mortar over a masonry surface to smooth or decorate it. Also called pargeting.

par lamp   Abbreviation for *parabolic aluminized reflector*, a type of bulb made of thick, heatproof glass and containing a shiny, curved reflective surface for greater light intensity. Available as floodlights or spotlights, for interior or exterior use. The bulb is rated by wattage and also by a number that indicates the bulb's diameter in 1/8-inch increments. For example, a par 38 bulb is 4 3/4 inches in diameter.

parquet floor   A finish floor covering of wooden blocks or strips arranged in various decorative patterns. See Fig. P-1.

particle board   See *building board*.

parting stop   A small, wooden piece used in the side and head jambs of a double-hung window to separate the upper and lower sash; also called a parting strip. See Fig. W-3.

parting strip   See *parting stop*.

partition   A wall that subdivides the space within a particular story of a building.

partition block   See *building block*.

parts per million   A measure of concentration, being the number of parts of one substance in one million parts of another substance. Abbreviated *ppm*.

party wall   1. A common wall separating two adjacent dwelling units, as in an apartment building. 2. A wall erected on the property line between two adjacent parcels of land and used jointly by both property owners.

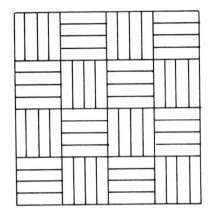

Fig. P-1. Parquet design

passage lock   See *lockset*.

passive solar system   A system for providing solar heat and/or cooling to a building without any mechanical means. Primarily a matter of architectural design, involving the materials and methods of building, placement of windows, use of berms and other landscaping, etc.

passive water preheater   A system of warming domestic water before it reaches the water heater whereby a black-painted tank is mounted in front of a bank of windows and tied into the cold-water supply line that enters the water heater such that sunlight can warm the water before it reaches the heater, thus reducing the load on the water heater itself.

patina   1. The glow or general appearance of age in the finish of a piece of wood. 2. An oxide film that forms on a metal such as copper, as it weathers.

patio door   See *door: sliding glass doors*.

Pavelock   A brand name for a system of decorative interlocking concrete paving blocks; used for driveways, walkways, etc.

paver   A brick, stone, or other natural or manufactured masonry unit used to pave a road or walkway, or as a stepping stone.

paving brick   See *brick*.

pavior   1. Alternate spelling of paver, a thin brick or stone often used as a floor covering. 2. An older term describing a person skilled in the installation of such materials.

payback period   The amount of time it will take an item, such as insulation or a solar heating system, to pay for its initial cost through fuel or other savings.

PB   See *plastic pipe*.

P board   Abbreviation for *particle board*.

PE   See *plastic pipe*.

pebble bed   A solar storage area consisting of a large bin filled with small, uniform-sized rocks.

peck   See *lumber defect*.

pediment   A broad, ornamental, triangular area similar to a low gable, located above a portico, door, or window.

peeling A defect in a coat of paint in which the paint film detaches and lifts off the underlying surface; usually caused by dirt, grease, moisture, or other foreign matter on the surface that was painted.

peel strength A measure of the strength of an adhesive, determined by the amount of force needed to separate the adhesive bond between two objects.

peen hammer See *hammer.*

peg 1. A wooden stake used to indicate foundation layout lines or surveying marks. 2. A wood dowel or pin used to fasten wood; often used instead of nails or screws. 3. Abbreviation for *polyethylene glycol*, a chemical compound sometimes used as a wood preservative.

pegboard See *building board.*

pellet stove A relatively new type of heating appliance, burning special wood pellets in place of traditional firewood.

pellets, wood Small pellets made from dry, compressed sawdust, intended for specific use in pellet stoves.

penalty clause A clause in a construction contract that imposes monetary or other types of penalties on a contractor if the project is not completed by the date set forth in the contract.

penetrating oil An oil used to penetrate and loosen threaded fittings that have become rusted or corroded together.

penetration 1. The depth to which a liquid, such as a stain, is absorbed into a piece of wood. 2. The depth to which the metal in a plate melts when being welded. 3. The depth to which a screw, nail, or other fastener enters an object.

penny Originally, the price of a given size of nails per hundred, but now a standard indicator of nail length; indicated by a number followed by the abbreviation *d.*

pentachlorophenol See *wood preservative.*

pentagon A polygon having five sides and five angles. In a regular or equilateral pentagon, each angle equals 72 degrees. See Fig. P-8.

percolation test A test to determine the ability of soil in a particular area to absorb liquid, the results of which are used to determine the size of the disposal field for a septic system; usually referred to as a perc test.

perc test See *percolation test.*

perforated pipe Plumbing pipe containing a series of small, evenly spaced holes, usually on opposite sides; used for drain tiles, septic disposal fields, and other drainage applications. See Fig. D-10.

perforated strap See *plumber's tape.*

performance bond A type of bond posted by a contractor to guarantee that contractor's performance of the contract specifications.

pergola An arbor or walk enclosed by trelliswork that is covered with vegetation.

perimeter The boundary line around any two-dimensional object, such as a piece of land.

perimeter insulation See *insulation, thermal.*

periodic duty Of or relating to an electrical system or component operating intermittently under loads that recur on a regular basis.

perlite See *insulation, thermal.*

perm A unit used to express the movement of water vapor through a material, equal to 1 grain of water vapor (1/7000 pound) penetrating 1 square foot of material in 1 hour under an air pressure difference of 1 inch of mercury (0.5 psi). Used in measuring and rating the effectiveness of a material as a vapor barrier. Abbreviation for *vapor permeance.*

permafrost Permanently frozen subsoil, occurring primarily in arctic and subarctic areas.

permeability The property of a material that allows or prohibits the passage of a liquid through it.

perpendicular 1. Of or relating to a line or plane that meets another line or plane at right angles. See Fig. P-8. 2. Being at right angles to the horizon; vertical.

perron In architecture, an exterior flight of steps, usually with a landing, that leads up to the entrance door of a building.

personnel door See *door.*

perspective drawing A pictorial drawing or sketch that depicts objects in a natural, three-dimensional relationship to each other; used to show a proposed room or building as it will actually look when finished.

pet door See *door.*

pH A chemical symbol for potential of hydrogen; used to express the relative acidity or alkalinity of a solution. A pH of 7 is regarded as neutral.

phase changing Of or relating to a material that readily changes from a liquid to a solid and back again. Phase-changing salts are used in some solar heating systems as a heat-storage material because such materials absorb large amounts of heat as they change phases from solid to liquid, and release the heat as they again become a solid.

phenolic resin A synthetic resin primarily having phenol and formaldehyde as its base.

phosphorescent paint Luminous paint that emits light in the absence of white light; the type of paint that glows in the dark.

photovoltaic power system A system of specially designed solar cells used to convert sunlight into dc electricity for various uses.

pi A mathematical symbol for the ratio of a circle's circumference to its diameter, approximately equal to 3.141592.

pick and dip A slang term relating to a bricklaying method in which, simultaneously, a brick is picked up with one hand and a trowel full of mortar is picked up with the other hand.

picket A narrow board, often pointed or otherwise shaped at the top, installed as a fence board; also called a pale.

pictorial drawing A drawing that represents an object in three dimensions, as though it were a photograph. See Fig. O-4.

picture-frame molding A molding with a decorative face and a flat or rabbeted back; used to frame pictures or as an accent molding.

picture molding A narrow molding attached to the walls of a room about 1/2 inch down from the ceiling, from which pictures are hung on thin wires. See Fig. M-4.

picture window See *window.*

pier 1. A pad or short column of masonry, usually cast in place; used to support another structural member. 2. The solid, load-bearing portion of a wall between window or door openings. 3. An upright support along a wall, similar to a buttress. 4. A platform or similar structure extending out into the water, primarily intended as an anchorage for boats.

pier and beam See *pier and post*.

pier and post Of or relating to a type of underfloor foundation and framing system that uses wooden posts, which are supported by piers or footings and in turn carry the girders on which the rest of the floor is framed; also called pier and beam or post and beam. See Fig. P-2.

pier block A pre-cast concrete block with a large square bottom tapering to a smaller square top. The top of the block can be solid or capped with an embedded piece of redwood lumber, usually 2 x 6 x 6 inches. Used to support other structural members, such as girders, and to allow an easy transition from masonry to wood. See Fig. P-2.

pigment See *paint*.

pigmented shellac Shellac to which a pigment has been added, usually white; often used as a primer to seal over stains before applying a finish coat of paint.

pigtail 1. An electrical wire extending out from a pigtail splice to connect with the wiring device it serves. 2. A short, heavy cord with a plug at one end and bare wires at the other used for connecting a stationary appliance, such as a clothes dryer, to its receptacle.

pigtail splice The joining of three or more electrical wires, at least one of which connects to the terminal of a wiring device such as a switch.

pilaster A rectangular column or section of wall that projects out vertically from the surface of the surrounding wall; used primarily to add strength, but also to decorate.

pile A heavy wooden, steel, or concrete column driven into the ground to support a building's foundation. Commonly used in loose or swampy soil or to support extreme loads such as multistory buildings. Also called a piling.

pile driver A machine using a steam-powered hammer, heavy weights, or similar methods to drive piles into the earth.

piling See *pile*.

pillar A freestanding column or shaft, slender in relation to its height, used as a support or as a base for another object.

pilot hole A hole drilled in a piece of wood to act as a guide for the insertion of a wood screw.

pilotless ignition Of or relating to an ignition system for gas appliances that uses a series of electrical sparks to ignite the incoming gas, eliminating the need for a pilot light; also called electric ignition.

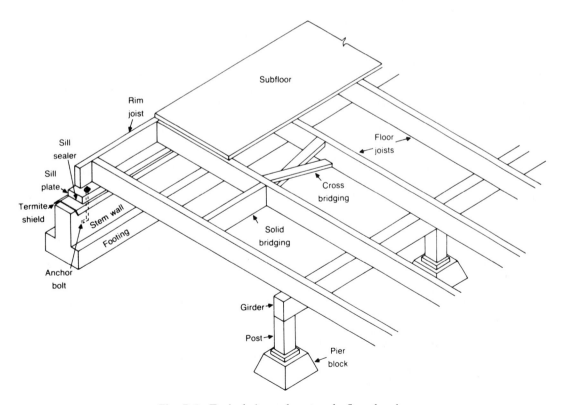

Fig. P-2. Typical pier and post underfloor framing

pilot light A small, continuously burning flame inside a gasfired appliance that serves to ignite the incoming gas when the appliance is turned on. See Fig. W-3.

pin 1. A round metal rod with a rounded head that is passed through the knuckles of a hinge to hold the two hinge leaves together. See Fig. B-12. 2. Any wood or metal piece that secures two objects together.

pine See *softwood*.

pin lock A removable steel pin device that is inserted into a sliding door or window for security.

pinnacle A tall, slender column, usually pointed; used as a decorative feature in some styles of architecture.

pipe clamp A type of woodworking clamp, used in combination with 1/2- or 3/4-inch threaded steel pipe and having two jaws: One contains a screw-activated block and is attached to one end of the pipe; the second is movable along the pipe's length to adjust to the size of the piece being clamped. Different lengths of pipe can be interchanged to provide a variety of clamp lengths.

pipe cutter A large, C-shaped hand tool designed for cutting metal pipe, consisting of a steel cutting wheel at the end of a threaded shaft. The wheel is tightened down against the pipe, then the tool is rotated around the pipe, gradually increasing pressure on the wheel until the cut is completed.

pipe dope See *pipe joint compound*.

pipe fitting 1. Any of a variety of fittings used to join lengths of pipe or to change their direction. 2. Installing a piping or duct system.

pipe insulation See *insulation, thermal*.

pipe joint compound A thick, nonhardening, liquid sealing material used on external pipe threads to help prevent leaks at the joints while still permitting easy disassembly.

pipe reamer A pointed, tapered device used to remove burrs from inside a metal pipe after cutting.

pipe repair clamp A two-piece tubular clamp, used in combination with a rubber pad, to repair small leaks in a pipe.

pipe strap A curved metal strap, available in various sizes, used to secure pipes against walls or joists.

pipe threader A tool consisting of a pipe-thread die set in a racheting or revolving handle; used to cut threads into a piece of pipe.

pipe vise A special-purpose vise having curved jaws with teeth; used for securing galvanized pipe during cutting and threading operations.

pitch 1. The inclined surface of a roof, equal to the rise divided by the span. For example, a roof with a 48-foot span, 8 feet high at the peak, would have a one-sixth pitch. 2. The ratio of rise to span. The example in 1. would have an 8:48 or 1:6 pitch. 3. A thick, natural resin occurring in some varieties of trees, particularly pines. 4. A thick, gummy substance derived from the refined residues of turpentine, oils, and other materials; used to seal and waterproof seams, particularly in wood.

pitch board A triangular template whose sides represent rise, run, and slope; used for the repetitive laying out of stairs or rafters. See *bevel board*.

pitched roof A roof that slopes as opposed to a flat roof.

pitch floor A type of floor consisting of small stones or rock pieces arranged in patterns.

pitch pocket 1. See *lumber defect*. 2. A container of tar or asphalt used to seal around an object that penetrates a roof.

pith See *wood*.

pivot window See *window*.

placing Putting wet concrete into forms. The term is commonly used instead of *pouring* because technically, concrete is too thick to pour.

plain-sawn Of or relating to lumber that has been cut from a log in continuous, parallel, vertical strips.

plan A drawing, photograph, template, written instructions, or other visual aid, often with a material list, used to describe and guide the construction of a building or other object.

plancier The bottom side of a cornice or eave. See Fig. C-4.

plane Any of a wide variety of hand-operated woodworking tools designed for smoothing, truing, and shaping wood. The basic plane consists of a wood or metal body, a handle, a base, and a cutting tool, called an iron. Various wedges and/or adjustment fittings are used to adjust the angle and depth of the iron with relation to the work. A few of the more common plane types follow:

BENCH PLANE A general-usage plane, designed for smoothing wood. The name is derived from its typical use at a woodworker's workbench.

BLOCK PLANE A short plane, usually 3 to 6 inches long, used for small work and for cutting across end grains.

CHAMFER PLANE A specialized plane with a base in the shape of an inverted V. Used for cutting a chamfer (beveled edge) on a piece of wood.

COMBINATION PLANE An intricate and precise plane using a number of interchangeable cutters, designed to create slots, dados, rabbets, moldings, and other cuts.

EDGE PLANE A short plane with the iron at the very front and exposed. Used to plane into corners and edges otherwise inaccessible to other planes.

JACK PLANE A relatively long plane, 12 to 18 inches, primarily used for rough planing prior to finishing and smoothing with other types of planes.

JOINTER PLANE Any of a variety of very long planes, ranging from 20 to 30 inches or more, used to true and straighten the long edge of a board prior to joining it against another one.

MOLDING PLANE A specialized plane designed to cut moldings. The base of the plane is shaped in the reverse of the desired molding shape, and one or more irons are ground and sharpened to match the molding profile. Alternate spelling: moulding plane.

RABBET PLANE A small, narrow plane used for cutting a two-sided, square-bottom groove along the edge of a piece of wood.

SMOOTHING PLANE A short bench plane designed for final smoothing, but also used for general shaping work.

plane iron The beveled, sharp cutting blade of a plane.

planer A stationary power tool having a long, flat table and a horizontally revolving cutter head, used to smooth rough wood or reduce its thickness to a desired size.

plank-textured plywood See *plywood siding*.

plan view A drawing that depicts an object in two-dimensional form as though viewed from directly above; also called a top view.

plaster A mixture of portland cement, lime, sand, and water, often with other additives; used over lath as a covering for walls. There are many types of plaster for different applications.

ACOUSTICAL PLASTER Various compositions of plaster containing pumice, vermiculite, asbestos, or rock wool, making a thick, porous plaster with high sound-absorbing qualities.

BOND-COAT GYPSUM PLASTER Specially prepared gypsum plaster with only water added; used as a base coat over concrete or other relatively smooth surfaces with low adhesion.

CEMENT PLASTER See *plaster, unfibered gypsum plaster.*

FIBERED GYPSUM PLASTER Gypsum plaster containing hair, wood fiber, or other fiber for strength.

FINISH PLASTER Plaster mixed with slaked lime and water to provide a smooth, easily worked final coat.

FIREPROOF PLASTER Plaster containing asbestos fiber and applied by spray.

HARD WALL PLASTER Gypsum plaster having sand and hair or other fibers for strength.

KEENE'S CEMENT PLASTER Calcined gypsum mixed with slaked lime and sand and providing a hard, smooth, moisture-resistant coating.

PLASTER OF PARIS A fast-setting and hard plaster made of calcined gypsum mixed with water or water and lime; used primarily for patching and ornamental work.

UNFIBERED GYPSUM PLASTER A dry mix with no additives, to which water and aggregate are added on the job; also called cement plaster.

plasterboard A wall covering in sheet form designed to take the place of the first two layers of plaster.

plaster box See *cut-in box.*

plaster ground Small strips of wood that are the same thickness as the finished plaster wall; used around windows, doors, and other openings when plastering to assist the plasterer in getting a straight wall and also to provide a surface for nailing the finish trimwork.

plaster of paris See *plaster.*

plaster ring A square metal plate that fits the front of a square outlet box and has a raised lip on the ring to bring it flush with the finished wall surface, and, depending on the type, to adapt the outlet box for use with one device, two devices, or a fixture.

plastic An organic synthetic material or a processed material that is capable of being molded, formed, shaped, or otherwise worked with.

plasticity The tendency of a material to be permanently deformed by stress.

plastic laminate A sheet material comprised of a decorative paper bonded to several layers of resin-treated kraft paper, then covered with a translucent, melamine-resin paper which becomes clear during pressing and curing. The resulting material is hard, durable, and resistant to water, alcohol, heat, and stains. Plastic laminates, of which the brand Formica

is probably the best known, are often bonded to particle board for use in furniture, countertops, wall coverings, and much more.

plastic pipe Plumbing pipe, tubing, and fittings made from various types of plastic and assembled with clamps, locknuts, or compatible adhesives. The main types of plastic pipe used in construction follow.

ABS PIPE A pipe made of acrylonitrile-butadiene-styrene, a form of plastic commonly used in the manufacture of drain, waste, and vent pipes. Common pipe diameters in this material are 1 1/2, 2, 3, and 4 inches.

CPVC PIPE A pipe made of chlorinated polyvinyl chloride, a type of plastic used in making plumbing pipe suitable for use with hot water.

PB PIPE A pipe made of polybutylene, a material used in the making of flexible plastic tubing for hot and cold water.

PE PIPE A pipe made of polyethylene; used in making flexible plastic tubing for cold water.

PP PIPE A pipe made of polypropylene; used primarily for making plumbing traps and drainpipes.

PVC PIPE A pipe made of polyvinyl chloride, a material commonly used in the manufacture of rigid plastic pipe for cold water.

SR PIPE A rigid pipe made of styrene rubber; used primarily underground.

plastic resin adhesive See *adhesive.*

plastic wood A mixture of wood and plastic in putty form, available in a variety of colors that are similar to different woods; used to fill and hide nail holes, dents, and other defects in wooden items.

plat A map, chart, or plan of a city or subdivision showing the property lines of individual parcels.

plate A horizontal framing member having three specific applications.

SILL PLATE A member bolted directly to a masonry wall; also called a mudsill or rat sill. See Figs. B-2, P-2, P-3, and P-10.

SOLE PLATE The bottom member of a framed wall; also called a base plate. See Figs. F-10, O-1, P-3, and P-10.

TOP PLATE The top member of a framed wall upon which the rafters and ceiling joists rest, often doubled so as to overlap at the corners and intersections for greater strength See Figs. C-4, F-10, J-2, O-2, P-3, P-10, R-3, R-5, and R-6.

plate cut See *seat cut.*

plate glass A thick, polished, high-grade glass with few or no defects; primarily used in large windows and other applications requiring high strength.

platen Any large flat plate used in a press, as in the manufacture of particle board.

plate rail A decorative wall shelf or molding having a shallow groove on the top to receive pictures or plates.

platform framing The most common method used for wood framing today, whereby the subfloor for the first story is laid up on the foundation, the walls are set on the subfloor, and the floor joists of each succeeding story are set on top of the preceding story. See Fig. P-3.

platform stairs See *stairs.*

plenum A sheet-metal distribution box located on top of a

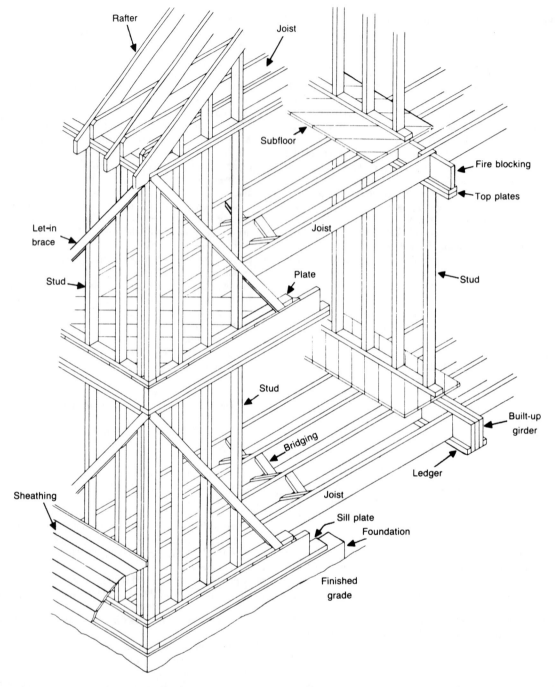

Fig. P-3. Main components of standard platform framing (Courtesy of the National Forest Products Association, Washington, DC)

furnace in which warm air is collected and routed into individual ducts.

plinth block A square or beveled block of wood at the base of a door casing to which the baseboard abuts. See Fig. B-4.

plot plan An architectural drawing depicting the boundaries of a parcel of land and showing the locations of buildings, driveways, walkways, easements, and other details of the current or intended use of the land. See Fig. P-4.

plough To cut a lengthwise groove in a piece of lumber. See Fig. D-1.

plow See *plough*.

plug 1. A small piece of wood used to hide a screw or nail head on a finished board. 2. A plumbing fitting having external threads and a square head, designed to be screwed into an internally threaded fitting to close it off. See Fig. T-4. 3. An anchor. See *anchor, masonry*.

plug cutter A cutting attachment used with a power drill to cut a small circular plug from a piece of wood. After a fastener is inserted into the resulting hole in the wood, the plug is replaced to conceal the fastener.

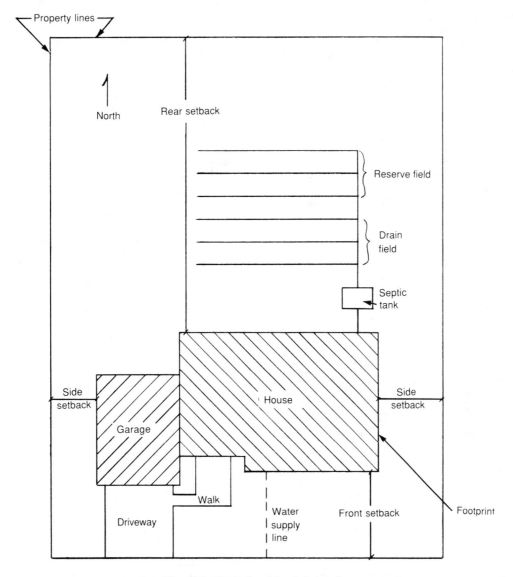

Fig. P-4. Typical residential plot plan

plumb 1. Exactly vertical; perpendicular to the floor or the horizon. See Fig. P-9. 2. To install all of a building's pipes, fixtures, and other plumbing fittings.

plumb bob A pointed weight attached to a line; used to determine or check if a surface is truly vertical. See Fig. S-19.

plumb cut 1. Any vertical cut, but mostly a vertical cut on an angled member, such as the end of a rafter. See Fig. R-5. 2. The vertical cut that is made when forming a bird mouth. See Fig. R-5.

plumber's helper See *plunger.*

plumber's putty A thick, gray or white putty used to seal the flange on a sink drain assembly to the sink, or for other plumbing fixture seals. See Fig. B-5.

plumber's tape Narrow, perforated metal tape, sold in rolls, used to strap and secure pipes and ductwork.

plunger A rubber suction-cup device attached to a long wooden handle and used to clear obstructions in sink traps and drain lines; also called a plumber's helper.

ply A layer or thickness of material, such as the veneers in a sheet of plywood or the layers in a built-up roof.

Plyedge The trade name for various narrow veneer strips, usually sold by the roll, designed to cover the exposed edge of a piece of plywood.

plywood A sheet material formed from layers of veneer or lumber that are glued and assembled under pressure. There are several types.

LUMBER-CORE PLYWOOD A sheet with a core of wood blocks, particle board, or other solid material covered on both faces

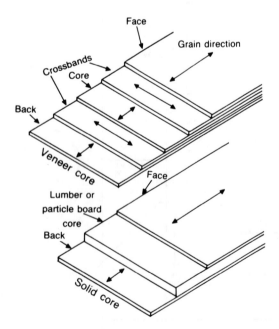

Fig. P-5. Plywood: veneer core (top), solid core (bottom)

with veneer, the grain of which runs lengthwise; also called *solid core plywood.* See Fig. P-5.

SOLID-CORE PLYWOOD See *lumber-core plywood.*

VENEER-CORE PLYWOOD The most common type of plywood, made from three or more layers of veneer, called crossbands, glued up at right angles to each other. An odd number of plies are always used so as to leave the grain of both face veneers running lengthwise. The most common sheet size is 4 x 8 feet, although longer lengths are available. Common thicknesses are 1/4, 3/8, 1/2, 5/8, 3/4, and 1 1/8 inches. See Fig. P-5.

Plywood is graded and specified by the type and quality of its face veneers and the type of glue used. Some common types follow.

EXTERIOR PLYWOOD Plywood having a totally waterproof glueline; suitable for any exterior use. Designated by the abbreviation X.

HARDWOOD PLYWOOD Plywood having hardwood face veneers, such as oak, walnut, birch, etc. Grades are #1—clear; #2—some pin hole knots; #3—good for painting. Face and back grades might differ, with the face grade listed first, i.e., 1-3.

INTERIOR PLYWOOD Plywood having lower-grade inner veneers and interior glue; able to resist occasional wetting, but not suitable for exterior or wet locations.

SOFTWOOD PLYWOOD Plywood having softwood face veneers, of which Douglas fir is probably the most common. Grades are N—clear with no repairs for natural finish; A—smooth with some neatly made repairs; B—smooth with more repairs and tight knots; C—knot holes to 1 inch, some limited splits; C plugged—improved C veneer with knotholes to 1/4 x 1/2 inch and splits to 1/8 inch wide, anything larger is plugged; D—knots to 2 1/2 inches, limited splits. The face and back grades might differ, with the face grade being listed first, i.e., A–C.

plywood blade A circular saw blade having numerous small teeth, used for ripping or crosscutting plywood, veneers, and plastic laminates without splintering or chipping.

plywood clip An H-shaped aluminum clip used between plywood panels in applications such as roof sheathing to provide structural support and reduce deflection between the unsupported edges of panels. See Fig. P-6.

plywood grades See *plywood.*

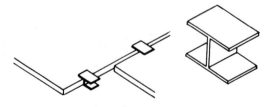

Fig. P-6. Plywood clips

plywood siding Softwood plywood sheets having a variety of decorative faces for use as interior wall covering or exterior siding. Common thicknesses are 3/8, 1/2, and 5/8 inch, and standard panel sizes are 4 feet wide by 8, 9, or 10 feet long. Some common American Plywood Association types follow.

BRUSHED PLYWOOD SIDING A plywood siding having a brushed surface to raise and accent natural grain and having square edges and no grooves.

CHANNEL GROOVE PLYWOOD SIDING A plywood siding having shiplap edges and 1/16-inch-deep, 3/8-inch-wide vertical grooves on 4-inch centers, cut into a lightly sanded face.

CORRUGATED PLYWOOD SIDING A plywood siding having deep, irregular, closely spaced corrugations and square edges.

FINE-LINE PLYWOOD SIDING A plywood siding having square edges and shallow, vertical grooves 1/32 inch wide and 1/4 inch o.c. cut in a smooth face.

MEDIUM DENSITY OVERLAID (MDO) PLYWOOD SIDING A plywood siding having a smooth resin-treated surface heat-fused to the plywood face, providing an excellent adhesion surface for paint. It is flat or grooved in different patterns, with square or shiplap edges.

PLANK-TEXTURED PLYWOOD SIDING A plywood siding having shiplap edges and 8-inch wide, rough-sawn sections that simulate individual, vertical boards.

REVERSE BOARD-AND-BATTEN PLYWOOD SIDING A plywood siding having shiplap edges and 1/4-inch-deep vertical grooves 1 or 1 1/2 inches wide and 8, 12, or 16 inches o.c. in a brushed, rough-sawn, or smooth face.

ROUGH-SAWN PLYWOOD SIDING A plywood siding having square edges and a rough-sawn surface, available flat or with narrow grooves 4 inches o.c.

STRIATED PLYWOOD SIDING A plywood siding having square edges and a combed effect with fine, random-width striations.

TEXTURE ONE-ELEVEN (T 1-11) PLYWOOD SIDING A plywood siding having shiplap edges, a rough, unsanded face, and deep, vertical grooves, usually 1/2 inch wide and 2, 4, 6, or 8 inches o.c.

pneumatic Of or relating to a tool or other device that is operated by compressed air.

pocket cut A section cut out of the face of a board, with none of the cuts reaching the board's edge. Pocket cuts are usually made by lowering a circular saw down onto the board, or by using a keyhole saw or jigsaw that is started in a predrilled hole.

pocket door See *door.*

pocket door latch A square unit having two side handles and one edge handle, set in a notch in the leading edge of a pocket door. Available in two types.

PRIVACY DOOR LATCH A pocket door latch that locks using a hook which engages a strike plate on the jamb. A door having such a latch can only be locked on one side of the door.

PASSAGE DOOR LATCH A pocket door latch that has no locking mechanism.

point The tip of a sawtooth. In general, the higher the number of points per inch of blade length, the slower and smoother the saw will cut. See Fig. P-7.

pointing 1. Smoothing or otherwise finishing off the mortar

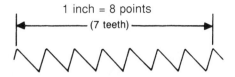

Fig. P-7. Points per inch on a blade of a handsaw

joints in a masonry wall. 2. Adding mortar to a joint with a trowel after the units are in place.

polarity 1. The state of electrification, either positive or negative. 2. The inherent quality or property of an object that exhibits opposite conditions or tendencies in opposite directions or toward opposite poles.

polarized Of or relating to a system or specially designed receptacles and plugs that have one leg larger than the other.

polarizing Of or relating to the use of color coding or other system intended to maintain an identified grounding conductor throughout an entire electrical system. See Fig. G-6.

polarizing leg One leg of the plug on an appliance or fixture cord that is larger than the other one, allowing it to be inserted into the receptacle in only one direction thus maintaining correct grounding throughout the system. See Fig. G-6.

poly Polyethylene plastic sheeting.

polychromatic Containing more than one color.

polyethylene expansion anchor See *anchor, hollow wall.*

polygon A closed figure, especially one having five or more straight sides. In a regular polygon, all the sides are of equal length. See Fig. P-8.

polygon miters Miters of matching angles that, when assembled, will form a regular polygon. For example, to construct an eight-sided polygon, all of the pieces would be mitered at 22 1/2 degrees.

polyisocyanurate insulation See *insulation, thermal.*

polystyrene insulation See *insulation, thermal.*

polysulfide rubber sealant See *sealant.*

polyurethane 1. An oil-modified urethane resin; used in varnish and many other finishing materials. 2. See *insulation, thermal.*

polyvinyl resin adhesive See *adhesive.*

pony wall A short wall used to support other members, such as floor joists or girders.

popcorn A nickname for the seamless acoustic material that is sprayed on some ceilings.

poplar See *hardwood.*

pop-off See *temperature and pressure relief valve.*

pop rivet A type of light-duty rivet installed with a pliers-like pop rivet gun by inserting the rivet in a predrilled hole, then squeezing the handles of the gun repeatedly until the rivet expands and locks the materials together.

pop-up An assembly consisting of a drain, a stopper, and various lifting rods and pivots; used to open or close the drain on a lavatory sink or bathtub. See Fig. P-9.

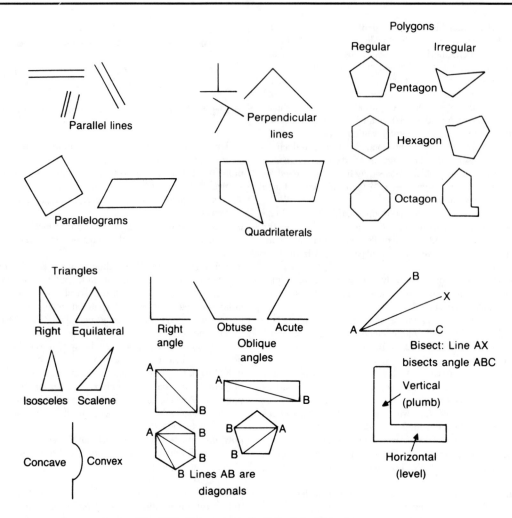

Fig. P-8. Polygons and other geometric shapes

porcelain   A type of hard, white, ceramic material, often glazed; used in the manufacture of bathroom fixtures, household accessories, and many other items.

porcelain enamel   A hard, smooth, waterproof glaze fused onto a metal surface under heat; commonly used over steel in the making of bathroom fixtures, appliances, and other objects. Available in various colors.

porch   An ornamental exterior entryway on the front, side, or rear of a building, usually covered.

pores   Comparatively large, open-ended wood cells set above one another to form continuous tubes. They appear as openings on the face of a piece of wood.

porosity   The percentage of a rock or other material that contains voids.

portable   Of or relating to any object that can be easily moved from one place to another for use.

portable appliance   See *appliance.*

portico   An open walk or porch covered with a roof that is supported by a continuous row of columns.

portland cement   A form of dry, hydraulic cement derived by combining approximately 4 parts limestone to 1 part clay, plus small amounts of iron oxide, silica, and other materials, then burning the mixture in a kiln, mixing the resulting pellets with small amounts of gypsum, then grinding it into a fine powder. Portland cement was developed in England, and gets its name from its resemblance to Portland stone, a widely used English building material. It is almost universally accepted by the construction industry for use in making concrete and other building materials. There are five basic types of cement, all of which are equal in strength and resistance to weather and aging.

TYPE 1 PORTLAND CEMENT   General-purpose portland cement.

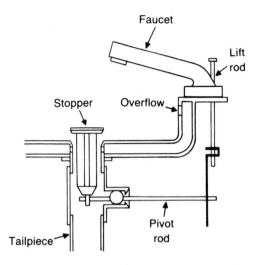

Fig. P-9. Typical bathroom sink pop-up assembly

TYPE 2 PORTLAND CEMENT A type of portland cement similar to type 1, but with a lower temperature of hydration (the heat created as concrete hardens); used for thick, heavy structures.

TYPE 3 PORTLAND CEMENT A portland cement with a high early strength, which hardens faster than the other types.

TYPE 4 PORTLAND CEMENT A type of portland cement giving off low heat, even lower than type 2; used in very massive structures, such as dams, etc.

TYPE 5 PORTLAND CEMENT An alkali-resistant portland cement; used where it may come into contact with water and soil having a high alkali content.

portland cement, white A type of portland cement made from materials specifically selected for their white color, such as chalk. It is a lower-strength cement; used primarily in decorative concrete facings.

positive pressure Air pressure within a building or device that is higher than the pressure outside.

post 1. A vertical structural member used to support a horizontal load. See Figs. O-1, P-2, and P-10. 2. A vertical structural member that carries a lateral load, as a fence post. 3. See *lumber.*

post anchor See *timber connector.*

post-and-beam construction 1. A method of framing using heavier lumber, such as posts, beams, and tongue-and groove planks, spaced farther apart than conventional framing. See Fig. P-10. 2. See *pier and post.*

post cap See *timber connector.*

potable water Clean water that is suitable for drinking, washing, and other common uses.

potential Electromotive force; electrical pressure.

potential energy See *energy.*

pot hook See *paint hook.*

powder-activated fastener A hardened steel nail, pin, or threaded rod that is driven into concrete or steel using a special gun and a small explosive charge.

powder room A general term applied to a small half bathroom, containing only sink and toilet.

power outlet An enclosed assembly of receptacles, circuit breakers, meters, and/or other equipment that supplies and controls power to mobile homes, recreational vehicles, or other mobile or temporary equipment.

PP See *plastic pipe.*

ppm See *parts per million.*

pre-cast concrete Concrete items that are factory poured in molds such as pier blocks, drain pipes, stepping stones, and decorative landscaping accessories; the opposite of cast in place.

precharged lines One of two copper tubes that are charged with refrigerant and sealed at the factory for use with a central air conditioning system to connect the inside evaporating unit with the outside compressor unit. One tube is a liquid line that carries the refrigerant in its liquid state from the compressor to the evaporator. The other tube is a larger, insulated suction line that returns the refrigerant to the compressor as a gas.

precut stud See *lumber.*

prefabricated Of or relating to an object or component of a system that is manufactured at a site other than where it will be used; opposite of site fabricated.

prefabricated construction A method of building where whole or partial walls, floors, and roofs are factory assembled, often with electrical wiring and plumbing in place, allowing rapid assembly of the building at the job site.

prefinished Any item, especially cabinets or woodwork, that is stained, painted, or otherwise finished at its point of manufacture, as opposed to being finished at its point of installation.

prefinished paneling See *paneling.*

prefurred stucco netting See *prepapered stucco netting.*

pre-hung door See *door.*

prepapered stucco netting Wire netting used for stucco that comes complete with line wire and a heavy kraft paper backing already attached, combining the otherwise separate steps of installing line wire, paper, and netting. Also called prefurred stucco netting.

prepasted wallpaper Wallpaper having a dry adhesive coating on the back side that is activated by dipping in water thus eliminating the need for a separate application of wet adhesive.

pressure connector See *solderless connector.*

pressure differential The difference in the pressure of a vapor between an area of high concentration and one of low concentration.

pressure regulator 1. A device used to control and adjust the pressure of air, gas, or other substances within a pressurized system. 2. A device located between a gas meter or tank and

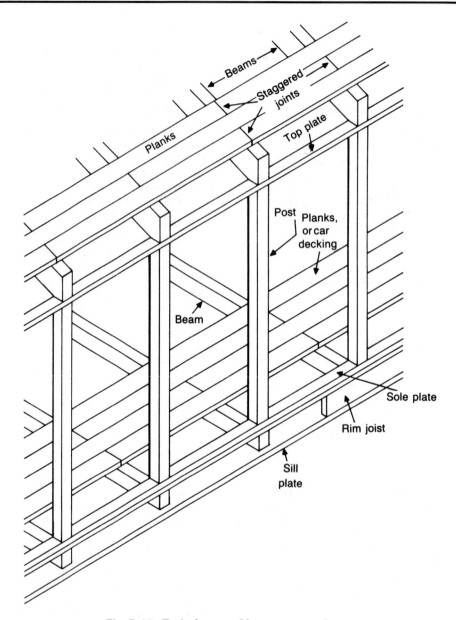

Fig. P-10. Typical post and beam construction

a gas furnace that is designed to maintain a slight positive pressure on the gas relative to the surrounding air.

pressure relief valve See *temperature and pressure relief valve.*

pressure-treated lumber See *lumber.*

prestressed concrete Pre-cast concrete units to which preliminary stresses have been introduced before placement, usually through the insertion of steel rods or mesh.

prevailing Of or relating to the direction from which the wind usually blows in a given area.

primary color A base color; a color that cannot be obtained by mixing other colors. In painting, the primary colors are red, blue, and yellow.

primary combustion See *combustion.*

primary conversion unit   See *furnace, combustion.*

prime   1. To apply the first coat of paint. 2. To displace air from a machine or liquid line so as to promote suction.

prime coat   The first coat of a two- or three-coat paint job, designed to seal the surface and prepare it for the subsequent coats. Different primers are available, depending on the material being painted. If a specialized primer is not used, this first coat is then usually called a base coat.

prime contractor   The contractor on a construction job who is in overall control of the project, including all subcontractors, schedules, and materials.

primer   1. Specially formulated paint used for the prime coat on a surface. 2. Any first coat of paint.

prime solar fraction   The ratio of the estimated amount of winter sunlight actually available at a solar collection site to that which would be available if the site had no obstructions.

prime window   The originally existing window to which a storm window was added.

privacy lock   See *lockset.*

private sewer   A main sewer line that receives the discharge of two or more building drains before joining a public sewer line.

projection   A structural or decorative member that protrudes out past the surrounding wall surface.

proof   The alcoholic strength of a mixture, denoted by a number that is twice the percent of alcohol by volume in the mixture. For example, a 100-proof mixture contains 50 percent water and 50 percent alcohol.

property line   One of the legal boundaries of a parcel of land; also called a lot line. See Fig. P-4.

protractor   A device used in drawing for measuring, laying out, and transferring angles.

provincial   Of or relating to furniture or architectural styles derived from and native to a specific region or province, usually bearing a descriptive geographical name, such as French provincial.

pry bar   See *wrecking bar.*

psi   Abbreviation for *pounds per square inch*, a measurement of pressure.

psychrometer   An instrument used to measure the amount of water vapor present in the air.

P-trap   A plumbing drainpipe shaped like a horizontal letter P, used as a fixture trap. See Fig. P-11.

P/T valve   See *temperature and pressure relief valve.*

public sewer   A system of sewage piping installed by a city, county, or other municipality to receive the raw sewage from individual buildings and transport it to a treatment site.

public water main   A water supply pipe administered by a city, county, or other authority, and intended for public use.

pull box   See *service pedestal.*

pull-chain fixture   A simple shaded or unshaded light fixture that is controlled by a pull-chain switch.

pull-chain toilet   A two-piece, gravity-fed toilet having a tank that is mounted approximately 6 feet off the floor and is activated by a long pull chain.

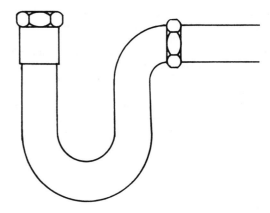

Fig. P-11. P-trap

pulley stile   One of the vertical side pieces of a double-hung window frame into which the pulleys for the sash weights are installed.

pulling elbow   A 90-degree conduit used to attach conduit to a panel or box. It has a removable cover to allow wires to be easily pulled and bent through the elbow for insertion into the box.

pullman   See *vanity.*

pull-out block   See *main disconnect switch.*

pull section   The area inside an underground service panel into which the service conductors are pulled for connection to the meter. See Fig. S-10.

pumice stone   A fine abrasive powder made from crushed volcanic lava and mixed with rubbing oil or water to rub out wood finishes. Similar to rottenstone, but more coarse and abrasive.

punch list   A list prepared at or near the end of a construction project by the owner, architect, or both, noting those things left to be completed or corrected by the contractor.

purchase   A mechanical hold or grip secured in raising or moving an object.

purlin   1. A horizontal framing member used to support roof rafters at a point approximately midway between the top plate and the ridge. See Fig. R-5. In flat roof framing, a horizontal member that intersects a main beam at right angles and supports rafters or prefabricated panels running parallel to the beam.

push block   See *push stick.*

push drill   A type of hand drill for small holes that is operated by pumping up and down on the drill's handle, which in turn causes the drill bit to rotate.

push plate   A decorative metal plate placed on a swinging door to protect the door's finish from constant handling.

push stick   A narrow, notched stick used to push or guide narrow lumber through a woodworking machine; also called a push block.

putlog A horizontal or diagonal crosspiece in a scaffolding having one end placed in a hole in the wall and the other end attached to the scaffolding.

putty A type of cement, commonly about 90 percent whiting (calcium carbonate) and about 10 percent raw linseed oil, that is kneaded to a doughlike consistency; used to set glass, fill small holes, and do other similar sealing.

putty stick A pigmented wood filler in stick form, similar to a crayon; used to repair scratches, nicks, nail holes, and other defects in finished wood.

PVA adhesive See *adhesive: polyvinyl resin adhesive.*

PVC See *plastic pipe.*

pyranometer An instrument that measures the total of direct, diffuse, and reflected solar radiation at a given spot; used in designing solar heating systems.

Pythagorean theorem One of the principles of a right triangle, stating that the square of the hypotenuse is equal to the sum of the squares of the two sides, and also that a right triangle will have sides whose lengths are a ratio of 3:4:5.

quad pane window See *window.*

quadrant 1. That section of a circle contained within an angle of 90 degrees; a quarter of a circle. 2. An arc of 90 degrees.

quadrilateral Of or relating to a closed figure having four straight sides. See Fig. P-8.

quarry tile A large, unglazed, machine-made tile; commonly used for outdoor paving and for interior and exterior floors.

1/4 bend Of or relating to a plumbing elbow having a radius of turn equal to 1/4 of a circle, or 90 degrees.

quarter round Of or relating to a small molding, available in various sizes, having a cross-section of a quarter circle. See Fig. M-4.

quarter sawn Of or relating to lumber that has been cut at approximately a 90-degree angle to the tree's growth rings. See Fig. G-3.

quartrefoil An architectural decoration in the form of a four-leaved or four-lobed pattern.

Queen Anne 1. Of or relating to a style of architecture dating to the eighteenth-century reign of England's Queen Anne that makes extensive use of red brick and emphasizes an uncomplicated form with simple ornamentation. 2. Of or relating to a furniture design of the same period, stressing soft curved lines, simple upholstery, and comfort.

queen post One of the vertical posts used in the construction of certain types of trusses.

quench To rapidly cool an object from a high temperature by immersing it in a liquid or gas.

quench-hardened nails See *nails, finishes.*

quick-closing valve A faucet or valve with an internal mechanism that automatically closes the valve and stops the water flow upon release of the handle.

quicklime See *lime.*

quill The steel cylinder in the head of a tool such as a drill press that contains the bearings and spindle and is geared to a handle or other device to allow it to be raised, lowered, or locked in position.

quirk A lengthwise groove that separates a flat surface from those which are curved or otherwise molded.

quoin A large, squared stone sometimes used to form the corners of a brick, stone, or other masonry building.

R

R In electrical calculations, the symbol for resistance in ohms.

rabbet A two-sided cut along the edge or end of a piece of wood; also called a rebate. See Fig. D-1.

rabbet plane A small, narrow hand plane used for cutting rabbets or other grooves in wood.

raceway Any of a variety of channels, usually metal, that are designed and used for the purpose of enclosing and supporting electrical wires, cables, or bars. Raceways can be surface mounted or buried underfloor in concrete slabs.

racking 1. In masonry, stepping back successive courses at a corner where two walls meet, then tying the corner units in to each other. 2. Pulling or pushing a framed wall or other structure into square. 3. The tendency of an unbraced rectangular frame to distort from square as a result of lateral stress.

radial arm saw A stationary power tool having a circular saw blade mounted on an overhead track; used for a variety of cutting operations on wood and other materials, especially for straight and angled crosscutting.

radial system A system of forced air heating ducts in which each individual duct runs directly from the plenum to the room it services.

radiant heating Heating a building, either by using hot water pipes in the walls, floor, and/or ceiling or by using electrically heated cables or panels to cause warm air to radiate into the building rather than be blown in.

radiation See *heat transfer.*

radius A straight line from the center of a circle to a point on the circle's circumference; equal to one half of the diameter. See Fig. C-3.

radon A colorless, radioactive gas produced from the decay of radium in certain rocks and soil. It is thought to pose a risk of lung cancer if it accumulates in poorly ventilated buildings in high enough concentrations.

radon daughter A charged metal atom that is a product of the radioactive decay of radon and can attach itself to dust. Both the attached and unattached particles can be inhaled and can lodge in the lungs.

rafter One of a series of structural members that make up the framing for a building's roof and roof overhang and that support the roof loads. See Figs. B-2, C-4, O-2, P-3, R-1, R-3, and R-5. The different types of rafters follow.

COMMON RAFTER A rafter running full length from wall plate to ridge at a right angle to both when viewed from above. See Fig. R-6.

HIP RAFTER A rafter running from the corner of a building to the end of the ridge at a 45-degree angle to both when viewed from above.

JACK RAFTER A shorter, intermediate rafter in one of three types. A *cripple jack rafter* is used in the area between a hip and a valley rafter. A *hip jack rafter* extends from the wall plate to a hip rafter. A *valley jack rafter* extends from the ridge to a valley rafter.

VALLEY RAFTER A rafter that extends diagonally from the plate to the ridge at the internal angle formed by the intersection of two roof slopes.

rafter length The distance a rafter must span, measured from the ridge to the outside edge of the wall plate, but not including any overhang. See Fig. R-6.

rafter level cut 1. The horizontal cut of a bird mouth, at right angles to the rafter plumb cut. See Fig. R-5. 2. A horizontal cut on the bottom of a rafter tail to provide a nailing surface for a closed cornice. See Fig. C-4.

rafter plumb cut See Fig. R-5. 1. The vertical cut on the end of a rafter where it meets the ridge. 2. The vertical cut of a bird mouth, at right angles to the rafter level cut. 3. A vertical cut on the end of a rafter to provide a nailing surface for the fascia.

rafter seat The horizontally cut portion of a rafter that sits on the wall plate. Also see *seat cut.*

rafter square See *framing square.*

rafter table A table of numbers printed on the blade of a framing square used in determining the length of various types of rafters at a variety of slopes.

rafter tail The end of a rafter that extends out past the wall plate. See Fig. R-5.

raggle A slot or groove that has been cut or formed into a masonry wall to receive a flashing.

rail 1. A horizontal member of a framework, such as a door, face frame, or sash. See Fig. D-3. 2. A horizontal member between two vertical supports, as in a fence or balustrade.

rail bolt A specialized type of bolt, pointed at both ends and having a wood screw thread on one end and a machine screw thread with a nut and washer on the other end. Used for blind-joining sections of stair rail. Also called a stair rail bolt. See Fig. R-2.

rain gutter See *gutter.*

raised grain A condition where the harder summerwood in a piece of lumber is raised above the softer springwood, leaving a rough surface.

rake 1. The extension of a gable roof beyond the end wall of the house. See Figs. C-13 and R-3. 2. A group of trim pieces applied to a gable end wall, forming the finish between the roof and the wall.

raked joint 1. A butt joint at the meeting of two inclined members deviating from a perfect plumb-cut joint. 2. A type

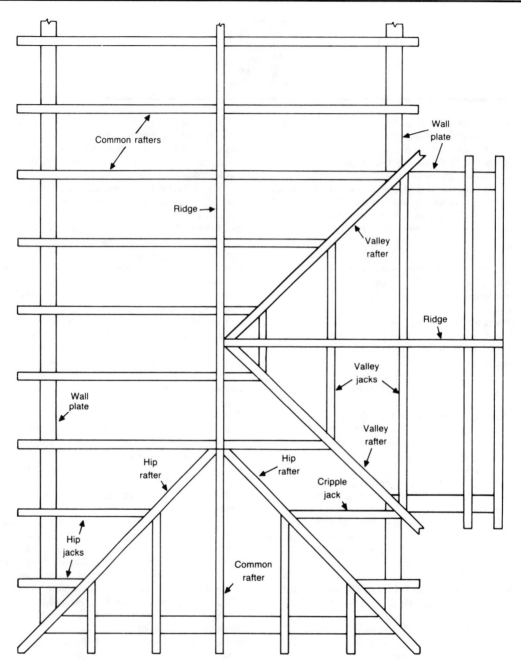

Fig. R-1. Rafter types

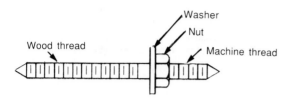

Fig. R-2. Rail bolt

of joint, used between bricks or concrete blocks in a masonry wall, in which mortar has been removed to a set depth.

ramp An inclined surface used to connect one level to another.

random lengths A quantity of lumber supplied in various lengths.

random widths A quantity of lumber supplied in various widths.

range hood See *hood*.

range outlet A 220-volt electrical outlet having a specific pattern of three slots in its face, designed to mate only with the plug on an electric range. The special outlet/plug combination is used as a precaution to prevent the range from being plugged into an outlet of incorrect voltage and/or amperage. A similar specific outlet/plug combination is also used for electric clothes dryers.

rasp A type of file having small, coarse, individual teeth; used for the rough shaping of wood.

ratcheting brace See *brace, ratcheting*.

ratio The relationship of two or more quantities to each other.

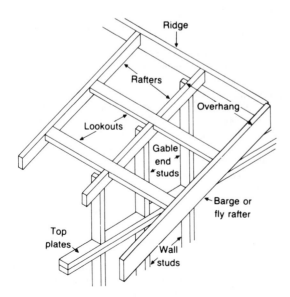

Fig. R-3. Components of a typical rake on a gable roof

rat sill See *plate: sill plate*.

rat tail file Any of a variety of round, tapered woodworking files, typically 4 to 20 inches in length, designed to shape and enlarge holes and cutouts. See Fig. R-4.

Fig. R-4. Rat tail file

raw linseed oil The crude product refined from flaxseed, with little or no other treatment.

raw sewage The liquid and solid waste from a building's soil pipes before treatment.

razor knife See *utility knife*.

readily accessible Of or relating to a fixture, connection, or piece of equipment to which direct access is provided without the need to remove a panel, door, or other covering.

ready-mix 1. Concrete that is premixed and delivered to the job site ready for use. 2. Premixed dry concrete, plaster, stucco, and other similar materials supplied in bags and needing only to be mixed with water to be ready for use.

ready-to-assemble Furniture, cabinets, and other items shipped in pieces and having all of the necessary slots, holes, and hardware needed for on-site assembly with a minimum of tools and skill. Abbreviated RTA.

ream To smooth and finish a drilled hole to an exact size.

reamer A tapered cutting tool used to grind off internal burrs in cut pipe, tubing, conduit, etc.

rebar See *reinforcing bar*.

rebate See *rabbet*.

receptacle A device installed at an outlet location to serve as an electrical contact for the connection of plug-in electrical appliances and equipment.

receptacle outlet An outlet box designed to receive one or more receptacles.

receptor A plumbing fixture or other device capable of adequately receiving the discharge of an indirect waste pipe.

recessed light fixture A lighting fixture that is set into a ceiling, soffit, cabinet, etc., so that it is flush with the surrounding surface.

reciprocating Of or relating to an object that moves alternately up and down or forward and backward.

reciprocating saw A portable power tool having a blade clamped to the end of a shaft, which moves out and back in a direct line with the motor; used primarily for cutting openings and for tear-out work.

recoat time The minimum length of time that a coat of paint or other finishing material must dry before a subsequent coat can be applied.

rectifier An electrical device that is used to convert alternating (bidirectional) current into direct (unidirectional) current.

redhead 1. A nickname for a wire nut that has a red color coding. 2. (*R*) A brand name for a type of wedge anchor used in masonry.

red-tag Informal term for a construction project that has been ordered temporarily stopped by building officials or other government agencies due to unsafe conditions or noncompliance with the building codes. Derived from the red "Violation Notice" or "Stop Work" tags used by many building officials.

reducer A fitting used to connect a pipe or duct of one diameter to one of a smaller diameter; also called a reducing coupling. See Fig. T-4.

reducing bushing A plumbing fitting having male threads of one diameter and female threads of a different, smaller diameter. It is screwed into a fitting to adapt it to a smaller pipe. See Figs. S-13 and T-4.

reducing coupling See *reducer*.

reducing elbow An elbow having two openings of different sizes. See Fig. S-13.

reducing tee A T-shaped plumbing fitting in which one outlet has a different diameter than the other two. See Fig. T-4.

redwood See *softwood*.

reeding A type of decoration, usually a series of rounded moldings running parallel and in contact with one another.

reflectance 1. A surface's ability to reflect light, expressed as a ratio of the light reflecting off a surface to the light striking the surface. 2. The ratio of solar radiation falling on an object to the amount being reflected off the object.

reflected radiation See *solar radiation*.

reflective film A shaded film applied to the interior surface of window glass to reflect sunlight away from the window and so minimize heat gain.

reflective glass A special type of glass that reflects sunlight back to the outside, reducing heat gain and glare.

reflective insulation See *insulation: thermal*.

reflectivity The tendency or ability of a material to direct light or sound.

reflector A device or surface coating used to direct light from a fixture in a desired direction.

refractory Any of a variety of solid, heat-resistant materials used to protect the interior of furnaces and other high heat areas.

refractory cement See *adhesive*.

refrigerant See *cooling system*.

refrigerator panel A wood panel with one faced edge, usually 24 x 84 inches in size, used in a run of cabinets to conceal and protect one or both sides of a refrigerator.

register A device, attached to the boot at the end of an air duct inside the building, having fins to direct the flow of air and adjustable louvers to regulate the amount of air flow or shut it off entirely. Some common types of registers follow.

BASEBOARD REGISTER A square or rectangular register that angles back from bottom to top, used at the juncture of floor and wall to direct air out and somewhat up.

CEILING REGISTER See *wall register*.

FLOOR REGISTER A rectangular, slip-in register, usually having fins that direct the air to both sides.

ROUND REGISTER A register used for ceiling installations where a 360-degree flow of air is desired.

WALL REGISTER A square or rectangular screw-on register designated as one-, two-, three- or four-way, depending on the number of directions in which the fin arrangement directs the air flow. This type of register is also used for ceiling mounting.

reglet A long, narrow slot or groove formed in a wall, designed to receive a flashing or to serve as an anchorage.

regular polygon See *polygon*.

reinforced concrete Concrete in which metal reinforcing bars have been embedded so that the two materials act as a unit for much greater strength in resisting forces.

reinforced concrete construction A method of building in which the walls, floor, and other structural members are constructed of reinforced concrete.

reinforcing Of or relating to metal bars, straps, or mesh placed in concrete for strength.

reinforcing bar A steel rod of various diameters, usually 10 feet in length, designed to be embedded in concrete. The rods have a deformed face to allow a better bond with the concrete. Commonly called rebar.

reinforcing wire Ten-gauge wire formed into sheets or rolls having a pattern of 6-x-6-inch squares; used for reinforcement in concrete slabs and other flatwork. Also called hogwire.

relative humidity The amount of water vapor in the atmosphere, expressed as a ratio of the actual amount of water vapor in the air to the amount of water vapor the air could potentially hold at the same temperature.

relief cut A gradual, preliminary cut made with a bandsaw or jigsaw, which then allows the saw to cut a sharper or deeper curve.

relief vent A plumbing vent used in certain circumstances to provide additional air circulation between drainage and vent systems.

remote-mounted collector A solar collector mounted on the ground, on an outbuilding, or any area other than the roof of the building it serves.

rendering A drawing of a proposed construction project, usually in color and well detailed, used to illustrate a completed building or to visually explain other design concepts.

renewable energy 1. Energy sources that are essentially in indefinite supply, such as solar, wind, and water. 2. Wood and certain similar energy sources that can be cultivated and grown to replace that which is used.

renewables See *renewable energy*.

repetitive usage The engineering principle that a combination of three or more structural members used together will be approximately 15 percent stronger than the sum of their individual strengths because any defect in one of the members will be supported by the other members.

reradiation Radiation that is reemitted from previously absorbed radiation. For example, sunlight that is absorbed in a storage mass, then given off as heat.

resaw To saw a thick piece of lumber into two or more thinner pieces.

reserve   1. On a large construction project, an amount of money held in a trust account or other holding account, used to guarantee to the contractor, lending institution, or other interested parties that sufficient funds are available to cover changes or cost overruns in the project. 2. See *reserve field.*

reserve field   Additional land area on a building site that is reserved for the future repair or expansion of an existing septic system's drain field. See Fig. P-4.

reset button   See *thermal cutout.*

resilient   Of or relating to a material that is able to withstand temporary deformation, springing back to its original shape when the deforming force is removed.

resin   A syrupy or solid secretion from various trees and plants (natural resin), or a synthetic organic chemical compound (synthetic resin), both of which are soluble in various organic solvents. Used in the manufacture of finishes, plastics, and many other materials. Also see *paint resin.*

resinous wood   Lumber from trees that contain fairly high amounts of resin, such as fir and pine.

resistance   The opposition to current flow. Electrical energy lost through resistance is given off as heat.

resistance heat   A form of electric heat in which electricity is passed through a metal element that has a high resistance, causing the element to heat up, giving off the heat to the surrounding air. Resistance heat is used in many types of heaters, furnaces, and cooking appliances.

resorcinol resin glue   See *adhesive.*

resteel   See *reinforcing bar.*

retaining wall   A wall constructed to contain a lateral force, such as a bank of dirt.

retarder   An ingredient that is added to paint, varnish, plaster, or other materials to extend the drying time.

retractable rule   See *rule.*

retrofit   To remodel or alter a structure to allow for the installation of a new or updated component.

return   1. A molding, cornice, or other finish that turns back from its original direction, usually ending against a wall or other surface. See Fig. T-7. 2. Commonly, all of the ducts and registers used to carry the air back to a furnace for recirculation.

return air   Cool air that is drawn back into a furnace for reheating.

return air duct   One of the ducts through which return air is routed to a furnace.

return nosing   The nosing of an open stair, mitered to return and extend past the balustrade.

reveal   1. That part of the edge of a door or window jamb not covered by the casing. See Fig. D-6. 2. The edge of any surface left partly uncovered by the trimwork.

revent   See *back vent.*

reverberation sound   A sound that reflects off of a wall, floor, or other surface, causing the sound to continue after its source has stopped; an echo.

reverse bevel   Any bevel, such as that on the edge of a door, which is at an angle opposite the normal bevel.

reverse board-and-batten   See *batten-and-board.*

reverse board-and-batten plywood   See *plywood siding.*

reversible base   A type of baseboard having a rounded edge on two opposite corners, allowing miter cuts to be reversed to save material and cutting time during installation.

revolutions per minute   The designation on a power tool or other powered device of the speed at which the blade or motor shaft revolves. Abbreviated RPM.

rheostat   A type of electrical resistor that can be adjusted to provide varying amounts of resistance, thereby changing current flow. Used for controlling motor speeds and lighting, and for other similar applications.

rib   1. One piece of a curved framework in a ship. 2. One of the arches in a vaulted ceiling or wall that meet and cross one another, dividing the vaulted space into triangles.

ribband   See *ribbon.*

ribbon   1. In balloon construction, a long piece of wood or metal set into the studs to support the floor joists; also called a ribband. See Fig. B-2. 2. See *ledger.*

rich lime   A quicklime formed from the calcining of pure or nearly pure limestone; also called fat lime.

ridge   The horizontal line at the junction of the top edges of two sloping roof surfaces. See Figs. O-1, R-1, R-3, R-5, and R-6.

ridge board   A board placed on edge at the ridge of a roof frame, to which the rafters are fastened. See Figs. R-3 and R-5.

ridge capping   See *hip and ridge.*

ridge pole   See *ridge board.*

ridge roll   A narrow row of mineral-surfaced asphalt roofing used to cover the ridge.

ridge ventilator   A raised, screened area along the ridge of a building, placed to allow hot air in the attic to escape.

riffler   Any of a variety of double-ended wood or metalworking files having curved, pointed ends. Used for carving, pattern-making, model-making, and similar applications.

right angle   An angle of exactly 90 degrees. See Fig. P-8.

right-hand door   See *door.*

right-hand reverse door   See *door.*

right triangle   See *triangle.*

rigid insulation   See *insulation, thermal.*

rim joist   A joist nailed perpendicular to the ends of the other floor joists at the sill level or where second-floor joists end on an exterior wall; also called a header joist or band joist. See Figs. J-2, P-2, and P-10.

rim lock   See *lock.*

ring-shank nail   See *nails, shank types.*

rip   To cut with or along the grain of a piece of wood; the opposite of crosscut.

ripping bar   See *wrecking bar.*

ripping hammer   See *hammer: claw hammer.*

ripsaw   See *handsaw.*

rise   1. The vertical height of a roof, measured from the wall's top plate to the centerline of the rafters at the ridge. See Fig. R-6. 2. The vertical height of a flight of stairs. See Fig. S-16. 3. The vertical height of a single step. See Fig. S-16.

rise per run   The increase in vertical distance in inches that a roof or other slope gains over every 12 inches of horizontal distance. See Fig. R-6.

riser 1. The vertical board placed between the treads of a stairway. See Fig. S-18. 2. A vertical run of pipe. 3. A short length of threaded pipe used in sprinkler systems to connect a sprinkler head to a riser fitting. 4. A pipe that connects an individual solar collector to the header.

riser elbow A fitting for use with plastic pipe in sprinkler systems, with one side being a slip fitting to attach to the pipe and the other side being threaded for a riser. Riser tees and side-outlet riser elbows are also available.

rivet A soft, unthreaded metal bolt with a head on one end; used to join two pieces of metal, leather, or various other materials. The rivet is inserted in a hole drilled through the two pieces being joined, then the straight end is flattened out with a rivet set to lock the pieces together. Rivet sizes

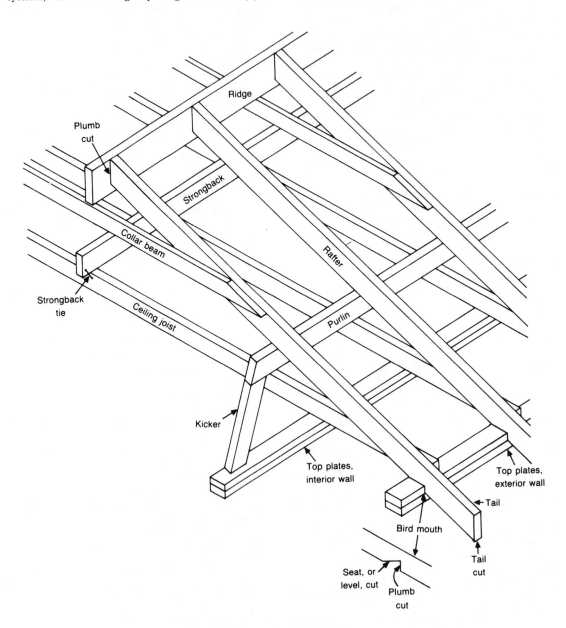

Fig. R-5. Parts of a typical roof frame

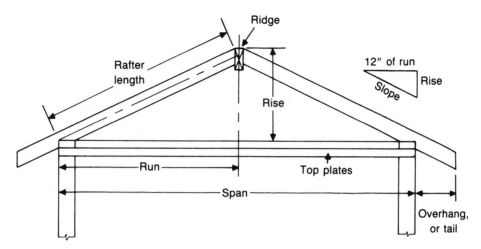

Fig. R-6. Roof framing terms

are usually specified by the rivet's weight per thousand; for example, 1000 two-pound rivets weigh two pounds.

R lamp   Abbreviation for *reflector lamp*, an incandescent bulb having an interior coating to help direct and reflect its light. It falls roughly between a standard light bulb and a par bulb in both cost and efficiency.

rock bed   A rock-filled enclosure, usually airtight, incorporated into a building's walls or foundation to store and reradiate solar heat.

Rockwell hardness rating   A rating of hardness given to a piece of metal after processing. Using a Rockwell Hardness Tester, a steel ball or diamond point is pressed into the metal under a predetermined load, and the penetration into the metal is indicated on a gauge.

rock wool   See *insulation, thermal*.

rococo   A style of architecture and decoration, following the Baroque period. It consists primarily of elaborate and delicate ornamentation in the form of soft curves, foliage, and other similar designs.

roller catch   See *friction catch*.

roller check   See *lumber defect*.

roll roofing   A roofing material composed of asphalt-impregnated felt, usually covered with crushed mineral; commonly available in 36-inch-wide rolls containing 108 square feet, and weighing 45 to 90 pounds per roll.

roll-up door   See *door*.

Roman brick   See *brick*.

Romex   A trade name for nonmetallic sheathed cable; extensively used in the construction industry for interior electrical wiring.

roof board   See *roof sheathing*.

roof decking   See *decking*.

roof drain   A receptacle on a flat roof in which rainwater runoff is collected. The drain is attached to a downspout, which

carries the water to the ground or to a disposal area.

roof frame   All of the various members that make up the rough framing of a roof structure. See Figs. R-5 and R-6.

roofing   Any of a wide variety of materials applied to a roof frame to make it waterproof.

roofing felt   See *felt paper*.

roofing nail   See *nails, common types*.

roofing tile   A terra cotta or other ceramic tile, available in various shapes and colors, used as a roof covering and having an extremely long lifespan and low heat gain. Main drawbacks are weight—as much as 1000 pounds per square—and high initial cost.

roof jack   1. A cone-shaped, sheet-metal flashing mounted on a flat base, designed to seal the area where a vent, pipe, duct, or conduit penetrates the roof. Some types have a self-sealing rubber ring that grips the pipe; others seal with a storm collar and/or a sheet-metal cap. Also called a roof safe. See Fig. R-7. 2. A sheet-metal can with an overhanging cap, mounted on a roof to provide ventilation for the attic; also called a roof vent. See Fig. R-7.

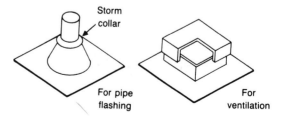

Fig. R-7. Two types of roof jacks

roof joist   A horizontal structural member of a flat roof that acts as both a roof rafter and a ceiling joist.

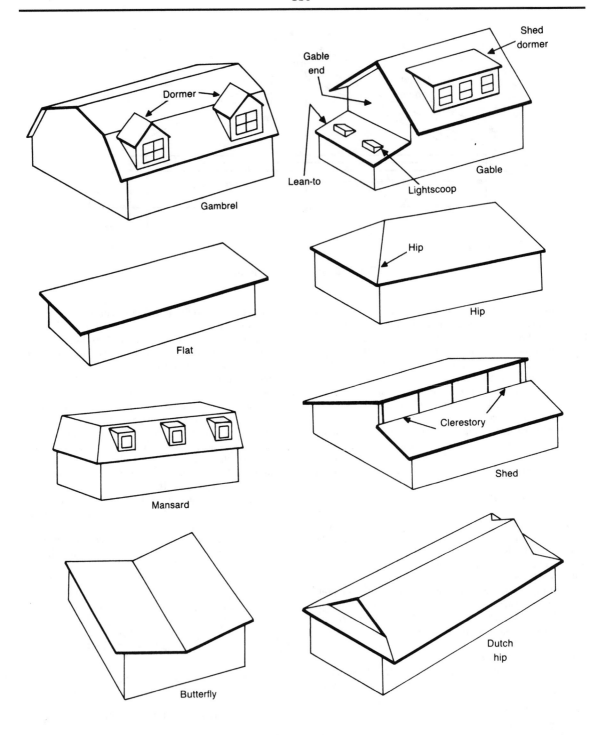

Fig. R-8. Common roof configurations

roof pond One of the open, water-filled troughs or plastic bags arranged on a building's roof to capture heat from the sun. They are covered at night, allowing the stored heat to radiate down into the building.

roof safe See *roof jack*.

roof sheathing Lumber, plywood, or other building boards fastened to the top of the rafters, over which the roofing is installed. See Fig. O-2.

roof tile See *structural clay tile*.

roof types See Fig. R-8.

roof vent See *roof jack*.

room divider Any object other than a wall that is used to break up a large, interior space, or to show the boundary between two rooms.

rose A decorative, metal trim plate used behind the knob on a lockset to cover the lock mechanism and the hole in the door; supplied with the lock in pairs, one for each side of the door.

rose nail A nail with a raised, decorative head used to decorate doors and other solid surfaces.

rose plate The steel retaining ring that holds a lockset in place on a door.

rosette 1. A round, flanged bracket used to support the end of a closet pole. 2. Any round, decorative feature resembling a rose.

rosin A type of resin obtained from the steam distillation of dead pinewood.

rotary-cut veneer A long, thin, continuous sheet of veneer peeled from a log that has been placed in a lathe and rotated against a broad cutting knife.

rotary hammer A type of portable power drill that combines a rotary drilling motion with an in-and-out hammering action; used to speed the drilling of holes in concrete and masonry.

rottenstone A powdered, abrasive material made from pulverized shale and limestone. It is mixed with rubbing oil and used to produce a smooth final finish on wood. Similar to pumice stone, but less abrasive.

rough-in To install the initial pipes, wires, ducts, etc., in a building's walls, ceiling, crawl space, or other area that will subsequently be covered.

rough lumber See *lumber*.

rough opening An opening constructed in a wall or other framework of the correct size to receive a door frame, window frame, or other finish unit. See Fig. F-10.

rough-sawn plywood See *plywood siding*.

rough sill The lowest member of a rough opening, upon which a window or door rests and over which the finished sill is applied. See Figs. F-10 and W-3.

route To remove wood using a power router or chisel, either for assembly of a joint or for decoration.

router A portable power tool with a high-speed vertical spindle, to which various cutters can be attached for making grooves, joints, inlays, or decorative edges in wood.

router plane A type of hand plane, usually with interchangeable blades, that functions similar to a power router for cutting dadoes, rabbets, and inlays.

rovings Small pieces of glass fiber, used in the making of fiberglass fabric. Also see *fiberglass fabric*.

rowlock 1. Bricks laid in such a manner that their ends are visible in a vertical position in the wall face. See Fig. B-10. 2. A row of bricks laid on edge.

RPM Abbreviation for *revolutions per minute*.

RTA Abbreviation for ready-to-assemble. See *ready-to-assemble*.

rubber cement See *adhesive*.

rubbing oil A light petroleum- or paraffin-based oil; used with pumice or rottenstone for polishing and rubbing out wood finishes.

rubble Blocks or pieces of stone in nonuniform sizes, laid up in a random pattern or in irregular horizontal courses as a decorative facing.

rubble masonry Uncut stone used for rough work, including fill, backing, fences, etc.

rule Any of a variety of metal, plastic, or wooden measuring devices. Common types include the following.

FLEXIBLE RULE A narrow graduated strip of metal or cloth that is contained in a metal or plastic case, or in an open reel. The rule is pulled out as needed, then rewound using a small crankstyle handle. Common lengths are 50, 100, and 200 feet.

FOLDING RULE Short lengths of wood, metal, plastic, ivory, or other materials, graduated in successive numbers. They are folded for easy transport and storage, and are partially or completely unfolded for measuring. Folded length is typically about 6 inches; unfolded lengths usually range from 2 feet to 6 feet or more.

RETRACTABLE RULE A thin, narrow, flexible, graduated strip of metal contained in a spring-loaded metal or plastic case. The rule is pulled out as needed, and can be held in that position using a small thumb latch that is part of the case. Releasing the latch allows the rule to retract automatically into the case. Common lengths are 6, 10, 12, 20, 25, and 30 feet.

STEEL RULE A thin, flat, rigid or semiflexible steel strip, graduated along one or both edges, usually in 1-, 2-, or 3-foot lengths.

run 1. The horizontal width of a roof, measured from the outside of the exterior wall to the center of the ridge; one half the span. See Fig. R-6. 2. The horizontal distance covered by a flight of stairs. See Fig. S-16. 3. The width of a step from the nose of the step to the riser of the next step. See Fig. S-16. 4. A horizontal or vertical series of pipes, ducts, or wires. 5. A sag, drip, or irregularity in a finished surface caused by an uneven flow of material. 6. A row of cabinets along any one wall. 7. To install or extend a pipe, duct, or electrical wire from one location to another.

runner 1. A guide used to align and assist the motion of a sliding part, such as a drawer. 2. A long, narrow strip of carpet used in a hallway or similar area.

running board One of the boards placed on either side of an electrical cable that has been run across the tops of the joists in an attic; used to keep anyone from stepping on or otherwise unintentionally contacting the cable.

running foot   See *linear foot*.

running splice   An electrical wire that is spliced into the run of another wire and wrapped with electrical tape, without being enclosed in a box. This was a common procedure with knob-and-tube wiring, but is no longer permitted by the electrical codes.

rush   The stems of various grasslike plants, dried and woven for use as chair seats.

rust   A form of corrosion in unprotected metal caused by a chemical reaction in which water and oxygen combine with the metal to form metallic oxide, a solid that resembles metal but is weak and brittle.

R-value   See *heat resistance*.

S

saber saw   See *jigsaw*.

sabin   A unit for the measurement of sound absorption, used in rating the acoustic properties of a material. The scale ranges from 0 (completely sound-reflective) to 1 (completely sound-absorptive).

saddle   1. A small, double-sloping roof, used where a vertical surface meets a sloping surface for the purpose of diverting rain or snow. Most commonly used behind a chimney where it passes through the roof. Also called a cricket. See Fig. F-7. 2. Wood or metal trim placed on the floor between two door jambs. See *threshold*.

saddle board   A V-shaped board or pair of boards sometimes used in place of ridge shingles to finish and protect the exterior ridge of a pitched roof. Also called a comb board.

saddle fitting   In plumbing, a tee or valve that is clamped over a predrilled hole in a piece of pipe, allowing for a new line to be taken off the existing pipe without disassembling it. See Fig. S-1.

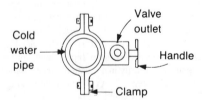

Fig. S-1. Saddle fitting

saddle hanger   See *timber connector*.

SAE   Abbreviation for *Society of Automotive Engineers*.

safety control value   A device on a gas appliance next to the pilot light that senses the presence of excess gas. If the pilot light is extinguished, it shuts the system down before an explosion can occur.

safety factor   See *factor of safety*.

safety glass   Any of a variety of reinforced types of glass, such as tempered or laminated, that offer greater impact resistance than regular glass and form small, rounded pieces when broken, instead of jagged slivers.

safety plate   See *timber connector*.

sag   The tendency of a liquid such as paint to run when applied to a vertical surface.

sailor   A brick laid so as to be standing on end, with its broad face exposed.

sail switch   A type of switch, used with forced-air heating and cooling systems to control an attachment such as a humidifier or an electronic air cleaner. A plastic panel, or sail, is attached to the switch and placed inside an air duct, where air movement in the duct will activate the switch. When the air movement ceases, the switch shuts off the accessory.

salt hydrate   See *eutectic salt*.

sander-grinder   A stationary power tool for sanding, grinding, and sharpening, having a long, narrow sandpaper belt that runs vertically over a set of pulleys. An adjustable table is mounted perpendicular to the belt to hold and guide the workpiece.

sand finish   1. A method of texturing the last coat of stucco, created by gently rubbing away the cement to expose the sand. 2. A rough texture for interior walls, created by adding sand or similar material to plaster, joint cement, or paint.

sand-float finish   Lime mixed with sand; used as a texture finish over plaster, stucco, or other wall surfaces.

sandpaper   Heavy paper coated with any of a variety of natural and artificial abrasives. Some of the terms associated with sandpaper follow.

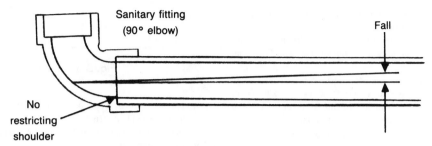

Fig. S-2. Sanitary fitting used with a soil pipe

ALUMINUM-OXIDE SANDPAPER A hard, long-lasting abrasive paper; used primarily for power-sanding wood.

CLOSED-COAT SANDPAPER Sandpaper having the abrasive particles spaced close together; the standard paper for most work.

FLINT The least durable and least expensive abrasive, often used as a "throw-away" sandpaper on painted areas that clog the paper quickly.

GARNET A natural, reddish abrasive; used for most all-around sanding of clean wood.

OPEN-COAT SANDPAPER Paper on which the particles are widely spaced; for use on materials that tend to clog the paper quickly, such as paint.

SILICON-CARBIDE PAPER A synthetic abrasive that is very hard and sharp, but somewhat brittle; used on hard plastic, ceramics, soft metal, and glass.

WET OR DRY ABRASIVE An abrasive mounted on tough, water-resistant cloth. Can be used dry, or with water for faster removal of the sanding residue.

sandpaper grades The method for determining and designating the coarseness of sandpaper. Sandpaper was originally graded by grit symbols, from 4 1/4 (coarsest) to 10/0 (finest). The method most common today uses mesh numbers, which represent the number of openings per square inch in a piece of screen through which the abrasive

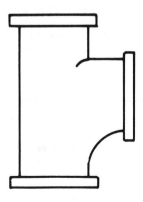

Fig. S-3. Sanitary tee

particles can pass, from number 12 (very coarse) to number 600 (very fine).

sandwich panel A building panel that is formed by facing a soft inner core with two rigid outer faces.

sanitary fitting A plumbing fitting used in joining DWV pipes and having no inside shoulder to obstruct the flow of waste; also called a drainage or DWV fitting. See Fig. S-2.

sanitary tee A T-fitting for waste and soil pipe systems, having a side opening that curves into the main body of the fitting, allowing waste products to flow in one direction only. See Fig. S-3.

sapwood See *wood*.

sash A single frame holding one or more panes of glass, that is set into the window frame and can be stationary or operable. See Fig. W-3.

sash balance A device, usually spring or tension operated, that counterbalances a double-hung window sash, eliminating the need for sash weights.

sash cord A thin rope that is attached to a vertically operating sash, is passed through a pulley in the window frame, and then is attached to a sash weight.

sash weight A cylindrical metal weight hung in a recess between the window frame and wall framing and connected to the sash with a sash cord; used to counterbalance the sash to keep it from falling when opened.

satin finish A paint or other finish material that reflects a small amount of light, but does not have a full gloss or luster.

saturated felt Felt that is impregnated with tar or asphalt.

saturation 1. Of or relating to the point at which the air cannot hold any additional water vapor; 100 percent relative humidity. 2. Of or relating to the point at which any material has absorbed all that it can.

sawhorse A portable structure formed by four braced, slanting legs fastened to a horizontal crosspiece; used to support lumber and other materials while they are being worked on.

Sawzall Brand name for a type of reciprocating saw; often used as a generic term to refer to any type of reciprocating saw. See *reciprocating saw*.

SBCCI Southern Building Code Congress International, Inc. Authors of *Standard Building Code*.

scab 1. A short length of lumber used to splice two other pieces together. 2. Any member that is added to lengthen or extend another member.

scaffold   A temporary system of poles, planks, and bracing erected around a building to support workers and material during construction.

scale   1. The relative and consistent proportion of a plan, map, or other drawing to the object it represents. For example, in 1/4-inch scale, 1/4 inch on the drawing would equal 1 foot of the actual object. 2. A special ruler marked in various scales; used for creating and reading scale drawings.

scalene triangle   See *triangle.*

scaling   1. A defect in a painted finish in which flakes of dried paint peel off, exposing the surface below. 2. Thin layers of flakes of dried concrete that chip away from the surface of a concrete slab or structure.

scantling   An older term for lumber having a cross-sectional dimension ranging from 2 x 4 inches to 4 x 4 inches.

scarf   To place an angled cut on the end of a board in order to create a scarfed joint.

scarf joint   A method of joining the ends of two boards using matching, overlapping bevel cuts. See Fig. S-4.

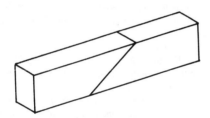

Fig. S-4. Scarf joint

schedule   A numbering system for various types of plastic pipe that indicates the pipe's pressure rating.

schedule of materials   A detailed listing and specification of the finish materials to be used in a construction project. These include windows, doors, moldings, cabinets, fixtures, and other finish materials, and are typically referred to by that name; i.e., "window schedule," "door schedule," etc.

schematic diagram   See *wiring diagram.*

scientific notation   A simplified method for working with large numbers, done by expressing the number as a power of 10. For example: $30,000 = 3 \times 10^4$ ($3 \times 10 \times 10 \times 10 \times 10$).

scissor truss   A type of V-shaped roof truss; used to give a room a sloped ceiling. See Fig. T-7.

score   To scribe, with a knife or other tool, a deep line in a material, along which the material can be creased or broken.

scored block   See *concrete block.*

Scotchlock   A brand name for a type of insulated, color-coded wire nut.

scotia   A type of molding in the form of an elliptical concave curve, similar to a cove molding but slightly flatter. See Fig. S-5.

scratch awl   See *awl.*

Fig. S-5. Scotia

scratch coat   The first coat of stucco or plaster applied to the wall, so named because it is scratched or grooved as it sets to form a good bonding surface for the next coat.

SCR brick   See *brick.*

screed   A straight strip of wood or metal drawn across the top of concrete forms or plaster grounds as a guide for maintaining an even thickness of concrete or plaster.

screeding   Using a screed to level fresh concrete or plaster; also called striking off.

screen bead   A thin, flat molding, usually with decorative grooves on one face, used to secure window screen material to a wooden frame. See Fig. M-4.

screen block   See *concrete block.*

screen rabbet   A rabbet cut into the outside edge of a window or door jamb to serve as a stop for the screen.

screw   Any of a variety of threaded fasteners used to join two objects. They include the following.

MACHINE SCREW A screw designed for insertion into a compatible nut or threaded hole.

SELF-TAPPING SCREW A type of screw that will cut its own threads when driven into a predrilled hole.

SHEET-METAL SCREW A screw with coarse threads and a sharp point, intended for light-to-medium-gauge sheet metal.

WOOD SCREW A screw with coarse threads and a tapered shank, designed to be driven into wood.

In addition to being described by their intended use, screws are also known by the shape of their head, which include fillister, flat, hex, oval, pan, and round configurations, and by the type of screwdriver they are compatible with, including allen (or hex), Phillips, square, standard, and torx. See Fig. S-6.

screw eye   A piece of hardware having a pointed screw thread at one end and a rounded loop at the other.

screw gun   A portable power tool used to drive and remove screws.

screw nail   See *nails, shank types: spiral threaded.*

screw starter   A small hand tool with a pointed, threaded

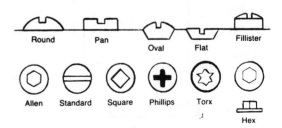

Fig. S-6. Common screw head configurations

shaft, used to tap shallow holes in wood to ease the starting of a wood screw.

screw terminal   A component having a tapped hole and a threaded screw; used to connect electrical wiring. See Fig. G-6.

scribe   1. To cut or plane a piece of wood so as to fit an irregular surface. 2. To cut the end of a piece of molding to fit the face contour of another piece, thus eliminating the need for a miter joint at an inside corner. 3. To scratch a line on the surface of a material, either as a reference mark or as a line along which the material can be bent or broken.

scroll saw   A stationary power tool with a very fine, vertically reciprocating blade that passes through a table; used for cutting intricate internal and external curves and other irregular shapes.

scrollwork   Decorative woodwork, usually cut with a scroll saw or jigsaw, most commonly used as exterior ornamentation. Also called spindlework or gingerbread.

scum   The layer of gas, grease, and small, solid particles at the top of a septic tank. See Fig. S-8.

scum stick   A folding measuring stick that can be inserted into a septic tank to check the depth of the liquid and scum layers. See Fig. S-8.

scuttle   An opening in a ceiling, having a hinged or fitted lid; used to provide access to an attic or roof.

seal   1. To make weathertight using gaskets, sealant, caulking, or other means. 2. A stamp or imprint from an architect or engineer, used to provide verification of the structural calculations and other design aspects of the plans for a construction project. 3. A similar stamp or imprint from a notary, used to provide verification of a person's signature.

sealant   Any of a variety of compounds used to seal joints in wood, metal, masonry, and various other materials. Sealants differ from caulking compounds primarily in their greater flexibility and effective temperature range, and generally have a substantially longer life span. Some common types of sealants follow.

BLACKTOP FIX   A flexible, waterproof compound that is black and generally not paintable, and is used to repair and seal joints in asphalt, concrete, flashings, and below-grade masonry.

CONCRETE AND MASONRY SEALANT   A light gray sealant for repairing cracks or sealing expansion joints in concrete, mortar, stucco, etc.

GUTTER AND LAP SEALANT   A sealant designed specifically for metal, and used to seal gutters, flashings, truck bodies, etc.

NEOPRENE SEALANT   A rubber sealant with good weathering capabilities, that is black in color for ultraviolet protection, has fair elasticity and shrinkage, and ages well, usually up to 20 years, but that stiffens in low temperatures.

POLYSULFIDE RUBBER SEALANT   A sealant that is similar to butyl rubber caulk, but with better elasticity and weathering properties, that does not become brittle in low temperatures, and that has a life expectancy of over 20 years.

SILICONE SEALANT   One of the more expensive of the sealants, but possessing a number of desirable characteristics, including a high degree of elasticity and resistance to weathering, suppleness at low temperatures, ability to be applied at any temperature, and ability to adhere to a wide variety of materials. Most types must be sealed with an oil-based primer before painting. There are three common types of silicone sealants. *Bathroom silicone sealant* is an extremely waterproof white sealant for use around tubs, sinks, etc. *Paintable silicone sealant* is a compound similar to regular silicone window and door sealant, but that can be painted over with most types of paint. *Window and door silicone sealant* is a waterproof, general-purpose compound with a life expectancy of around 50 years, and the ability to handle temperatures down to – 40° F.

seal-down shingle   See *seal-tab shingle*.

sealer   A clear or pigmented finishing material applied directly to uncoated wood or other material to seal the material's surface. Some types of transparent, waterproof sealers are used as the final finish coat; other types simply seal the wood as a base for the application of lacquer, varnish, or other finishes.

seal-tab shingles   A composition shingle that has a strip of dry asphalt adhesive running horizontally above the line of exposure, which when installed, is softened by heat from the sun, causing the shingle to adhere to the underside of the shingle above it; used to increase the shingle's wind resistance. Also called self-sealing shingles.

Seal-tite   The trade name for a type of flexible conduit wrapped with a heavy water- and abrasion-resistant coating; suitable for exterior applications.

seamer   A hand-held or pole-mounted steel tool consisting of a flat plate with a rounded ridge on the bottom side; used to form expansion or control joints in new concrete flatwork.

season   To remove excess moisture from a piece of wood, rendering it less likely to warp or change dimension.

seasonal efficiency   1. In a heating system, the ratio of usable heat produced by the system to the amount of energy the system uses to operate, calculated over an entire heating season. 2. A rating of a solar collector's performance, being the ratio of solar energy collected to solar energy striking the collector, spread over an entire season.

seat   The tapered, machined area of a valve or faucet that controls the flow of water.

seat cut   The horizontal cut that is made when forming a bird mouth or similar notch. See Fig. R-5.

secondary colors   Colors made by combining equal amounts of two primary colors. In painting, the secondary colors are orange (red and yellow), green (blue and yellow), and purple (red and blue).

secondary combustion   See *combustion*.

secondary venting   Additional vent stacks used to vent fixtures that are too far away to be tied into the main soil stack. See Fig. D-14.

second growth   New trees that have grown in an area where a large portion of the previous stand has been removed, usually by logging or forest fire.

section A drawing of an object as if it were cut by an intersecting plane, showing the shape and inner detail of the object at that point. See Fig. O-4.

sector That section of a circle bounded between two radii and the connecting arc. See Fig. C-3.

seepage pit A large, perforated pit that receives the effluent from a septic tank; used in areas that are too small, hilly, or otherwise unsuited for a normal disposal field.

segment That section of a circle bounded by an arc and its chord. See Fig. C-3.

selective surface A thin coating on a solar collector's absorber plate that is highly absorbent of the sun's rays while allowing very little heat to be lost through reradiation; used to increase collection efficiency.

self-closing hinge Any of a variety of spring-loaded hinges designed to close a door automatically.

self-drilling anchor See *anchor, masonry*.

self-rimming sink A type of sink, particularly one used in a kitchen or bathroom, that has a wide, rolled lip around its outer perimeter, which is sealed down to the countertop in which the sink is installed, thus eliminating the need for a separate sink rim.

self-sealing shingle See *seal-tab shingle*.

semicircle One-half of a circle; an arc of 180 degrees. See Fig. C-2.

semi-gloss Of or relating to a paint or other finish having a moderate luster, approximately midway between flat and gloss.

semihoused stringer An open, cutout stringer that is attached to a solid board called a backing stringer; a simpler version of the housed stringer.

semi-indirect lighting A fixture that combines direct and indirect lighting.

sensor See *termistor*.

septic system A system for the private disposal of waste from a building not having access to a public sewer system, consisting of a septic tank, one or more distribution boxes, and a series of drainage lines laid out in a disposal field. See Fig. S-7.

septic tank An underground watertight receptacle for the waste material discharged from a building's sewage system, in which solid waste settles to the bottom of the tank where it is disintegrated by bacterial action, and liquid waste is discharged into the earch through a series of perforated pipes. See Figs. S-7 and S-8.

series connection 1. In a solar heating system, a method of connecting a group of individual collectors so that the heat-transfer fluid runs from the outlet of one collector to the inlet of the next; normally limited to three or four collectors. 2. In electrical wiring, a series of connections that allows the electricity only one path to follow. Rarely used in home wiring, because only the hot wire extends from fixture to fixture, with the neutral wire returning from the last fixture in line. If one fixture burns out, the circuit is opened and all the fixtures go out.

serrated Cut into alternating pointed hills and valleys, like the teeth on a saw.

serrated trowel See *notched trowel*.

service 1. One of two types of electrical supply coming into a building from the utility. Three-wire service supplies one neutral and two hot conductors, giving a voltage potential of 240 volts at the service panel. See Fig. S-9.

service conductor A heavy-gauge electrical wire that connects the service equipment with the power company's main lines. See Fig. S-9.

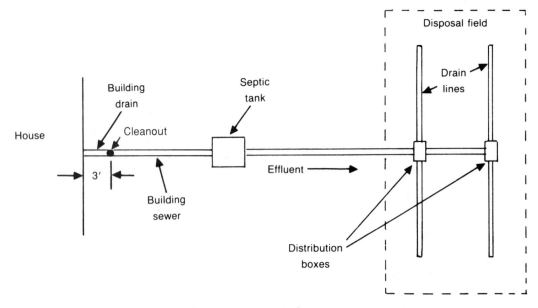

Fig. S-7. Standard septic system

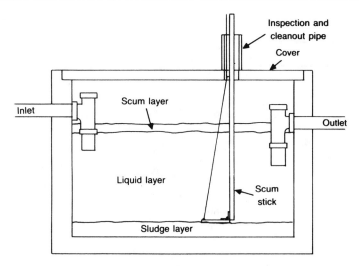

Fig. S-8. Typical septic tank

service drop   The overhead service conductors, provided and installed by the power company, that connect the transmission lines on the utility pole to the service-entrance conductors of an individual building. See Fig. S-9.

service-entrance conductor   One of the conductors used between the electric meter on the service panel and the point of attachment of the service drop. See Figs. S-9 and S-10.

service-entrance conduit   A large-diameter conduit, usually threaded on one end where it enters the hub; used to enclose and protect the service-entrance conductor and often to provide a point of attachment for the service drop. See Figs. S-9 and S-10.

service equipment   All of the conductors and mechanical equipment that receive incoming electricity from the power company and transfer it, through a main shutoff, to the branch-circuit overcurrent protection devices. See Fig. S-10.

service lateral   The underground service conductors between the street main or transformer and the point of connection to the service-entrance conductors. See Fig. S-9.

service panel   A box or cabinet that is the main power distribution center for a building, receiving incoming electricity from its source and distributing it out to the branch circuits. It contains the main disconnect switch, the grounding and neutral connections for the entire system, and the individual circuit breakers or fuses. See Figs. S-9 and S-10.

service pedestal   A covered, waterproof box used with underground electrical service in which the incoming service conductors from the power company are connected to the service-entrance conductors from the service panel; also called a pull box. See Fig. S-9.

service sink   A special-purpose sink designed for use in an area other than the kitchen or bathroom. They include laundry trays, bar sinks, and surgeon's sinks.

service stairs   See *stairs*.

set   The distance that the teeth on a saw blade are bent so as to project past the side of the blade. Setting the teeth in this manner creates a kerf that is wider than the thickness of the blade, preventing the saw from binding. See Fig. K-1.

setback   1. The distance a building is located from the front, side, and rear property lines. See Fig. P-4. 2. A method of constructing buildings of two or more stories in which the face of each succeeding story is set back from the one below it.

setback thermostat   See *clock thermostat*.

set nail   See *nails, head types*.

setting up   The initial drying of a material to the point that it will no longer flow.

settling   The unequal sinking of a structure or parts of a structure, resulting in cracks, stuck doors, and other problems.

sewer brick   See *brick*.

sewer, private   See *private sewer*.

sewer, public   See *public sewer*.

S4S   Abbreviation for *surfaced four sides*, a designation for a piece of lumber having all its sides surfaced smooth. Other combinations include S2S (surfaced two sides), S2S1E (surfaced two sides and one edge), and S2E (surfaced on two edges only).

shade   1. To darken a paint color by adding black. 2. A graduation of a color, as a shade of blue. 3. To provide a structure with protection from the sun.

shaft   The central and primary supporting portion of a column, between the base and the capital. See Fig. C-8.

shake   1. A type of cedar roofing material similar to a wood shingle, but larger, thicker, and hand-split for more surface texture. Medium shakes are approximately 3/4 inch thick at the butt end; heavy shakes are about 1 1/4 inches thick. 2. See *lumber defect*.

shall   A term commonly used in building code definitions and provisions, indicating that the code requirement is mandatory.

shallow-trap toilet A toilet similar to a conventional toilet but

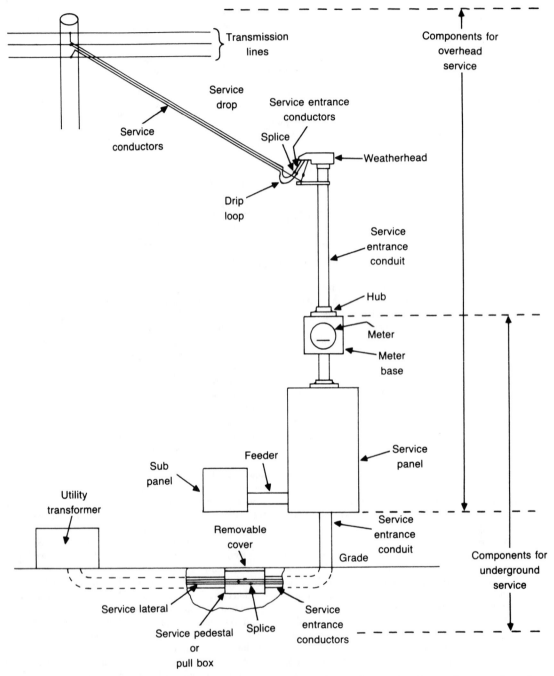

Fig. S-9. Three-wire electrical service, showing a typical installation for overhead or underground service

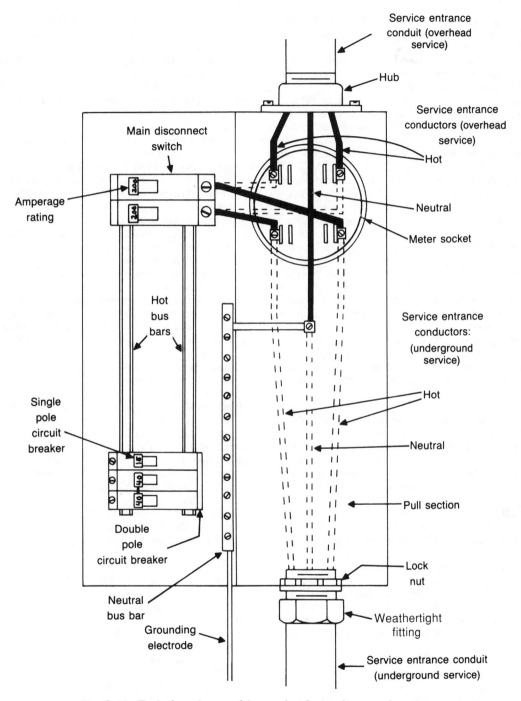

Fig. S-10. Typical service panel for overhead or underground service

having a shallower tank and a smaller passageway in the bowl, reducing the amount of water needed for each flush.

shallow well   See *well*.

shallow-well pump   A nonsubmersible pump designed for extracting water from shallow wells, usually capable of 20 to 50 psi at depths up to 25 feet.

shaper   A stationary power tool having a large, flat table, a guide fence, and a high-speed vertical spindle to which various cutter heads are attached; used for cutting decorative or glue-joint edges on lumber pieces.

shear   1. Of or relating to stress imposed on an object by the force of two adjacent objects sliding over each other. For example, when a piece of plywood is nailed to a stud, the nails are subjected to shear forces. 2. Of or relating to stress imposed on an object by a twisting force, such as the force acting on a bolt head while it is being turned by a wrench. See Fig. S-22.

shear strength   The maximum amount of shear force that an object can withstand before it fails.

sheathing   A structural covering of lumber, plywood, or certain building boards fastened to the outside of wall studs or roof rafters, and to which the roofing or finished siding is applied; also called sheeting. See Figs. B-2, C-4, O-2, and P-3.

sheathing paper   A building paper, either plain kraft paper or paper that is asphalt impregnated, used in wall, floor, and roof construction to protect against the passage of air and, to a lesser degree, moisture. Common weight is 4 to 10 pounds per square in 36-inch widths.

shed dormer   A shed-roofed projection from a building's sloping roof; used to house windows and provide additional upperfloor headroom. See Fig. R-8.

shed roof   A simple form of pitched roof, having only one slope. It can stand alone on certain types of smaller houses or be attached to the side of a larger structure, as in the case of a room addition, when it is also referred to as a lean-to roof. See Fig. R-8.

sheet flooring   See *sheet goods*.

sheet goods   Floor coverings manufactured and installed in large sheets, such as linoleum, as opposed to individual tiles; also called sheet flooring.

sheeting   See *sheathing*.

sheet metal   Metal that has been formed into flat or corrugated sheets, usually by rolling. Available in a variety of sheet sizes, styles, gauges, and surface treatments.

sheet-metal roofing   Corrugated steel or aluminum sheets, in various sizes, shapes, and colors, used as a roofing material.

Sheetrock   A trademark for a type of plasterboard used as an interior wall covering. See *drywall*.

shell   A building with its interior and exterior walls framed, and all windows, exterior doors, siding, and roofing installed so as to be weathertight. Homes are sometimes sold as shells, allowing the homeowner to finish the interior.

shellac   A finishing material for wood that is obtained from secretions of the lac bug, an insect found primarily in India. It is refined and dissolved in denatured alcohol, and produces a smooth, fairly hard finish, but is not waterproof. Shellac is available in two colors: orange, used as a primer and an additive for other finishes, and white, which has been bleached and is used as a finish or sealer under varnish.

shellac cutting   Thinning shellac with denatured alcohol to the proper consistency for various applications; designated by the number of pounds of shellac that are dissolved in one gallon of alcohol, such as a one-pound cut, a two-pound cut, etc.

shellac sticks   A dry shellac resin in stick form, that is tinted in a variety of colors to match most woods, applied with a hot knife, and rubbed out with naphtha; used to fill and repair defects in finished wood.

Sheraton   Of or relating to a furniture style originated by English furniture maker Thomas Sheraton in the late 1700s and characterized by simplicity and graceful, straight lines.

shim   1. A thin, flat- or wedge-shaped piece of wood used to level or plumb an object during installation, as with a cabinet or door frame. See Fig. D-6. 2. A thin piece of metal or other material used between various parts in a piece of machinery to remove excessive play or otherwise adjust the way the parts fit and interact. 3. To use shims to move an object until it is plumb, level, or otherwise properly positioned and adjusted.

shingle hatchet   A type of hammer containing a thin hatchet blade on its head, instead of a claw; used primarily for installing wood roofing shakes and shingles.

shingle mold   A trim piece used with wood shakes or shingles that is applied to the barge rafter directly under the roof covering; also called gable mold. See Fig. C-13.

shingle nail   See *nails, common types*.

shingle panel   A backing board in a 4- or 8-foot length having individual shingles bonded to it; used primarily to speed installation of siding, but some types are available for roofing applications.

shingle   A piece of asbestos, asphalt, fiberglass, slate, tile, wood, or other material, cut or formed into a stock length, width, and thickness for use as a roof or wall covering. See Fig. F-7.

shiplap   Of or relating to lumber or plywood having two rabbeted edges: one edge rabbeted on the face, the opposite edge rabbeted on the back. Used primarily as siding, with the rabbeted edges overlapping. See Fig. S-11.

shoe   See *base shoe*.

S hook   A small, S-shaped metal rod used to connect chains, fixtures, and other objects.

shoring   Lumber and timbers used as temporary bracing for a building or excavation.

short circuit   1. An improper, low-resistance connection occurring between two current carrying (hot) wires or between a hot wire and a grounded wire or object. 2. (hyphenated) to create such a connection.

short length   Generally, lumber in a length of less than 8 feet.

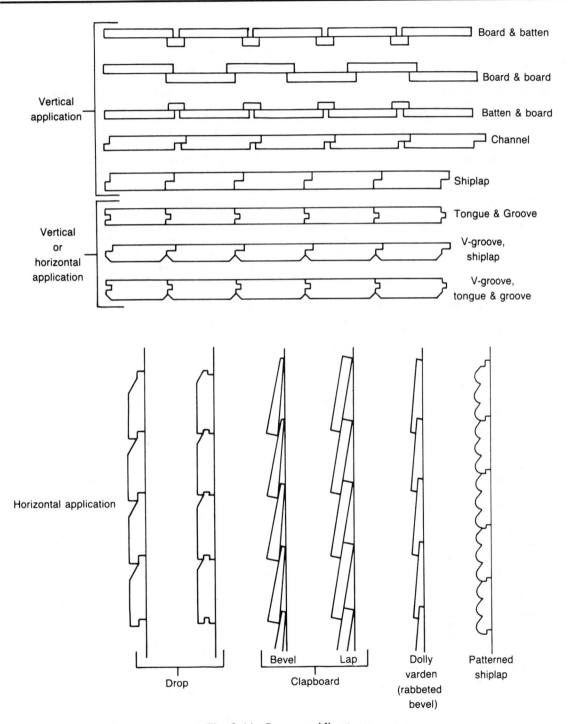

Fig. S-11. Common siding types

short oil resin   See *oil resin.*

shortwave radiation   See *solar radiation and greenhouse effect.*

Shotcrete   See *Gunite.*

shower head   Any of a variety of devices that act as the outlet for water entering a shower, designed to convert the flow into a spray, intermittent pulse, or other pattern.

shower pan   A waterproof, panlike enclosure that forms the bottom of a stall shower and is made of plastic, fiberglass, or metal in standard sizes, or is custom made from hot tar and felt and covered with ceramic tile.

shrink film   A thin, clear plastic film that shrinks when heated; sometimes used on the inside of window frames as a temporary storm window.

shutter   1. A louvered or flush frame of wood or other material, usually fixed to the wall next to a window as a decoration or hinged to close across the window for protection. 2. Adjustable louvers or plates used to regulate air flow, as with the burners in a furnace.

side cut   The angled saw cut made on the side of a hip or valley rafter to enable it to meet a ridge or common rafter.

side lap   The distance that the side of a shingle or the end of a row of roofing felt or roll roofing overlaps the end of the one preceding it.

side light   A fixed glass panel, normally obscure, set on one or both sides of an entry door. See Fig. F-2.

side matched   See *tongue and groove.*

side-outlet cross   A cross fitting for plumbing having five or six outlets instead of four. The fifth (and sixth) outlet extends out from one or both sides, at a 90-degree angle to each of the other four.

side-outlet elbow   A plumbing fitting with three outlets, each of which is at a 90-degree angle to the other two.

side point nail   See *nails, point types.*

siding   The exterior finished wall covering of a frame building, consisting of horizontal or vertical boards, shingles, plywood sheets, or other materials. See Figs. C-4, O-2, S-11.

siding nails   See *nails, common types.*

siding shingles   One of various kinds of shingles, usually wood, slate, or asbestos-cement, used over sheathing and sheathing paper as an exterior wall covering.

sieve   A screen or screen-bottomed container; used for separating sand from gravel or for other sifting applications.

silent switch   A type of electrical switch having a mechanism designed to eliminate any audible click when operated.

silex   A form of silica that does not expand or contract in moisture and that is chemically inert; often used in the making of paste wood fillers.

silicon-carbide paper   See *sandpaper.*

silicone acrylic caulk   See *caulking compound.*

silicone sealant   See *sealant.*

sill   1. The lowest wood member of a frame house, bolted to the foundation; also called a mudsill or sill plate. See Figs. B-2, P-2, P-3, and P-10. 2. The member making up the lower side of a rough or finished opening, as a door sill or windowsill.

See Fig. W-3.

sill beam   See *rim joist.*

sill plate   See *plate.*

sill sealer   Any of a variety of resilient, waterproof materials used between the top of a foundation and a sill plate or between a concrete slab and a sole plate to seal against air, dirt, and insects; also called a bedding strip. See Fig. P-2.

silver steel   See *steel.*

single-cut file   A file having parallel rows of teeth or ridges that run diagonally across the file's face in one direction only.

single-family dwelling   A building that is designed for occupancy and use as a home by one person or one family, and that usually is the only dwelling located on a parcel of land.

single-handle faucet   A type of faucet, typically for a kitchen or bathroom sink or for a shower or bathtub, that allows the hot and cold water to be mixed and regulated with one handle instead of two.

single-hung window   See *window.*

single-pole circuit breaker   See *circuit breaker.*

single-pole switch   See *switch.*

single roll   Of or relating to a roll of wallpaper containing 36 square feet.

single-strength glass   Common window glass, measuring just under 1/8 inch thick.

sinker nail   See *nails, head types, flat countersunk.*

sink front   A cabinet face frame with doors and a toe kick, but with no floor, sides, or back; used between two adjacent cabinets in some kitchen installations where undersink plumbing makes installation of a conventional cabinet impractical.

sink rim   A stainless steel ring that fits around the outside edge of certain types of sinks to attach them to countertops.

sink strainer   See *basket strainer.*

siphon   To move liquid through a pipe or a tube from an upper level to a lower level, passing over an intermediate point that is higher than either of the two levels. Pressure at the lower point is greater than pressure at the upper point, and draws the liquid into the pipe.

site   1. The parcel of land upon which a structure is constructed. 2. The location and working area of a construction project.

site fabricated   Of or relating to any component of a building, such as a cabinet or door frame, that is cut out and assembled on the job site, as opposed to being prefabricated.

site finished   Any item, especially cabinets and woodwork, that is stained, painted, or otherwise finished at the point of use, as opposed to being done at the point of manufacture.

site plan   An architectural drawing that shows the perimeter of the building site, the outline of the proposed building and other major improvements as they will sit on the site, the site's elevation and contours, and the location and outline of the site's major natural features.

site-specific   Anything, but especially an architectural design or set of construction plans, that is designed and intended to fit the natural features of a particular building site.

1/16 bend A plumbing elbow having a radius of turn equal to 1/16 of a circle, or 22 1/2 degrees.

size To coat a wall with sizing compound prior to wall-papering.

sizing compound A solution of resin, glue, starch, or similar materials used to seal the pores in a surface material such as plaster or plasterboard; often done preparatory to hanging wallpaper to prevent the wall surface from prematurely absorbing liquids from the wallpaper paste.

skeleton 1. The bare frame of a building before the installation of sheathing, siding, or other enclosing materials. 2. The bare frame of a piece of furniture before the application of padding or fabric.

skewback An abutment having an angled face to receive the thrust of an arch.

skewed Angled; asymmetrical; turned or extended in a direction away from the original or intended path.

skewed hanger See *timber connector.*

skidding In logging, moving logs from where they are cut to where they will be loaded, usually by dragging.

Skilsaw The brand name for a type of electric circular saw; often used as a generic term to refer to any type of circular saw.

skin 1. A thin veneer panel used for appearance over other woods, as on cabinets and doors. See Fig. D-3. 2. A tough layer of dried paint or varnish that forms in a container, caused by exposure to the air. 3. The overall outer covering of a structure.

skinning a wire Removing the outer jacket or individual insulation from an electrical wire.

skip See *lumber defect.*

skip sheathing 1. Boards, usually 1 x 4 inches, that are laid perpendicular to the rafters with a space between them to act as a base for the installation of wood shakes or shingles; also called open sheathing or spaced sheathing. 2. Boards applied in a similar manner over wall studs to act as a base for sidewall shingles.

skip trowel Of or relating to a method of applying a textured finish to a plasterboard or plaster wall, usually by mixing sand with thinned joint cement, then applying the mix with a trowel or blade that "skips" as it moves, leaving a random pattern of cement.

skirting 1. Metal or wood panels placed around the base of a mobile home or similar raised structure to conceal the underpinnings. 2. See *baseboard.*

skylight A roof-mounted window of glass, plastic, acrylic, or other material set in a wood or metal frame and used to admit natural light into a building. Skylights can be fixed, or operable to also provide ventilation.

slab 1. Concrete flatwork, especially that which serves as a base for a building or other object. 2. A slice of a material, such as from a stone. 3. The outer piece cut off a log preparatory to milling.

slab insulation See *insulation, thermal.*

slab on grade A concrete slab that is poured directly on the ground, including footings and wire mesh or other reinforcement. See Fig. D-10.

slag Waste material deposited in a furnace during the process of refining metal from ore. It is recovered and reprocessed for use in other building materials.

slag wool See *insulation, thermal.*

slake To mix with water or moist air.

slaked lime Quicklime to which water has been added, forming a crumbly mass.

slamming stile The jamb leg in a door frame against which the door closes.

slat A narrow, thin piece of wood or other material; commonly used in making louvers, trellises, lattices, etc.

slate A type of natural rock that is easily split into flat, thin sheets, also called slates; often used for roofing, siding, and floor covering.

sledge hammer See *hammer.*

sleeper 1. A piece of lumber embedded in concrete to serve as a point of attachment, as for a door threshold, etc. 2. A wooden member laid on the ground or on a concrete slab as a support for joists or decking. See Fig. S-12.

sleeper clip An H-shaped clip embedded in concrete and used to secure a sleeper. See Fig. S-12.

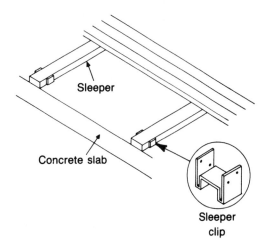

Fig. S-12. Sleepers and sleeper clips

sleeper wall See *pony wall.*

sleeve 1. The hollow metal tube inside a lockset that works with the cylinder to move the latch. 2. A tube or tubular opening formed into concrete to allow for the passage of wiring or plumbing.

sleeve anchor See *anchor, masonry.*

sleeve bearing A precisely milled tube, usually of bronze, used as a bearing or bushing around a revolving shaft, as in an electric motor.

sliding door See *door.*

sliding glass door   See *door.*

sliding T-bevel   An adjustable metal blade in a handle; used to lay off and duplicate angles. Also called a T-bevel.

sliding window   See *window.*

slip fitting   An unthreaded plumbing fitting for use with copper or plastic pipe into which the pipe slips and is joined with solder (for copper fittings); solvent cement (for plastic fittings); or a compression nut, friction ring, and compression washer (for some metal and plastic fittings). See Fig. S-13.

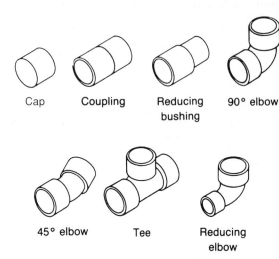

Cap   Coupling   Reducing bushing   90° elbow

45° elbow   Tee   Reducing elbow

Fig. S-13. Common slip fittings

sliphead window   See *window.*

slope   1. The ratio of the rise of a roof to its run. For example, a roof that is 8 feet high at the ridge and has a 24-foot run would have an 8:24, or 4:12, slope. See Fig. R-6. 2. A similar ratio applied to any inclined surface, such as a pipe or a grade.

sludge   The solid waste material that settles to the bottom of a septic tank. See Fig. S-8.

slump stone   See *concrete block.*

slump test   A rough measure of the consistency of freshly mixed concrete, done by noting how well a shovelful will form a mound without slumping.

slurry   A dry, powdered material, such as cement, that is mixed with large amounts of water until it is easily pourable.

slushed joint   In masonry, a vertical joint that is filled by pushing mortar into it with a trowel after the units are in place. Slushed joints are considered inferior to normal joint methods.

smoke chamber   The wide, tapering area within a masonry fireplace that creates the air currents necessary for proper draft. See Fig. F-4.

smoke sample   A test of the smoke in the flue of an oil-fired furnace that uses a special smoke-testing instrument to take a sample of the smoke, which is then compared to the colors

on a smoke scale chart to evaluate the condition and efficiency of the burner.

smoke shelf   A projection within the smoke chamber of a masonry fireplace that prevents descending air from blowing smoke back into the room. See Fig. F-4.

smoothing plane   See *plane.*

smooth nail   See *nails, shank types.*

snake   See *drain auger.*

snap lock   A slot-and-tongue assembly formed into opposite edges of a sheet-metal duct whereby the edges are interlocked and crimped or hammered flat, forming a solid pipe.

snow board   A low board placed on the roof just above the eaves to interrupt the force of sliding snow or to divert water from over doors and windows when gutters are not used.

socket   The internally threaded receptacle in a light fixture designed to receive the light bulb. Sockets are described by the type of bulb base they accept, which differ primarily in diameter and thread size. Some common socket types are (from small to large); miniature, candelabra, medium or Edison, and mogul. Special three-way sockets are also available in medium and mogul sizes for use with three-way bulbs.

socle   A square block at the base of a column or pedestal that is similar to a plinth block, but larger.

sodium light   See *gaseous discharge light.*

soffit   1. The underside of a part of a building, such as a cornice, staircase, or arch. See Figs. C-4 and C-13. 2. A decorative, boxed-in area, as over kitchen cabinets.

soffit lighting   Light fixtures that are recessed into interior or exterior soffits, primarily for accent lighting; also called cornice lighting.

soffit vent   A screened opening placed in an exterior soffit to allow air to enter the attic. See Fig. C-4.

soft face hammer   See *hammer.*

soft jaws   Lead wood, or other soft material used over the steel jaws of a wrench or vise to protect the finish of the object being worked on.

soft steel   See *steel: mild steel.*

softwood   A general term referring to lumber produced from needle-bearing conifer trees, but having no bearing on the actual softness of the wood. Some of the more common softwoods follow.

CEDAR, RED A light, soft, and straight-grained softwood that splits easily and is red to cream in color; commonly used for closet and chest liners, fencing, shingles, and pencils.

CYPRESS A straight-grained and easily worked softwood that is yellow to light brown in color; used in general construction, boats, trim, and paneling.

FIR, DOUGLAS A strong, durable, and somewhat hard softwood that is tan-yellow in color; used extensively for construction lumber and plywood. Abbreviated doug fir.

HEMLOCK A straight-grained, durable, strong, and easily worked softwood that is light brown in color and is very similar to Douglas fir; used almost exclusively for construction.

KNOTTY PINE Pine lumber containing numerous solid knots

and white to light yellow in color. Most often milled into 3/4-inch thick boards for use as a decorative wall and ceiling covering.

PINE   A soft, straight-grained, and durable softwood that is white to light yellow in color and is very easily worked, with little tendency to warp. Commercial varieties include ponderosa, lodgepole, white, and sugar. Used for furniture, trim, shelving, construction, and general woodworking.

REDWOOD   A straight-grained, soft, and easily worked, but brittle and not overly strong softwood that is moisture resistant and red-brown in color; used extensively for exterior structures such as decks, barns, fencing, and roof and sidewall shingles.

SPRUCE   A light, soft, and straight-grained softwood that works easily and is pale reddish-brown in color; used for construction, fencing, and boats.

softwood plywood   See *plywood.*

soil cover   See *ground cover.*

soil pipe   Any pipe used to carry liquid or solid waste material. See Fig. D-14.

soil-pipe cutter   See *chain cutter.*

soil stack   The main vertical waste pipe into which toilet and branch drains are connected, having a lower end that ties into the main house sewage line and an upper end that extends through the roof to act as a vent. See Fig. D-14.

solar access   The legal right to have access to sunlight. Many communities have passed laws placing restrictions on the size and placement of buildings, trees, and other objects that might block the sunlight falling on another person's property.

solar degradation   The process by which a material is deteriorated by the ultraviolet portion of sunlight.

solar house   A building using active, passive, or hybrid solar heating systems from which it receives at least 50 percent of its annual heat.

solarium   See *sunspace.*

solar orientation   The positioning of a home or solar collector as close to true south as possible, so as to gain maximum benefit from the winter sun.

solar radiation   The energy from the sun; sunlight, which comes to earth as radiation waves of different lengths, ranging from shorter waves in the ultraviolet range, through intermediate waves that produce visible light, to the longer waves in the infrared range, which are felt as heat. Solar radiation strikes the ground or a collection surface in one of three ways. See Figs. G-4 and S-14.

DIFFUSE RADIATION   Sunlight that is scattered by air molecules, dust, or water as it passes through the atmosphere, as on a cloudy or hazy day. This type of radiation does not cast a shadow and cannot be focused.

DIRECT RADIATION   Sunlight that comes in parallel rays directly from the sun, has the ability to cast a shadow, and is the most intense; also called beam radiation.

REFLECTED RADIATION   Sunlight that is bounced or reflected off trees, snow, buildings, and other objects.

solar savings   A calculation used in determining the mon-

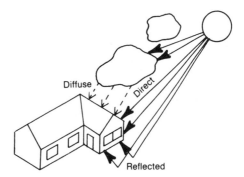

Fig. S-14. The three types of solar radiation

etary savings derived from the installation of a solar water or space heating system, equal to the life-cycle cost of a solar heating system subtracted from the life-cycle cost of a conventional heating system.

solar savings fraction   That portion of a building's space or water heating needs which are provided by solar energy.

solar screen   1. A mesh material in its own frame that is added to the outside of a window to reduce sunlight intensity by diffusing the light before it strikes the glass. 2. A perforated masonry wall, primarily used to block sunlight, but also used for privacy or to block noise.

solar screen tile   See *structural clay tile.*

solar storage capacity   The maximum amount of heat that can be stored in a solar heating system for later use.

solder   1. A soft wire of various metals and alloys that melt between 300° F and 800° F, used to join copper pipe or wire. 2. To join two metal objects with solder.

soldered joint   A joining of metal parts using solder.

solderless connector   A small brass or copper tube crimped onto the bare ends of two wires in order to form a connection without soldering; also called a pressure connector.

soldier   A brick laid so as to be standing on end, with the narrow edge exposed in the wall face.

solenoid   An electromagnetic coil that surrounds a movable rod. When current is induced into the coil, the rod moves out, returning when the current stops. Often used to activate or control various types of machinery.

solenoid valve   A valve activated by a solenoid for controlling the flow of gas or liquid in a pipe.

sole plate   See *plate.*

solid bridging   See *bridging.*

solid-core door   See *door*

solid fuel   Combustible fuel such as wood, coal, and a variety of other materials, as opposed to gaseous or liquid fuels.

solid-top block   See *building block.*

solid wire   Wire that is solid in cross-section and composed of only one conductor, as opposed to stranded wire; for use as an electrical conductor or for other purposes.

soluble   Of or relating to a material that is capable of being dissolved in a liquid.

solvent Any of a variety of liquids that are capable of dissolving a particular material.

solvent cement Liquid cement used to fuse various plastics, such as ABS and PVC; primarily used in joining plastic plumbing pipes and fittings.

solvent-weld To fuse materials together using an adhesive.

soot A soft, black material composed primarily of carbon particles that results from incomplete combustion.

sound 1. Solid; free from decay, defect, or flaw. 2. The wave motion of molecules in the air—typically in the range of 20 to 20,000 cycles per second—that stimulates the auditory organs.

sound absorption The ability of a material to absorb sound waves; used in rating acoustical insulation materials.

sound-transmission class The measure of a material's ability to stop the transmission of sound waves at all frequencies. The higher the sound-transmission class number, the more efficient the material. Abbreviated STC.

sound-transmission loss The number of decibels that a sound loses as it passes through various materials; abbreviated TL.

space design A general term used to refer to the layout and decoration of a building's interior areas; interior design.

spaced sheathing See *skip sheathing*.

space heat Heat that is provided for use in an enclosed area, such as a building, as opposed to heat for water or other uses.

space heater 1. Any of a variety of fixed or portable heaters, usually electric or kerosene-fired, that are used to heat an individual room as needed. 2. Any heating system that supplies heat for an enclosed space, whether one room or an entire building.

spackling compound A type of plaster paste used to fill holes, cracks, and other surface irregularities in a variety of materials prior to painting. Available as a powder for mixing with water, or premixed.

spall 1. A fragment or piece of broken stone or masonry. 2. A broken brick.

spalling The deterioration and chipping away of masonry or concrete, usually caused by moisture and freezing.

span 1. The distance between structural supports, as for a joist, rafter, etc. 2. In roof framing, the overall distance between the outside faces of two opposing exterior walls; twice the run. See Fig. R-6.

span chart See *span table*.

span table A table listing allowable spans for various sizes and types of lumber, based on such factors as spacing, load, and intended area of use; also called a span chart.

spandrel The triangular area between the exterior curve of an arch and the right angle of its surrounding framework.

spandrel beam A beam supported by two columns and concealed in an exterior wall, used to carry the weight of the curtain walls above it.

spark arrester 1. A screen placed over or around a chimney top to prevent sparks from escaping. 2. A screen placed over the muffler of a gas-powered tool, such as a generator or chain saw, to serve the same purpose.

spar varnish A durable, long-lasting, heat- and weather-resistant varnish; designed for service under severe conditions on exterior surfaces. Originally named for its suitability for use on the spars of ships.

spatter finish Dark paint spattered on a surface for a decorative, speckled effect.

spec 1. Abbreviation for specification. 2. Abbreviation for specify. 3. Abbreviation for speculation: see *spec house*.

spec house Abbreviated term for a house constructed for sale, as opposed to one being built for a specific owner. Short for speculation house.

specialty contractor A contractor engaged in a particular trade or construction field that requires specialized knowledge, ability, and equipment.

specification A written list or document defining the details of a construction project, usually including the methods of workmanship and the type, quality, quantity, and size of materials.

specific gravity The ratio of the weight of 1 cubic foot of a material to the weight of 1 cubic foot of cold water.

spigot 1. The lipped end of a piece of cast-iron pipe that fits into the hub end of a similar pipe. 2. A hose bibb, especially one used for washing machine connections.

spike A steel nail 4 inches or greater in length.

spike knot See *lumber defect: knot*.

spindle A rotating horizontal or vertical shaft or rod to which a cutter head or other bit can be fastened.

spindlework See *scrollwork*.

spiral ratchet screwdriver A type of screwdriver that is operated by pressing down on the handle, which in turn engages a spiral groove on the shank and causes the screwdriver bit to rotate. A spring chuck enables the screwdriver tips to be changed, and the direction of rotation can be reversed for removing screws.

spiral stairs See *stairs*.

spiral-threaded nail See *nails, shank types*.

spire The tapering, cone-shaped roof on top of a tower.

spirit level A bar of wood, steel, or aluminum containing vials of spirits with a trapped bubble of air that will center between marks on the vial when the bar is held perfectly horizontal or vertical; used to check plumb and level.

splash block A masonry block or trough set at the bottom of a downspout to receive and drain off water from the roof.

splayed 1. Of or relating to a slanted surface or beveled edge. 2. Of or relating to a poorly cut joint forming a wedge-shaped gap between the two boards.

splayed reveal A reveal formed at an angle other than, and usually greater than, 90 degrees.

splice 1. A connection between two ropes, cables, or wires. 2. A connection between electrical conductors. See Fig. S-9.

spline A small piece of wood fit into slots in two adjoining members to reinforce the joint.

split See *lumber defect*.

split block See *concrete block*.

split circuit See *circuit, three wire.*

split complementary colors A grouping of three complementary colors obtained from the color wheel by first selecting one color, then taking the two colors that flank the color opposite on the wheel to the first color chosen. See Fig. C-7.

split level A building constructed on three or more interconnected levels.

split system A type of air conditioning system in which the evaporator coil is mounted inside the house in the furnace plenum, the condensing unit is mounted outside the building, and the two units are connected by pipes containing refrigerant that circulates under pressure.

splitting maul A heavy, long-handled tool, usually having a head with a broad hammer face on one side and a tapered blade on the other side; used for general driving and hammering work, and for splitting wood.

spokeshave A two-handled plane having an adjustable blade; pulled across a piece of wood for finishing inside and outside curves and other irregular shapes.

spotlight An incandescent bulb designed to provide a narrow, concentrated beam of light; used primarily to illuminate or accent a particular area or object.

spray adhesive See *adhesive.*

spread rate The manufacturer's recommendation of the thickness at which a liquid or semiliquid should be applied; usually expressed as the maximum number of square feet a certain quantity of the product will cover.

spring bender A long, tightly wrapped spring with one flared end; used when bending copper tubing to prevent kinking it.

spring clamp A light- to medium-duty clamp with spring-operated jaws. Pressure on the handles opens the clamp, and when released, the spring draws the two jaws tight against the pieces being clamped.

spring steel Tempered steel with a high carbon content, used in making springs.

springwood See *wood.*

sprinkler system 1. A system of interior ceiling- or wall-mounted pipes with heat- or smoke-activated sprinkler heads, designed to spray water over the building's interior in the event of a fire. 2. A manually or electrically operated system of pipes, valves, and sprinklers designed to water exterior or interior landscaping.

spruce See *softwood.*

spud One of the small nozzles that are attached to the gas manifold in a gas furnace and used to supply gas to the burners.

spud wrench An adjustable, wide-jawed, toothless wrench designed to fit the large locknuts on plumbing drainpipes and traps. See Fig. S-15.

spur bit A type of wood-boring bit having a small, brad point for centering and cutting lips around the outside edge of a hole.

square 1. A unit of measure for roofing and other building materials equal to 100 square feet. 2. To form a right, or 90-degree, angle.

Fig. S-15. Spud wrench

square foot A measurement of area equal to 1 foot in length and 1 foot in width, regardless of thickness. One square foot contains 144 square inches. See Fig. C-16.

square footage The size of a room, building or other area in square feet, arrived at by multiplying the area's total length by its total width. See Fig. C-16.

square head A bolt having a four-sided square head, as opposed to a hex head.

square inch A measurement of area equal to 1 inch in length and 1 inch in width, regardless of thickness. See Fig. C-16.

square measurement A measurement of area equal to the area's width multiplied by its length. See Fig. C-16.

square nail See *nail.*

square nut A nut having four sides, as opposed to a hex nut.

square to round A type of sheet-metal transition fitting used to adapt rectangular ducts or fittings to round ones.

square yard A measurement of area equal to 1 yard in length and 1 yard in width, regardless of thickness. One square yard contains 9 square feet or 1,296 square inches. See Fig. C-16.

SR See *plastic pipe.*

Stabilite The trade name for hardwood veneers impregnated with chemicals, then laminated under high pressure. The veneer layers are bonded with their grains parallel, so the resulting material is easier to work than cross-grained plywood. Mainly used for model and pattern making.

stability 1. A measure of the ability of a material to resist freezing, cracking, drying, or other changes in its original condition when stored or in use. 2. The rigidness of a structure; its ability to resist movement.

stack 1. A flue or chimney. 2. A group of flue pipes arranged in one chimney or chase. 3. A main vertical DWV pipe extending one or more stories to its vent.

stack dried Of or relating to newly milled lumber that has been stacked with air spaces between the layers and left to dry. No other form of heat is used in the drying process. Abbreviated S-dried. Also called air dried.

stack effect The natural movement of warm air from a low place to a high place, both by virtue of its lighter, less dense mass, and by the effects of wind-created low-pressure areas, particularly above a building.

stack temperature The temperature of the air and other gases passing through the flue of a furnace, chimney, or other such device.

staggered joints End joints between sheathing or siding boards that are alternated between their supporting members to prevent all of the joints from falling on the same support and to more evenly distribute load and stress. See Fig. P-10.

staging Temporary scaffolds or work platforms used during construction.

stagnation A condition in which a gas or fluid is not stirred by a current or flow.

stagnation temperature The temperature reached within a solar collector during periods when the air or liquid-transfer material is not circulating.

stain A liquid used to color wood, either to enhance its own natural color or to imitate the color of other woods. There are three basic wood stains.

ALCOHOL-BASE STAIN A stain that produces good color with little effort; available in a wide range of colors.

OIL-BASE STAIN A pre-mixed stain with good color selection, that is easy to use but prone to fading with time. End grains should be sealed with linseed oil to prevent exaggerated darkening.

WATER-BASE STAIN Powdered dyes that must be mixed with water. It produces very good color but tends to raise the grain of certain woods.

stainless steel A metal alloy made from very low carbon steel that contains varying amounts of chromium. It is strong, ductile, and extremely corrosion resistant.

300 SERIES STAINLESS STEEL Stainless steel that contains 12 to 20 percent chromium.

400 SERIES STAINLESS STEEL Stainless steel that contains at least 12 percent chromium and is magnetic.

staircase See *stairway*.

stair fitting Any of a variety of parts, usually prefabricated, used in the construction of some types of stair rail systems.

stair rail See *handrail*.

stair rail bolt See *rail bolt*.

stairs A series of uniform steps that connect and provide access from one level to another. Stairs are primarily classified as the following. See Fig. S-17.

CIRCULAR STAIRS See *stairs: winding stairs*.

CLOSED STAIRS Stairs having walls along both sides.

COMBINATION STAIRS Stairs having a combination of both open and closed stairs within the same run.

MAIN STAIRS Stairs that provide the primary access between two inhabited floors of a building.

OPEN STAIRS Stairs having no walls on either side, or a wall on one side only.

PLATFORM STAIRS Stairs that are broken up by platforms, called landings, at which the direction of the stairs usually changes; also called L- or U-type stairs.

SERVICE STAIRS Stairs that provide access to a secondary or uninhabited area of a building, such as a basement, roof, or storage area.

SPIRAL STAIRS Stairs that ascend in a tight circle around a central post.

STRAIGHT STAIRS Stairs that directly connect two levels, with no turns or landings.

WINDING STAIRS Stairs that gradually change direction as they ascend and are either circular or elliptical in shape; also called circular stairs.

stairway A flight of stairs, especially one that connects two floors of a building, including the steps, railing, and all other structural components; also called a staircase. See Fig. S-18.

stairwell A framed opening in a building within which a stairway is constructed. See Fig. S-16.

staking out Measuring and marking out an area of proposed excavation, using stakes driven into the ground. See Fig. S-19.

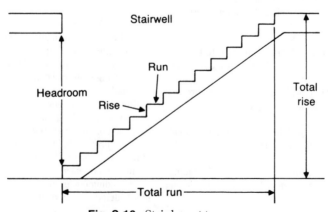

Fig. S-16. Stair layout terms

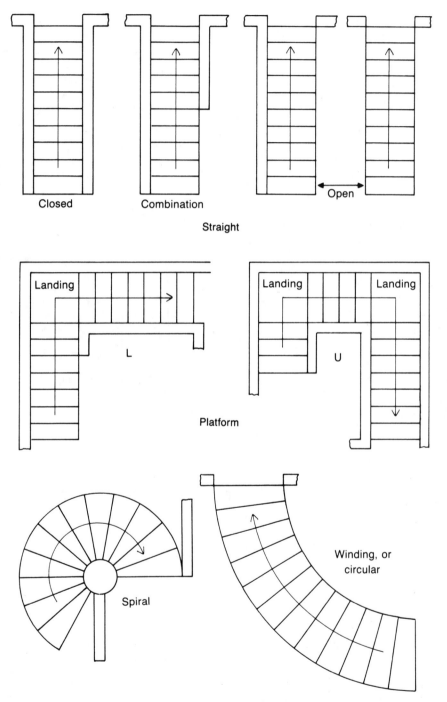

Fig. S-17. Common stair types

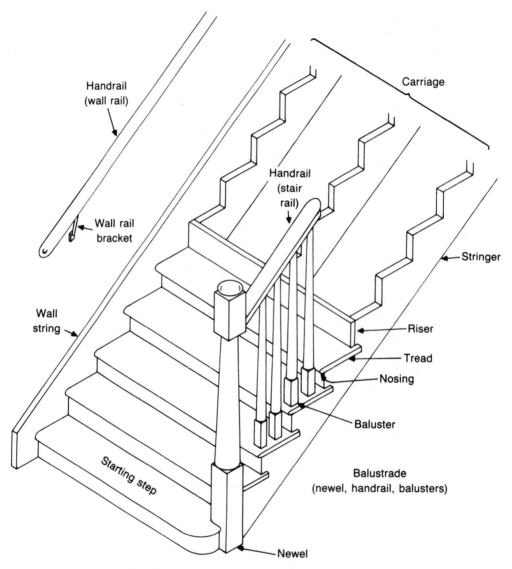

Fig. S-18. Main components of a typical stairway

standing finish An older term for the finish materials used around doors, windows, baseboards, and other finished areas of a home's interior.

stand-off Of or relating to a type of screw eye having a long shaft and an insulated disk in the eye; commonly used to secure television antenna wires to prevent them from rubbing against a roof or wall.

standpipe A vertical pipe extending from a trap; used to secure uniform pressure in a water-supply system or as a drain for a clothes washer. See Fig. D-14.

staple 1. A U-shaped fastener formed from a loop of round or flattened wire with pointed or sharpened legs. Available in a variety of widths, leg lengths, and wire gauges. 2. See *nails, common types.*

staple gun A hand-, electric-, or air-operated tool used to drive staples.

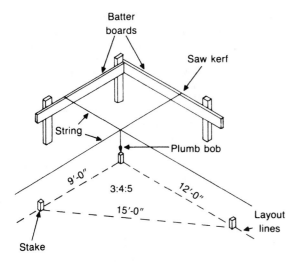

Fig. S-19. Staking out a foundation

star drill A hardened steel drill with a star-shaped face, driven into masonry with a hammer to form a hole.

starter collar A round, open-ended, sheet-metal fitting, used with a radial forced-air heating system, that is crimped directly onto a hole cut in the plenum, forming a point of attachment for a run of ducting; also called a straight collar.

starter strip A roll of 9-to-12-inch-wide mineral-surfaced asphalt roofing installed at the lower edge of roof sheathing. It forms the first course of a composition roofing installation, over which the first course of shingles is laid.

starting current The amount of current that an electric motor initially requires in order to bring it up from stopped to full rotation.

starting easement A simple, upward-curving stair fitting having a rounded end, designed to attach over the top of the newel post and provide a transition to the handrail. See Fig. S-20.

starting fitting A type of stair fitting that attaches to the top of

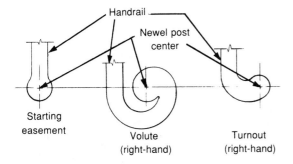

Fig. S-20. Starting fittings

the newel post and provides the starting transition to the handrail. See Fig. S-20. Also see *starting easement; turnout; volute.*

starting step The first and lowest step in a stairway, often wider and more decorative than the other steps. See Fig. S-18.

static electricity Electricity that is stationary or at rest, as opposed to current electricity, which is flowing.

static head In a plumbing system, the pressure required to overcome gravity in order to push water up a pipe. In a vertical pipe, 0.434 pounds per square inch are required to push water up 1 foot (1 psi will raise water about 2.3 feet).

stationary appliance See *appliance.*

statute of limitations A legal time limit placed on the filing and enforcement of a lien or other legal action.

STC See *sound-transmission class.*

steady-state efficiency See *heating efficiency.*

steam bending Using wet steam to bend wooden members; used in the making of laminated beams, and for furniture or other decorative work.

steam dome An area above a steam boiler that allows for expansion of the steam.

steam-heating system A method of providing heat to a building in which water is heated in a central boiler until it turns into steam, then the steam is passed through pipes to individual radiators in each room, where it gives off its heat and condenses back into water, which flows back to the furnace for reheating. There are two types of systems.

ONE-PIPE SYSTEM A steam-heating system that uses one pipe to and from the boiler, with steam traveling up the pipe while water returns down it.

TWO-PIPE SYSTEM A steam-heating system that uses one pipe to supply the steam and a second pipe to return water to the boiler.

steam trap A device used to regulate and control the flow of steam and returning water in a two-pipe steam-heating system.

steel A type of hard, strong metal, made from iron that is alloyed with up to 1.5% carbon and often small amounts of other metals, including chromium, nickel, manganese, and others. There are a variety of special steel allows intended for specific uses; common types include the following.

CARBON STEEL Steel having a relatively high carbon content, used in the manufacture of tools. Increased carbon makes the steel harder and better able to accept and hold a sharp edge, but increases its brittleness.

CHROME-VANADIUM STEEL Carbon steel to which chrome, vanadium, and other alloys have been added, producing steel that is strong and tough, but that does not hold a sharp edge. Used in the manufacture of screwdrivers, wrenches, and similar tools.

HIGH-SPEED STEEL Carbon steel to which other alloy materials have been added, including tungsten, chrome, and vanadium. Used primarily for high-speed, high-friction cutting tools such as drill and router bits.

MILD STEEL A relatively soft steel containing approximately 0.3%

carbon. It is easy to bend and shape, but does not contain enough carbon to be hardened. Commonly used for structural applications instead of toolmaking. Also called low or soft steel.

SILVER STEEL Highly polished carbon steel, often used in the manufacture of small tools and instruments.

TEMPERED STEEL Hardened steel that has been reheated and air- or oil-cooled. This process creates steel that is less brittle, but that retains its hardness and its ability to achieve and maintain a sharp cutting edge.

TOOL STEEL Any of a variety of hard, durable steel alloys with a medium to high percentage of carbon, used in the manufacture of tools.

steel-frame construction A method of building in which the structural components are steel or steel supported.

steel pipe Pipe manufactured from steel for use in plumbing systems, having threaded ends that are used to join the pipe to compatible threaded fittings, although some types can be soldered or welded. Sometimes called iron pipe.

steel rule See *rule.*

steel wool An abrasive made of fine steel threads matted together and formed into pads. Grits range from 4 (coarsest) to 0000 (finest).

stem wall The vertical wall of a foundation, between the footing and the sill plate. See Fig. P-2.

stencil 1. To brush or spray paint over a precut template, producing a design on the surface underneath. 2. Such a template.

step-down transformer See *transformer.*

step-up transformer See *transformer.*

step flashing Short, L-shaped sheet-metal pieces used as roof flashing. They are applied in an overlapping row, one per course of shingles, where the roof meets a vertical surface, such as a second-story wall. See Fig. F-7.

stepped foundation A foundation formed in a series of steplike increments; used when placing a building on sloped ground.

sticker 1. Strips of wood placed between layers or stacks of lumber to aid in drying. 2. A machine that produces moldings.

stick frame 1. A building constructed using individual pieces of lumber, as opposed to modular, prefab, masonry, or steel. 2. To construct a stick frame building.

stile A vertical member of a framework, such as a door, face frame, or sash. See Fig. D-3.

stipple A heavy, decorative paint finish, usually created by rolling thick paint onto a surface using a carpet-covered paint roller.

stone panel A plywood panel to which natural rock aggregate has been epoxy bonded; used as a decorative exterior facing.

stool A wide, flat molding, usually with one rounded edge, used as a finish over the window sill framing. It fits between the side jambs and against the bottom rail of the sash or sash frame. See Figs. M-4 and W-3.

stoop A small porch, platform, or set of stairs at the entrance to a building, usually not roofed.

stop 1. Thin strips of molding attached to the side and headjambs of a door or window, against which the door or window closes. See Figs. D-6 and W-3. 2. A water shutoff valve, normally used to control an individual water line, as under a sink. 3. Any device that acts to limit or control the action of a particular object.

stopper 1. A rubber device that covers the flush valve opening at the bottom of a toilet tank; also called a flapper. See Fig. F-9. Any of a variety of devices for closing off the drain in a sink or bathtub. See Figs. B-5, P-9, and W-1.

storm collar A sheet-metal strip that is attached around a pipe where it passes through a flashing and that flares out from top to bottom, shedding water out and away from the pipe/flashing joint. See Fig. R-7.

storm door A one-panel wood or metal frame door, containing a full or half panel of glass, installed outside an existing door for additional weather and security protection.

storm window 1. An additional fixed or operable window placed against the exterior or interior frame of an existing window to provide extra thermal insulation against outside cold and help to block air infiltration. 2. Plastic, plexiglass, or other material used inside or outside of an existing window for thermal insulation, usually removed at the end of winter.

story The habitable portion of a building, extending from the top surface of one floor to the top surface of the next floor, or in the case of a building's highest story, to the ceiling or underside of the roof.

story pole A rod or strip of wood used in laying out and transferring measurements for stairways, shingle and siding courses, or wall openings; also called a story rod.

story rod See *story pole.*

stove mat See *hearth pad.*

straight collar See *starter collar.*

straightedge A length of wood, metal, or other material with a true edge; used for laying out lines, checking accuracy, or guiding tools.

straight peen hammer See *hammer: peen hammer.*

straight stairs See *stairs.*

straight stop A plumbing shutoff valve in which the inlet and outlet sides are in line with each other.

strain The deformation of an object by an imposed stress.

stranded wire Wire comprised of a quantity of smaller wires wrapped or otherwise grouped together; for use as an electrical conductor or other purpose.

S-trap A type of fixture trap shaped like a horizontal S; designed for draining down through the floor.

strap See *timber connector.*

strap anchor See *timber connector.*

strap hinge A heavy-duty hinge having both leaves formed from thick steel straps. See Fig. S-21.

strapped wall A type of wall construction in which additional framing members are installed horizontally over the exterior wall studs to deepen the wall cavities and allow for the installation of greater amounts of wall insulation.

stratification The tendency of air within a building to form

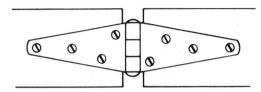

Fig. S-21. Strap hinge

temperature layers, with cool air staying near the floor and warm air staying near the ceiling.

strawboard See *building board*.

street elbow A plumbing elbow having external threads on one end and internal threads on the other. See Fig. T-4.

street tee A T-fitting in which one or two of the outlets has external threads.

stress The load per square inch that acts against an object. See Fig. S-22.

stressed skin Plywood, veneer, or other facings glued to opposite sides of an inner, supporting framework to form panels, doors, etc.

stretcher A brick or other masonry unit laid so that its longest dimension is horizontal and its face is parallel with the wall's face. See Fig. B-10.

stretcher block See *building block*.

striated Of or relating to a material that is marked or deco-

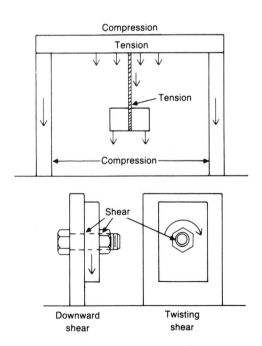

Fig. S-22. Common forms of stress

rated with a series of small parallel grooves or recessed stripes.

striated plywood See *plywood siding*.

strike plate A metal plate that is screwed into a mortise on a door jamb and engages the latch bolt of a lockset.

striking Using a trowel to cut away excess mortar in a freshly laid masonry wall.

striking off See *screeding*.

string See *stringer*.

stringboard A board that covers the ends of a step and hides the true string; used in stair building.

stringer 1. In stairs, an inclined side of a stair that supports the stair treads and risers. See Fig. S-18. 2. One of the horizontal members between two posts in a fence to which the fence boards are attached. 3. A support for cross members in a floor or ceiling.

strip See *lumber*.

strip flooring See *hardwood floor*.

strip gauge A molded recess or mark on the back of a switch, receptacle, or other device used to gauge the amount of insulation that should be stripped off the end of a wire for proper connection to the device; used primarily on backwired devices. See Fig. G-6.

strip shingle Any of a variety of composition shingles in widths of 12 to 15 inches and lengths of 36 to 40 inches, usually having one or more tabs, or cutouts, to give the appearance of smaller, individual shingles.

strongback A piece of lumber set on edge on top of the ceiling joists and at a right angle to them. It is secured to the joists with special ties to lessen the possibility of sagging. See Fig. R-5.

strongback tie A double-pointed, Z-shaped nail used to secure ceiling joists to a strongback. See Fig. R-5.

struck joint A mortar joint between masonry units that has been finished with a trowel.

structural clay tile Hollow or partially solid masonry building units made from burned clay, fire clay shale, or a mixture of these materials. Some types of structural clay tiles follow.

FACING TILE Tile for interior and exterior use with an exposed face.

FIREPROOFING TILE Tile that has been treated to provide fire protection for structural members.

FLOOR TILE Tile of sufficient wear and abrasion resistance for use on floors.

FURRING TILE Tile used as a non-loadbearing lining for the inside of exterior walls, over which other materials are usually installed.

HEADER TILE Tile constructed with recesses for receiving brick headers when constructing a wall with a masonry facing.

LOADBEARING TILE Tile of sufficient strength for use in loadbearing walls.

ROOF TILE Tile made specifically for use as a roofing material.

SOLAR SCREEN TILE Tile made specifically for use in the construction of solar screens.

structural lighting Permanent light fixtures that are installed during construction and become part of the finished building.

structural lumber See *lumber.*

structural panel A sheet manufactured from wood veneers or chips, such as plywood, waferboard, OSB, or others, that must meet the performance standards of governing bodies such as the American Plywood Association to qualify for use in structural applications. See *building board.*

structural slab A poured concrete floor or, occasionally, a roof, having embedded steel or other reinforcement to support a concentrated load at that point.

stub joist A short ceiling joist that extends at right angles from the last regular ceiling joist to the sidewall.

stub out 1. The end of a waste or water pipe that extends through a wall or floor for connection to a fixture. 2. To install pipes or ducts in such a way, prior to applying finished wall and floor coverings.

stub wall A wall that is shorter than the other interior walls in a building; used as a partition, as backing for cabinets in exposed areas, or for other such uses.

stucco An outside wall covering, usually a mix of portland cement, sand, and water, that is applied wet over wire netting.

stucco mold A decorative, redwood molding having a U-shaped recess along one edge; used as a casing around exterior doors and windows when stucco is used as the wall covering. Stucco is forced back into the recess, helping to prevent cracking along the molding.

stucco netting Wire netting, similar to chicken wire, attached to the exterior wall framing as a base for stucco.

stud A wooden or metal structural member installed vertically in a series to make up a wall or partition. See Figs. B-2, F-10, P-3, and R-3.

stud anchor See *anchor, masonry.*

stud brace See *timber connector.*

studding The framework of a wall.

studio window See *window.*

Styrofoam The brand name for a variety of polystyrene products; often used incorrectly as a generic term for any type of foam insulation or foam product.

sub- Below, beneath, or secondary.

subcontract 1. A contract for the supply of labor and/or materials, usually between a contractor and a subcontractor. 2. Any construction contract not made directly with the building's owner.

subcontractor An individual or firm that contracts for work with the prime contractor instead of with the owner of the project.

subfloor Lumber or plywood laid directly over the floor joists or girders, and over which tyment and finishes installed. See Figs. B-2, P-2, and P-3.

subgrade See *below grade.*

submersible pump An electrically operated, waterproof pump capable of being completely submerged in water; used in well shafts, sumps, pools, fountains, etc.

subpanel A secondary service panel connected to the main

service panel by a feeder cable; used to house overcurrent protection devices in a remote area such as a detached shop or garage. See Fig. S-9.

substantially completed A construction contract term generally accepted to refer to that stage of a construction project when the building is finished to a point where the owner can take occupancy.

substrate The backing or subsurface to which another material is applied or attached.

substructure The lower, supporting portion of a building, usually including the footings, foundation, and joists.

summer 1. A heavy, horizontal timber placed over a wide opening, acting as a header to support the structure above it. 2. A large stone atop a column that supports an arch or similar structure.

summer solstice The day, on or about June 21, when the sun rises and sets farthest to the north and is at its highest elevation of the year, appearing almost directly overhead; traditionally, the start of summer. See Fig. S-23.

summerwood See *wood.*

sump pit A pit or receptacle in the ground where water from heavy rain, basement drains, or below-grade fixtures is collected and pumped out.

sump pump A special pump used to empty a sump pit. When water in the pit reaches a certain point, a float device activates the pump motor, shutting it off again when the pit is drained.

sunset hour angle The angle of the sun as it sets. See *hour angle.*

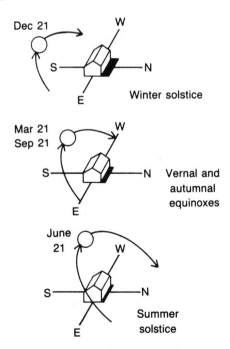

Fig. S-23. Sun's yearly path

sunspace A south-facing room or area that is attached or incorporated into a building, is heavily glazed, and contains internal thermal mass; used to provide solar heat to the building through convection or radiation. Also called a solarium.

sun tempered Of or relating to a building that derives some of its heat from passive solar gain, as through a large number of south-facing windows, but that has no specific storage system.

superinsulated Of or relating to an energy-efficient method of house construction involving three basic design and construction considerations: substantially increased insulation amounts; limited glazing on the north, east, and west walls; and substantially decreased air infiltration.

superstructure That portion of a building above the foundation and other supporting members.

supply line A flexible tube or hose used to connect a water shutoff valve to a faucet, toilet, or other fixture.

surfaced lumber See *lumber*.

survey 1. The accurate measurement of an area of land, usually by a licensed surveyor, that establishes the land's boundaries and the location of significant improvements and natural features. 2. The drawn and/or written description of an area of land that results from a surveyor's measurements and observations.

suspended ceiling A grid of inverted, T-shaped bars hung on wires from overhead support framing, into which removable panels are inserted to form a finished ceiling that allows easy access to the area above it.

swaging Expanding the end of soft pipe, such as copper, to receive another pipe, thus eliminating the need for a coupling. A special punch called a swaging tool is used, which is driven into the pipe with a hammer.

swamp cooler See *evaporative cooler*.

sweating 1. Moisture that collects on plumbing pipes and fixtures, caused when water vapor in the air condenses on the colder surface of the plumbing. 2. Soldering copper pipes together.

swelling The expansion of the cells in a piece of wood as a result of exposure to moisture, causing the wood to increase in dimension.

swinging door See *door*.

switch An electrical device installed in a circuit to open and close the path of current flowing through that circuit. Common switch types follow.

SINGLE-POLE SWITCH A switch having two screw terminals, used for controlling a fixture from one location.

3-WAY SWITCH A switch having three screw terminals, used for controlling a fixture from two different locations.

4-WAY SWITCH A switch having four screw terminals, used in conjunction with two 3-way switches to control a fixture from three different locations.

switch box See *electrical box: wall box*.

symmetrical 1. Of or relating to an object having exactly equal features, contours, and shapes on either side of a centrally dividing axis. 2. Of or denoting a structure or arrangement that is well balanced or well proportioned.

synthetic Produced or manufactured; artificial; not occurring or obtainable naturally.

T

table saw A stationary power tool having a large, flat table and a rotating, circular saw blade; used for ripping, crosscutting, dadoing, and performing many other operations on wood and other materials.

tab 1. A cutout in a shingle. 2. See *timber connector*.

tachometer A device used to measure and indicate revolutions per minute.

tack 1. A short nail having a broad, flat head and a sharp chisel point; used primarily in upholstery work. 2. The stickiness, or adhesion ability, of a paint, glue, or similar material.

tack hammer See *hammer*.

tackle A set of ropes and pulleys used for lifting or moving heavy objects; also called block and tackle.

tack rag A porous cloth impregnated with varnish so as to be relatively sticky; used to remove surface dust and sanding residue before finishing materials are applied.

tack strip A narrow wooden strip that has numerous small nails driven through it such that their points protrude through the top and that is nailed to the floor around the perimeter of a room to hold wall-to-wall carpeting in place.

tacky Of or relating to a partially dried coating of paint, glue, or other material that is still sticky and is not easily removed from the underlying surface.

tail 1. The end of a rafter or truss that extends out past a sidewall. See Figs. R-5, R-6, and T-7. 2. The end of a floor joist that extends out past the foundation.

tail cut   The plumb cut made on the overhanging end of a rafter. See Fig. R-5.

tail joist   A relatively short beam or joist attached to a floor or ceiling header at one end and resting on the foundation or wall plate at the other end. See Fig. J-2.

tailpiece   A drain pipe that connects a basket strainer or sink pop-up assembly with a trap. See Figs. B-5 and P-9.

tailpiece extension   A pipe and locknut assembly used to extend the length of a tailpiece in certain installations. See Fig. B-5.

take off   To form a material list, cutting list, order sheet, or other set of specifications from the details given on a set of plans.

takeoff collar   A sheet-metal duct fitting having a rectangular base attached to a round, flexible elbow; attached to a hole cut in the main duct of an extended plenum duct system as a starting point for a branch duct.

tall unit   Any kitchen cabinet that extends from the floor to a height even with the top of wall cabinets, such as an oven cabinet. Common heights are 84 and 96 inches.

tambour   A rolling top or front made of thin wood strips attached to a cloth backing, such as that used on a rolltop desk.

tamp   To pack down with repeated impact, as for loose soil.

T and G   See *tongue and groove*.

tang   The pointed shank of a tool such as a file or chisel that is opposite from the cutting edge and is inserted into a handle. See Figure R-4.

tangent   1. A line that touches a curve at one point. See Fig. C-3. 2. A line or curve that touches another curve at one or more consecutive points.

tangential   Coincident with or parallel to a tangent line.

tangential sawing   In lumber milling, cutting a log lengthwise, perpendicular to a radius.

tankless water heater   A type of water heater that supplies hot water as the demand for it occurs, rather than heating it in advance and storing it in a tank.

tap   1. An externally threaded tool used for cutting internal threads into a predrilled hole. 2. A faucet or spigot for drawing water from a pipe.

tape   See *measuring tape*.

taper   1. To narrow gradually and regularly toward one end. 2. A person skilled in the taping and finishing of plasterboard.

tapered edge   The long edge of a sheet of plasterboard that is beveled down slightly to allow easier recessing of the joint tape.

Tarmac   The trade name for a paving material having coal tar as its principal ingredient.

task lighting   Using a fixed or portable light fixture to illuminate a particular work space, rather than for general lighting of an entire area.

taut   Pulled or drawn tight, as a rope.

T-bar ceiling   See *suspended ceiling*.

T-bevel   See *sliding T-bevel*.

tee   A T-shaped plumbing or duct fitting having three outlets, one of which is perpendicular to the other two. See Figs. S-13 and T-4.

tee nut   A fastening device consisting of an internally threaded steel disk with four spiked legs that is driven into a predrilled hole in a piece of wood, providing a threaded insert into which a machine screw or bolt can be inserted.

tee plate   A flat, T-shaped, predrilled metal plate used for joining and reinforcing two pieces of wood or other material at the perpendicular intersection. See Fig. T-1.

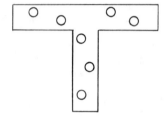

Fig. T-1. Tee plate

Teflon tape   A trademark for thin, white, nonadhesive tape used to wrap pipe threads as a sealant.

temp   Abbreviation for *temporary*.

temper   1. To reheat and oil-quench a piece of metal that has been hardened, in order to eliminate brittleness. 2. To mix plaster or mortar with water to the proper consistency.

temperature-pressure relief valve   1. A safety valve on top of a water heater that allows dangerous buildups of steam or hot water to escape. See Fig. W-2. 2. A similar valve for solar collectors, designed to allow heat-transfer fluids to escape from a solar heating system in the event that the maximum working temperature or pressure of the system is exceeded.

temperature relief valve   A valve designed to be activated by temperature alone, which will open and discharge at a predetermined temperature.

temperature swing   A fluctuation in temperature that occurs inside a building, caused by the on-and-off cycling of heating equipment, the opening and closing of doors and windows, etc.

temperature zoning   A concept used in solar heating design to separate rooms into zones, such that rooms needing more heat (living room, study, etc.) are grouped on the home's south side for best use of winter sunlight, while rooms needing less heat (kitchen, bedroom) are placed on the north side.

tempered glass   A strong, impact-resistant type of glass that is manufactured through a reheating and cooling process to reduce stress in the glass.

tempered hardboard   See *building board: hardboard*.

tempered steel   See *steel*.

tempering valve   A valve in which cold water is mixed with small amounts of hot water before being supplied to a fixture; used primarily to prevent condensation on toilet tanks.

template A pattern made of plastic, wood, metal, or other material that is superimposed on an object as a guide for repetitive marking or cutting. Templates of standard shapes and symbols are used extensively in drafting.

temporary Of or relating to braces, blocking, covers, or anything else used for short-term support or protection while a structure is being built, repaired, or altered; often referred to as temps.

tenon A projection, usually square or rectangular in section, that remains on the end of a piece of wood after the surrounding stock has been cut away and that is inserted into a matching mortise on another piece to form a mortise-and-tenon joint.

tensile strength The maximum amount of tension that can be imposed upon an object before it fails.

tension Stress imposed upon an object by stretching forces, which attempt to elongate the object. For example, when a weight is lifted with a rope, the rope is in tension. Opposite of compression. See Fig. S-22.

tension wood See *lumber defect.*

termite A small, antlike insect that enters the wood structure of a building from a ground nest and eats the inside of wood members while leaving a shell of sound wood which often conceals the damage. There are about 56 species of termites in the United States.

termite shield A shield, usually of galvanized sheet metal, placed in or on a foundation wall, around pipes, or in other underfloor areas to prevent the passage of termites. See Fig. P-2.

terrace A portion of land, especially one near a house, that is raised above the surrounding grade and slopes down, usually lawned or otherwise landscaped.

terra cotta Literally, cooked earth; a burned, glazed or unglazed clay product; used in making decorative veneers, plumbing pipe, roofing tiles, and many other products.

terrazzo A type of finished floor created by pouring a wet mixture of marble or granite chips and portland cement over a concrete subfloor. When dry, the surface is ground and polished.

tertiary color A color created by combining adjacent colors on a color wheel, such as red and orange, or by combining two secondary colors, such as orange and purple.

tessellated 1. Of or relating to an object constructed of squares or cubes, such as tiles or blocks. 2. Of or related to anything arranged in a checkered pattern

tessellated floor An older decorative style of flooring consisting of small regular blocks of stone, marble, or other material, arranged in symmetrical patterns and set in cement.

texture 1. Any of a variety of decorative treatments applied to a finished wall surface, such as plasterboard, plaster, or stucco. 2. The grain of a piece of wood.

textured block See *concrete block.*

texture one-eleven plywood See *plywood siding.*

texture paint Thick paint, often with additives such as sand, that can be manipulated with a brush, roller, or other tool to create a variety of surface textures.

therm A unit of heat, typically used in the measurement of natural gas consumption, equal to 100,000 Btus.

thermal 1. Pertaining to or determined by heat. 2. Of or relating to anything that causes, uses, or produces heat.

thermal break A material or space used to separate two other materials that are high in thermal conductance; for example, a piece of rubber used to separate the interior and exterior aluminum frames of a window, thus reducing the amount of heat which can be conducted through the metal. See Fig. D-3.

thermal bridging A material having a relatively low R-value that allows heat to be conducted through it, thus bypassing the building's insulation. Wall studs are an example.

thermal chimney A vertical passageway through which warm air rises by virtue of the stack effect; used to passively distribute heat and to induce ventilation in a building.

thermal circulation See *gravity circulation.*

thermal cutout A type of overcurrent protection device, usually located directly on a piece of equipment, that reacts to the increase in heat accompanying an increase in electrical current, as in a motor which has jammed or overloaded. The cutout opens the circuit to stop the current flow to the equipment. Some types burn out and must be replaced; others can be reactivated by pressing a reset button when the source of heat buildup has been removed.

thermal door A door and door frame combination that has an R-value of at least 2, and that is completely weatherstripped.

thermally broken Of or relating to an object such as a door or window that is constructed with a thermal break.

thermally improved metal Any metal object, such as an aluminum window frame, that has been manufactured with a thermal break to reduce conductance.

thermally protected Of or relating to a notation on the nameplate of an electric motor indicating that the motor is protected by a thermal cutout.

thermal mass Any of a variety of materials, such as rock, water, concrete, etc., that is used to absorb and store solar radiation for later reradiation in the form of heat.

thermal window See *window: double-pane window.*

thermistor A device that reads temperature changes in a solar heating system and then signals the information to the differential thermostat; also called a sensor. See Figs. C-5 and O-3.

thermocouple A device consisting primarily of two dissimilar metals that are joined together and that will conduct heat at different rates, producing a weak electrical current; commonly used as a safety device on gas appliances. As long as the appliance's pilot light is on, the thermocouple stays warm and generates current, which keeps the gas valve open. If the pilot light goes out, the current stops, closing the valve that supplies the gas to the appliance. See Fig. W-2.

thermoplastic Of or relating to a material that softens and becomes plastic when exposed to a sufficient amount of heat.

thermosetting Of or relating to a material that is not affected by heat once it has cured.

thermosiphoning In a passive solar heating system, a method of using the principles of convection to allow heat from the sun, which is collected by various means at the home's lowest point, to rise naturally through the building, then, as the air gradually cools, to fall and be routed through ducts back to the collector, where it is warmed and recirculated. See Fig. T-2.

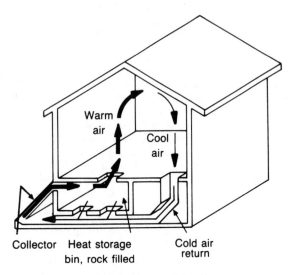

Fig. T-2. Thermosiphoning

thermosiphon water heater A passive solar heating system for domestic water that uses a collector at ground level to heat the water, which rises naturally into an overhead storage tank before entering the building's hot water plumbing system. See Fig. C-6.

thermostat An instrument used to automatically control the operation of various heating, cooling, or pumping devices by reacting to changes in air temperature.

thermostat anticipator An adjustable control inside of a furnace thermostat that is normally set to the amperage of the furnace and is used to adjust the thermostat's response time between actual temperature and set temperature, helping to eliminate wide swings in temperature range.

thermostat cable A type of electrical cable for low-voltage wiring that consists of two or more small-diameter, color-coded conductors contained in one common, outside jacket.

thimble A type of sleeve made from fire clay, mortar, or insulated metal, placed in a combustible wall as an opening through which a flue pipe can pass.

T-hinge A type of heavy hinge having one regular leaf and one long strap leaf; used on gates and similar structures. See Fig. T-3.

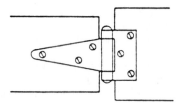

Fig. T-3. T-hinge

thinner A volatile liquid used to thin and regulate the consistency of various paints and varnishes.

thin-set adhesive A powdered, cement-based adhesive used for applying ceramic tile to various subsurfaces.

thin-wall See *EMT*.

threaded fitting A pipe fitting having internal and/or external pipe threads for connection to threaded pipes. See Fig. T-4.

3:4:5 A method of establishing or checking square, based on

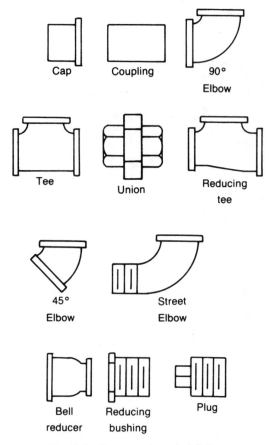

Fig. T-4. Common threaded fittings

the ratio of the legs of a right triangle. One side of the object being checked is marked three units (inches, feet, etc.) out from the corner. The adjacent side is marked four similar units from the same corner. If the corner is square (90 degrees), a measurement diagonally between the two marked points will equal 5 units. Any ratio of 3:4:5 (6:8:10, 12:16:20, etc.) will work the same way. See Fig. S-19.

three-phase   Of or relating to a type of electrical power that is transmitted to a building from three transformers which are synchronized so as to provide steady voltage in overlapping waves, without the power swings associated with single-phase power, thus allowing for more efficient operation of large electric motors.

three-ply   1. Of or relating to any material, such as plywood, that is composed of three separate plies or layers. 2. Of or relating to a structure or surface formed from three separate layers of applications, as with a built-up roof.

three-prong adapter   See *grounding adapter*.

three-prong plug   See *grounded plug*.

three-tab   Of or relating to a commonly used type of composition shingle having cutouts that divide the lower half of the shingle into thirds. When installed the tabs give the appearance of three individual shingles.

three-way bulb   A special type of incandescent light bulb containing two filaments of different wattages, for example 50 and 100 watts. Each filament can be used independently, providing 50 or 100 watts of light, or together, achieving the equivalent of 150 watts of light.

three-way socket   A type of light socket having two contacts in its base instead of one, designed for use with a three-way bulb. A special switch attached to the socket allows selection of the bulb's low- or high-wattage filament, or a combination of both.

three-way switch   See *switch*.

three-wire service   See *service*.

threshold   A strip of wood or aluminum used over the floor or door sill to close and insulate the space between the floor and the door bottom. See *saddle*.

throat   The narrow area beneath the smoke chamber in a masonry fireplace that houses the damper; designed to increase the velocity of the upward current of smoke into the flue. See Fig. F-4.

through check   See *lumber defect*.

through stone   A stone, brick, or block that extends through a wall to form a bond with another parallel or intersecting wall.

thumbscrew   A machine screw with a flattened end that can be gripped and turned by hand.

tie beam   See *collar beam*.

tie-dated nail   See *nails, head types*.

tier   A continuous vertical section of masonry, one unit thick; also called a wythe or withe. See Fig. B-10.

tie rod   A metal rod in tension, usually with a turnbuckle or other adjustable fitting, used to tie together and stabilize a building's walls. Tie rods may be temporary, as when a building is being repaired, or may be a permanent repair in their own right, in which case the exposed portions of the rod may be ornamental. See Fig. T-5.

tight knot   See *lumber defect: knot*.

tile   Any of a variety of building units made from fired clay, cement, glass, or other material.

tile-backing board   See *building board: cement board*.

tile decking   See *decking*.

tilt angle   The angle above the horizontal at which a solar collector is placed for maximum exposure to the sun. This angle varies depending on the angle of the sun at a given latitude.

tilt-up construction   A method of construction in which wall sections of concrete or wood are prefabricated elsewhere, then brought to the site and tilted up into position.

timber   See *lumber*.

timber connector   Any of a variety of metal devices widely

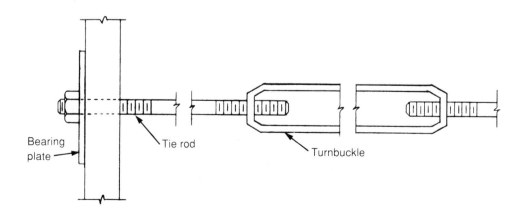

Fig. T-5. Tie rod

used to connect, support, or reinforce structural wood joints in construction. Some of the more common types follow. See Fig. T-6.

ANCHORS AND CLIPS Various connectors used to tie and anchor wood members to each other or to concrete in different applications.

BEAM CAP See *timber connector: post cap.*

BEAM SEAT A plate having two or three prongs, which are embedded in concrete, and a flange, which is connected to the beam.

GIRDER HANGER A hanger used in certain applications to attach a girder to the foundation.

GUSSET An L bracket with a 45-degree reinforcing web used to add support where two members meet at right angles.

HEADER HANGER A hanger used to attach a header to the king

studs, thus eliminating the need for a cripple.

HINGE CONNECTOR A special connector used to splice and support two beams of different heights.

JOIST HANGER A U-shaped bracket of stamped sheet metal or welded steel that is secured to a girder or ledger to receive and support the ends of a joist.

KNEE BRACE A U-shaped strap that spans the bottom of a beam for additional stability between the beam and the intersecting purlins or joists.

ORNAMENTAL CONNECTOR A hanger, strap, cap or other timber connector having a notched, hammered, painted, or otherwise decorated surface.

PANELIZED ROOF HANGER A small, top-flange hanger for use with prefabricated roof panels.

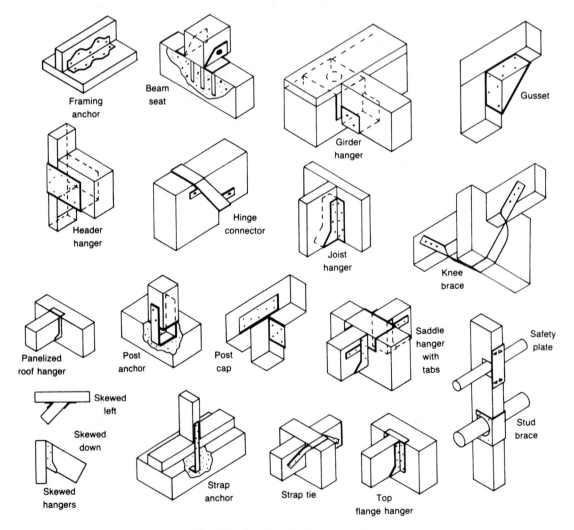

Fig. T-6. Standard timber connectors

POST ANCHOR One of a variety of connectors that are embedded in concrete to receive and secure a post. Some types can also be secured to wood, as on a deck.

POST CAP Any of a variety of T-, L-, or U-shaped devices for attaching a post to a beam or a concrete support. Larger sizes are often referred to as beam caps or splice caps.

SADDLE HANGER A double joist hanger with a flat or channel shaped top, designed to be hung over the top of a beam or other supporting member.

SKEWED HANGER A joist or saddle hanger designed to angle down or to the side; used for connecting two members at angles other than 90 degrees.

SPLICE CAP See *timber connector: post cap*.

STRAP ANCHOR A hooked strap used in various applications to tie wood framing to concrete or masonry.

STRAP TIE Metal strips that span the top of a beam to tie together the purlins on each side.

STUD BRACE AND SAFETY PLATE A plate used to reinforce studs that have been cut for the passage of pipes, or to protect pipes and wires from damage.

TAB One of the horizontal straps attached to the sides of various hanger flanges to improve seismic stability.

TOP-FLANGE HANGER A joist hanger having two flanges at a 90-degree angle to the back of the hanger that are secured over the top of the supporting member for additional strength.

timer See *time switch*.

time switch An electrical switch that can be set for a given period of time, after which it will shut off the electrical fixture or unit it controls; also called a timer.

tin shingle A small piece of tin or sheet metal used as a flashing or to repair a roofing shingle.

tin snips Heavy-duty, short-bladed shears used primarily for cutting sheet metal.

tint To lighten a paint color by adding white.

TL Abbreviation for *transmission loss*. See *sound transmission loss*.

toe kick A recessed area between the bottom of a cabinet and the floor, usually about 4 inches high by 4 inches deep, permitting a person to stand close to the cabinet without contacting it with their toes; also called a toe space.

toenailing Nailing two boards together by driving a nail down at an angle through the edge or face of one board into the edge or face of the second board.

toe space See *toe kick*.

toggle bolt See *anchor, hollow wall*.

toilet auger A short, curved, flexible cable with a small blade at one end, specifically designed for hand operation to clear obstructions from the trap inside a toilet or from the closet bend.

toilet dam Pieces of plastic or rubber-coated metal that are inserted vertically into a toilet tank to close off part of the tank's area and reduce the amount of water required for flushing, thus saving from 1/2 to 2 gallons of water per flush.

tolerance The amount of deviation allowable from a given dimension or size.

ton 1. A measurement of weight equal to 2000 pounds. 2. A rating of the cooling capacity of an air conditioner equal to approximately 12,000 Btuh, derived from the amount of cooling furnished by the melting of a metric ton (2,204 pounds) of ice in 1 hour.

tone To alter a paint color by adding both black and white, or by adding gray.

T-1-11 plywood See *plywood siding*.

tongue 1. A projecting edge on a board; used in forming a tongue-and-groove joint. 2. The shorter of the two arms of a framing square, at right angles to the blade and usually 16 inches long and 1 1/2 inches wide.

tongue and groove Of or relating to lumber, plywood, or other panels that have been milled to have a tongue along the center of one edge and a groove along the center of the other edge. The tongue of one board fits into the groove of the next, providing greater strength over wide spans. Also called dressed and matched or side matched.

tool apron Any of a variety of multipocketed aprons, usually leather, used to hold tools, nails, and other supplies while working.

tool crib A box or other enclosure for holding tools; especially a large, truck-mounted box.

tooling Compressing and finishing off the mortar joints in a masonry wall with a tool other than a trowel.

tool steel See *steel*.

toother One of the projecting masonry units in a wall that have been temporarily ended by toothing.

toothing Constructing a masonry wall with a temporary end such that the last unit of every other course projects to allow an intersecting wall to be interlocked with the wall, or to allow the wall to be extended at a later date.

top coat The final coat of plaster or stucco applied to a wall; also called a finish coat.

top-flange hanger See *timber connector*.

topographical map A drawing that depicts the topography of a particular land area, showing elevations, slopes, locations of water, and often soil and foliage conditions; commonly called a topo map.

topography The physical features of an area or piece of land, including elevations, depressions, lakes, streams, etc.

topo map See *topographical map*.

topping mud A type of joint compound that is smooth and easily sanded; used for the final topping coat.

top plate See *plate*.

top view A drawing that depicts an object in two-dimensional form as though viewed from directly above; also called a plan view.

torpedo level A small spirit level, usually 1 foot long and tapered in at both ends.

torque The rotary or twisting force of an object, such as a shaft that is spinning or a wrench that is tightening a bolt.

torque wrench A special type of wrench having a dial or scale to indicate the amount of torque being applied to the object it is tightening.

total rise The distance from the floor to the top of the last rise in a set of stairs. See Fig. S-16.

total run The distance from the beginning of the first run, on the first stair, to the end of the last run, on the last stair in a set of stairs. See Fig. S-16.

to the weather See *exposure.*

towel warmer A bathroom accessory that circulates the home's hot water through a series of pipes which have been formed into a towel rack. Electrically operated, oil-filled models are also available.

toxic Caused by or pertaining to a poison.

tracery In architecture, any delicately formed design or pattern of fine lines.

tracing An original drawing done in pen or pencil on transparent paper or cloth from which copies can be made by various methods, such as blueprinting.

track lighting Ceiling- or wall-mounted bars into which adjustable light fixtures can be inserted to provide light as the need arises.

traffic paint High-wear, abrasion-resistant paint used for curb marking, road stripes, etc.

trammel points Adjustable metal points placed on a bar or tube, used to scribe circles and arcs having a large radius. One of the points may be replaced with a pencil holder for drawing instead of scribing.

transformer A device that allows the characteristics of incoming current to be changed into a different outgoing current.

STEP-DOWN TRANSFORMER A transformer used to reduce the incoming voltage to a lower outgoing voltage.

STEP-UP TRANSFORMER A transformer used to increase the incoming voltage to a higher outgoing voltage.

transit See *level transit.*

Transite 1. The trade name for a fireproofing material made from asbestos and portland cement, which is molded under pressure into sheets. 2. Of or relating to a type of soil pipe made from the same materials.

transition 1. Of or relating to a fitting that allows two different sizes, shapes, or types of pipe or duct to be connected; also called an adapter. 2. Of or relating to a fitting that allows two dissimilar materials to be joined.

translucent Allowing the passage of light, but obscuring a clear view of objects, as a translucent window.

transmission heat loss See *heat loss.*

transmission line One or more conductors used to carry electricity from its source, such as a generator, to the load it supplies, such as a house. See Fig. S-9.

transmittance 1. A ratio of the amount of solar radiation passing through a surface to the amount striking it. 2. A term sometimes used to describe U-value. See *heat resistance: U-value.*

transom A small window sash above a door, hinged at the bottom to allow it to open inward for ventilation.

transom bar A horizontal bar separating the top of a door from a transom, and to which the transom is hinged.

transparent Admitting the passage of light and permitting a clear view of objects beyond, as a pane of clear glass.

transverse Lying across or attached to two supports, as a beam suspended between two walls. See Fig. O-1.

transverse support A horizontal or inclined support lying across two or more points, which in turn supports other loads along its length; for example, a girder lying across two piers, which in turn supports the floor joists laid across it. See Fig. O-1.

trap A plumbing fitting connected to a fixture, such as a sink or bathtub, and consisting primarily of a curved section of pipe with one inlet and one outlet. Water trapped in the bottom of the curve forms a seal that prevents sewer gases from entering the building through the fixture's drainpipe. See Figs. B-5, D-14, and P-11.

trap arm That portion of a drainpipe or waste pipe between a fixture's trap and its vent. See Fig. B-5.

trap door A small, hinged or removable door that covers an access opening to an attic, crawl space, or other concealed area.

trap primer A plumbing device designed to deliver water into a trap at a slow, controlled rate, so as to maintain a trap seal with the trap. Primarily used with floor drains to prevent evaporation of the seal from infrequent use.

trap seal The amount or level of water that is held in the curve of a trap to form the seal. See Fig. H-6.

traverse rod A drapery rod that is operated by a cord at one end and is used to open the draperies from one side to the other, or from the middle to both sides.

traverse window See *window: sliding window.*

tray base A narrow base cabinet, usually with one or more vertical dividers, used to store trays, cookie sheets, etc.

tread The horizontal part of a step onto which the foot is placed. See Fig. S-18.

treenail *also* **trenail** A heavy wooden dowel used to fasten timbers; also called a trunnel.

trefoil An architectural decoration that resembles a three-lobed plant, such as clover.

trestle A heavy sawhorse or frame formed by braced, slanting legs attached to a horizontal crossbar; used to support planking to form a low scaffold.

triad color harmony Color compatibility obtained by using colors from three equidistant points on the color wheel, such as red, yellow, and blue. See Fig. C-7.

triangle A three-sided figure that is the most structurally solid of all shapes, in that it cannot shift without changing the length of one of its sides. A triangle is described by the length of its sides or by the relationship of its angles, which always total 180 degrees. See Fig. P-8. Common triangles follow.

EQUILATERAL TRIANGLE A triangle having three sides of equal length.

ISOSCELES TRIANGLE A triangle having two sides of equal length.

RIGHT TRIANGLE A triangle having two legs that form a 90-degree angle.

SCALENE TRIANGLE A triangle having no sides of equal length.

trickle-type collector A solar collector for liquid systems in which water is pumped into the top of the collector, where it then flows by gravity along a channeled, black metal plate to the bottom, absorbing heat from the plate as it goes.

trim The finished moldings and other materials applied around doors, windows, floors, ceilings, or other interior and exterior areas where two surfaces meet.

trimmer 1. A beam or doubled joist to which a header is nailed when framing for a chimney, stairway, or similar opening in a floor or ceiling. See Fig. J-2. 2. A cripple stud used to support the header when framing a wall opening; also called a trimmer stud or jack stud. See Figs. D-6, F-10, and W-3.

trimmer stud See *trimmer (2)*.

trimming Dressed and polished stone used for sills, moldings, and other ornamental purposes.

triple-pane window See *window: double-pane window*.

triple-wall Having three separate walls, layers, or sections, each separated by an airspace to prevent heat buildup and transfer; commonly found in the form of three concentric pipes used for fireplace flues.

tripped Of or relating to a circuit breaker that has automatically opened to shut off electricity to a circuit. The tripped position of the breaker's handle is different from the off position, to indicate that the breaker shut down automatically, as opposed to having been shut manually.

Trombe wall A thick masonry or concrete wall that is heated by exposure to banks of windows on one side, and in turn conducts its heat to the home's living area on the other side; named for its developer, Dr. Felix Trombe.

trowel Any of a variety of flat metal tools used for applying or finishing concrete, plaster, mortar, or stucco.

true Accurately formed, laid out, or constructed; straight, flat, or level.

true direction A direction according to the stars, as opposed to a magnetic compass. True north is indicated by the polestar, or North Star, also called Polaris. True direction is an important consideration in the placement of solar collectors.

trunnel See *treenail*.

truss A structural unit consisting of a variety of members made up into a series of triangles. The units can span long distances with a minimum of material, and are often used for constructing floors and roofs. See Fig. T-7.

truss plate See *gang-nail plate*.

try square A fixed, L-shaped tool used for laying out perpendicular lines and checking squareness. See Fig. T-8.

T-square A T-shaped device consisting of a short head attached at precise right angles to a long blade. The head is placed against the edge of a drawing board or sheet of material, allowing the blade to be used as a guide for marking or cutting.

tubeholder Any of a variety of plastic receptacles that hold and provide contacts for fluorescent tubes.

tubing cutter An adjustable, C-shaped tool with a hardened

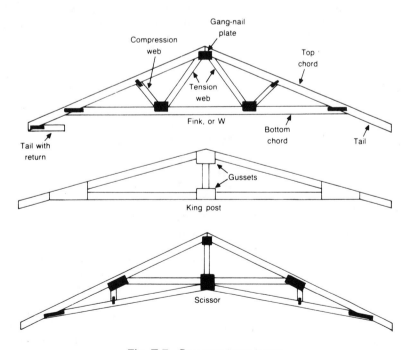

Fig. T-7. Common truss types

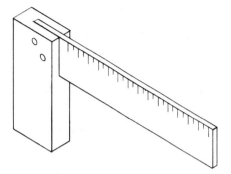

Fig. T-8. Try square

steel cutting wheel that is rotated around a piece of pipe or tubing to make a square cut.

tub spout diverter A type of bathtub spout containing a knob-activated diverter such that when the knob is down, water flows through the spout into the tub and when the knob is up, the water is diverted to a shower head.

tubular lock See *lock*.

tuck pointing Using fresh mortar to fill in defective or missing mortar joints in a masonry surface.

tung oil An oil obtained from the seeds and nuts of the tung tree, varieties of which are cultivated worldwide; used alone as a wood finish, or as an additive in the manufacture of other products, such as varnish and linoleum.

tungsten carbide See *carbide*.

turnbuckle 1. A hollow metal device threaded on each end to receive screw eyes, to which ropes or cables are attached. Rotating the turnbuckle draws the screw eyes together equally, tightening the cable. 2. A similar adjustable device used to tighten and tension a tie rod. See Fig. T-5.

turning One of various cylindrically shaped objects produced on a lathe, such as a spindle or baluster.

turning vane One of the curved sheet-metal pieces used inside a duct elbow to better direct the flow of air around a turn.

turnkey A construction method in which the architect and/or the contractor handles virtually all of the project's details from design to finishing, requiring the owner simply to "turn the key" in the front door lock.

turnout A curved section of handrail, attached to the top of a newel post and providing a starting point for the stair rail. See Fig. S-20.

turpentine A liquid distilled from the sap of pine trees; used as a thinner in paints, enamels, and varnishes, and also as a brush cleaner.

twist See *lumber defect: warp*.

twist bit A bit similar to a twist drill but having a sharper angle at the point; used for drilling in wood.

twist drill A bit having a tapered point and helical grooves; used for drilling in metal.

2 x Symbol for two by, referring to any lumber having a nominal thickness of 2 inches, regardless of width or length.

two-gang Of or relating to an electrical outlet box that holds two switches and/or receptacles. One "gang" will hold one device.

two-pipe system See *steam heating system*.

two-wire service See *service*.

type NM cable See *nonmetallic sheathed cable*.

type X plasterboard 5/8-inch thick plasterboard to which asbestos fiber and/or other materials have been added for fire resistance.

U

UBC Abbreviation for *Uniform Building Code*.

U-bolt A curved steel rod, shaped like the letter U, threaded at each end to receive nuts; primarily used with a plate behind the nuts to fasten a flat object to a round one.

UCC Abbreviation for *Uniform Commercial Code*.

ufer ground A 10-foot length of steel rebar that is embedded in a building's foundation and bent so as to leave about 12 inches above the concrete; used instead of a cold-water pipe for grounding the building's electrical system.

UL See *Underwriter's Laboratories*.

UL approved See *Underwriter's Laboratories*.

ultrasonic Of or relating to sound frequencies that are beyond the audible range.

ultraviolet Of or relating to light that has a wavelength less than that of the visible light at the violet end of the spectrum, and in the range between visible violet light and X-rays.

umbrella nail See *nails, head types*.

unconditioned space Any area, either within a building or outside of it, that is not served by a heating or cooling system. Examples might include attics, underfloor areas, basements, etc.

undercabinet light A small light fixture, usually fluorescent, mounted on the underside of a wall cabinet to light the surface of the countertop.

undercoat 1. In a three-coat paint job, the second coat, between the prime and finish coats. 2. In repainting, the first of two coats. Sometimes called a body coat.

undercut A member cut at less than 90 degrees in order to provide a tight fit against a beveled surface.

underfloor Referring to those spaces located beneath the floor of the building's first story, exclusive of the basement. See Fig. C-15.

underfloor insulation Thermal insulation placed underneath and in contact with the underside of a floor. See Fig. C-15.

underfloor wiring system A system of underfloor or in-slab conduits or other electrical chases, allowing simplified rewiring for future changes; used primarily in commercial applications.

underground house A type of passive solar home that is partially or completely covered with earth on the roof and three sides. The south side is usually left open for access to solar heat, and the north side is often partially uncovered for cross ventilation.

underground service Electrical service that is provided to a building from underground, carried from the power company to the service panel through a trench or buried conduit. See Figs. S-9 and S-10.

underlayment Building boards, usually 3/8 inch thick particle board, that are applied over the subfloor as a smooth surface for the application of finish flooring.

underpinning Posts or other shoring, usually temporary, that are used to support a building during repair or construction.

Underwriter's knot A type of square knot used to secure two small-gauge electrical wires into a plug or fixture base.

Underwriter's Laboratories An independent organization that conducts safety tests of electrical appliances, fixtures, and other devices, which, if meeting their standards, are labeled as *UL approved*, an important source of consumer protection.

unfibered gypsum plaster See *plaster.*

unfinished Any item, especially cabinets and other woodwork, sold in its natural state, without any prior application of stain, paint, or other finish.

unglazed collector A solar collector without a glass top or other covering.

unified threads A standardized series of screw threads for some types of machinery that are in use in the United States, Canada, and Great Britain.

Uniform Building Code A book of standardized requirements for the methods and materials used in construction, designed to provide greater safety and uniformity of building laws. First published by the International Conference of Building Officials (ICBO) in 1927 and revised and republished approximately every three years, it forms the basis for almost all building codes in use today. Abbreviated UBC.

Uniform Plumbing Code A book published by the International Association of Plumbing and Mechanical Officials (IAPMO) that establishes standard plumbing practices and materials and is the basis for most of today's plumbing codes. Abbreviated UPC.

union A fitting that allows two lengths of pipe to be assembled and disassembled without the need to take apart the entire run. See Fig. T-4.

UPC See Uniform Plumbing Code.

upeasing An upward-curving stair fitting, used at transition points between handrails where they pass over a landing or balcony.

upfeed system A water supply system in which water flows into the building under pressure at ground level and is forced upward to service fixtures on upper floors. Building height with an upfeed system is limited to approximately 60 feet.

up-flush toilet A special type of toilet for installation below the level of the building's sewer line, as in a basement, that when flushed uses a jet of water to break up solids, followed by a second jet to force the liquified waste up a pipe into an overhead sewer line. Maximum lift is about 10 feet at 40 psi of water pressure.

up-flush sink A sink below the level of a building's lowest sewer line that features a motor-powered centrifugal pump into which waste water from the sink drains, and which then discharges the liquid up to the sewer pipe.

uplight A light fixture so designed as to direct all of its light upward.

up-vent heater A type of gas-fired room heater designed to be recessed into the space between studs and vented up through the roof. It can be single-wall, facing into only one room; or double-wall, facing into two adjoining rooms.

urea-formaldehyde See *insulation, thermal.*

urethane resin See *paint resin.*

used brick See *brick.*

useful energy gain The net amount of solar energy absorbed by a collector for use in heating, equal to the gross amount collected minus that which is lost back to the surroundings.

U-shaped Of or relating to anything having three distinct arms, wings, sections, or other areas that are at right angles to each other in a shape resembling the letter U.

U-shaped kitchen A kitchen in which the appliances and cabinets are arranged on three adjacent walls in a U shape, generally considered to be the most efficient kitchen layout. See Fig. K-2.

utility cabinet A tall, relatively narrow cabinet used for general storage purposes or as a broom closet.

utility knife A general-purpose, multiuse knife consisting of a metal handle that holds a razorlike, replaceable blade. A variety of blade styles are available for different applications. Also called a razor knife.

U-type stairs See *stairs: platform stairs.*

UV Abbreviation for *ultraviolet.*

U value See *heat resistance.*

UV resistant Of or relating to materials, such as plastic sheeting, that have been treated to withstand the damaging effects of ultraviolet solar radiation.

V

vacuum breaker A device installed on the end of a hose bibb or other plumbing valve to prevent the backflow of contaminated water into the house water supply.

vacuum relief valve 1. A safety valve designed to prevent the buildup of excessive vacuum within a pressure tank. 2. A valve installed at the high point of a liquid solar system above the collectors in order to allow air to enter into the return piping to relieve vacuum pressure and allow the system to drain by gravity.

valance 1. A decorative, wall-mounted shield or covering; used to hide curtain rods, light fixtures, and other objects from view. 2. A decorative board used to connect two wall cabinets on either side of a window or obstruction. 3. A molding, board, or manufactured unit serving as a valance.

valley The internal angle formed by the intersection of two sloping roofs.

valley flashing A sheet-metal flashing, usually in 10-foot-long sections, used as weatherproofing under the shingles at a roof valley.

valley rafter See *rafter*.

valve A device that starts, stops, and regulates the flow of water, gas, steam, etc.

vanity A cabinet installed in a bathroom or dressing area, either with or without a sink; also called a pullman.

vapor barrier Any material used to prevent condensation and retard the movement of water vapor, usually considered as having a perm value of no more than 1.0. Vapor barriers may form the facing of batt or blanket insulation, or be applied separately to the heated side of floors, ceilings, and exterior walls.

vapor-barrier paint Special paint with a very low permeance that is used as a primer over drywall, plaster, or masonry to allow the coated wall to act as its own vapor barrier.

vapor permeable Of or relating to any material that will allow water vapor to pass through it.

variance 1. An exception to an existing zoning law, sometimes granted to a property owner under special circumstances. 2. Any deviation from true, or from a given dimension.

varnish A clear liquid finishing material that has a base of fossil gum or synthetic resin and that dries to a hard, high-gloss finish which is waterproof and heat resistant. Common varnish resins include acrylic, alkyd, epoxy, polyurethane, and urethane.

VA tile See *vinyl asbestos tile*.

vault An arched or curving structure that forms a roof or ceiling.

vaulted ceiling A single- or double-sloping ceiling that rises to a height greater than that of the sidewalls. See Fig. V-1.

V-block A block of wood or metal having a 90-degree, V-shaped slot running horizontally through its center; used to hold a cylindrical object such as a pipe while cutting or drilling. See Fig. V-2.

vegetable glue See *adhesive*.

vehicle See *paint*.

velocity The distance traveled by an object in a specified time; often used to describe the movement of air.

veneer 1. A thin strip or sheet of wood; used as a facing material or as one of the layers making up a sheet of plywood. It can be produced by peeling off long sheets from a log

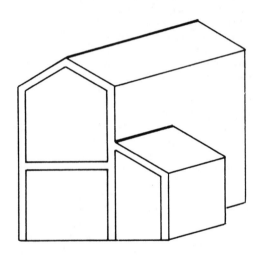

Fig. V-1. Vaulted ceilings

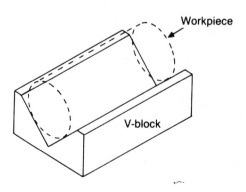

Fig. V-2. V-block for holding round stock

rotating in a lathe (rotary-cut veneer), or in strips cut lengthwise from the log with a saw (sawn veneer) or a knife (sliced veneer). See Fig. P-5. 2. Any thin material used to cover a core or base of another, less expensive material. See Fig. D-4.

veneer inlay A handmade or prefabricated design that is inlaid into a veneered surface.

veneer punch A punch for repairing veneered surfaces that removes an irregular shape from around a defect, then cuts an identical patch from scrap veneer to fill the hole.

veneered wall A framed wall that is covered with a layer of stone, nonstructural brick, or other decorative facing material, usually only one unit thick; also called a faced wall.

vent 1. Any opening that allows outside air to enter a building, or allows inside air to leave it. See Fig. I-1. 2. That portion of a window that opens, allowing for ventilation.

ventilation The circulation of air within an enclosed space, such as a room or building, to replace fouled or stale air with fresh air.

ventilation, mechanical Ventilation that is provided through the use of fans or other mechanical means.

ventilation, natural Ventilation that is not otherwise assisted by fans or other mechanical means; the natural flow of air through a building. See Fig. I-1.

vent pipe 1. In plumbing, a pipe that connects a plumbing fixture to a main or secondary vent stack. See Fig. D-14. 2. Any flue, duct, or pipe that connects an interior space with the outside air for ventilation purposes.

vent shaft An open-topped exterior shaft formed by the intersection of four or more walls of a building, such as a hotel or apartment, and serving to allow light and air into rooms that open onto it, typically bathrooms or service areas.

vent stack A vertical pipe that extends through the roof of a building, into which the vent pipes from various plumbing fixtures are connected to provide an outlet for sewer gases and fouled air, and to maintain the necessary atmospheric pressure to prevent siphoning of contaminated water through the fixture traps.

veranda A long, open-sided roofed porch, one or two stories high, on the front and/or sides of a building.

verge The edge of shingles or other roofing that overhangs the end rafter of a gable roof.

verge board See *barge board.*

vermiculite See *installation, thermal.*

vernal equinox The day, on or about March 21, on which the sun has traveled half the distance north on its yearly north-south path, reaching a point halfway between the northern and southern limits; traditionally, the first day of spring. See Fig. S-23.

vertical Perpendicular to the plane of the horizon; plumb; extending up and down. The opposite of horizontal. See Fig. P-8.

vertical section A drawing of a building or other object cut vertically top to bottom, showing interior details.

vestibule A small entrance hall or lobby outside a building, having a door at each end; used to minimize heat loss or gain

when people are entering and leaving the building. Also called an air lock.

V-groove Of or relating to tongue-and-groove or shiplap lumber having beveled edges on one face producing a V-shaped channel when joined together. Usually available in 3/4- or 1 1/2-inch thickness, and in 6- or 8-inch widths. Commonly used for siding or as roof decking with an open beam ceiling.

vibrator A machine placed on or in fresh concrete to agitate it. Vibrating concrete causes it to act as a liquid, flowing and leveling itself more readily, which helps eliminate honeycombs.

vice A bench-mounted clamping tool with movable jaws; used for holding an object while working on it.

Victorian Of or relating to architectural or furniture styles originating during the reign of Queen Victoria of Great Britain, from 1837 to 1901.

vinyl asbestos tile A type of floor tile that combines polyvinyl chloride with other chemicals. It can be used indoors below grade and is relatively inexpensive. Commonly referred to as VA tile.

vinyl-coated nail See *nails, finishes.*

vinyl siding Rigid, factory-colored polyvinyl chloride compounds produced in various shapes and sizes for use as a siding material. It is durable and suitable for new work, but is used primarily for re-siding.

vinyl tile Tile that is similar to vinyl asbestos tile, but lacking the asphalt or asphalt-substituting resins. It is somewhat more attractive than vinyl asbestos tile but is also more expensive, its cost being dependent on the amount of vinyl in its makeup.

viscosimeter A device used to determine the viscosity of a liquid by measuring its rate of flow.

viscosity The resistance of a liquid to flow; also the thickness, or body, of a liquid.

visible transmittance See *daylighting.*

Visqueen A commonly used trade name for plastic sheeting.

vitreous Material that has been vitrified.

vitrified Of or relating to ceramic material that has been made glassy and waterproof by exposure to heat. Vitrified material is used in the production of toilets and other waterproof ceramic objects.

void 1. An open or hollow space. 2. Having no legal effect; canceled or nullified.

volatile That which, at ordinary temperatures, evaporates rapidly upon exposure to air.

volatile thinner A rapidly evaporating liquid used to thin finishes without changing the ratio of pigment to nonvolatile vehicle.

volt A unit of measurement used to describe electrical potential, equal to the difference of potential produced by a current of one ampere across a resistance of one ohm.

voltage The difference of potential, equal to current times resistance. Symbolized by the letter *E, e,* or *V.*

voltage drop A decrease in electrical current caused by

wires that are undersized for the current they are carrying, especially over long distances. The noticeable slowing of an electric motor or loss of brightness from a light bulb are common indicators of voltage drop.

volute   A semicircular stair fitting, spiraling out from a center cap. Used on top of a newel post to provide a starting point for the handrail. See Fig. S-20.

V-rustic   Of or relating to a type of V-groove lumber, usually 4 to 8 inches wide and having a rough or resawn face; primarily used as an exterior siding material. See Fig. S-11.

W   In electrical calculations, the abbreviation for *watts*.

waferboard   See *building board*.

wainscot   1. A low interior wall covering of boards, plywood, or paneling extending from the floor to a height of 3 to 4 feet and usually contrasting with the wall surface above. 2. Any decorative surface that extends a quarter way to halfway up a wall, as with brick wainscoting. 3. To line with boards or paneling.

wainscot cap   A molding having a rabbet on the back of its bottom edge of a depth matching the thickness of the wainscot material, applied horizontally to the top of wainscotting as a finishing piece. See Fig. M-4.

wainscoting   The boards or paneling used to wainscot a surface.

wale   A horizontal member used to stiffen or support forms for concrete.

wallboard   See *drywall*.

wall box   See *electrical box*.

wall cabinet   Any cabinet in a kitchen or other area that is totally suspended from the wall.

wall cavity   The space within a wall between the interior wall covering and the wall's exterior siding or sheathing. See Fig. F-10.

wall insulation   See *insulation, thermal*.

wallpaper   Any of a variety of decorative paper, cloth, vinyl, or other materials, used as a wall or ceiling covering. American wallpaper is commonly measured in units called single rolls, which contain 36 square feet, but are usually sold in double or triple roll quantities, which contain 72 and 108 square feet respectively. Common wallpaper types include the following.

FLOCKED   Decorative patterns of fluffy fabric or other material applied over a cloth, paper, or vinyl backing.

FOIL   Any of a variety of metallic or simulated metallic wallpapers, usually of thin, shiny, brightly colored aluminum laminated to a paper or cloth backing.

GRASS CLOTH   A type of wallpaper made from natural plant fibers such as hemp or jute, laminated to a cloth or paper backing. May be dyed or left in its natural color.

PAPER   Relatively inexpensive treated or untreated paper, with or without a backing, intended for low to moderate wear.

PREPASTED   Wallpaper with a dry adhesive already applied to the back. Dipping or otherwise wetting the paper with water will activate the adhesive.

TEXTURED   A general term for wall coverings made of silk, grass cloth, hemp, burlap, or other relatively course and textured materials. Designed for low to medium wear areas, and generally among the most expensive types of wallpaper.

VINYL   Vinyl that is laminated to paper or cloth, creating a more washable and stain-and-moisture-resistant wall covering. Intended for heavy wear and high moisture areas.

wallpaper lining paper   A special, nonprinted paper used to cover walls and smooth out irregularities prior to applying wallpaper.

wallpaper steamer   A tool used to remove wallpaper from the wall. It consists of a small electric tank that heats water to create steam. The steam is passed through a hose to a flat, perforated metal plate that is pressed against the wall, allowing the steam to soften the wallpaper and adhesive.

wall rail   See *handrail*.

wall-rail bracket   A two-piece bracket designed to attach a wooden handrail to the wall. See Fig. S-18.

wall string   In stair construction, the inside trim attached to the wall along the steps. See Fig S-18.

wall tie   1. Galvanized metal straps or wires used between tiers of masonry veneer to bind them to a wood wall. See Fig. B-11. 2. Metal strips or wires used between the tiers of a masonry wall for reinforcement, or to tie together intersecting walls.

wall washer   A light fixture installed so as to provide an even amount of light over a partial or entire wall surface.

walnut   See *hardwood*.

wane   See *lumber defect*.

wardrobe door   See *door*.

warm colors Colors in the red, yellow, and orange ranges, so named for the impression of warmth they give to a room or building.

warm white See *fluorescent lamp.*

warp See *lumber defect.*

wash The top portion of a material or surface that has been given a slope to shed water.

wash coat A very thin coat of varnish, shellac, or other finishing material that is designed to seal the surface before the application of other finishes.

washerless faucet A single-handled faucet that does not contain the normal valve washer and seat found in a two-handled faucet, but instead water is mixed and controlled by a ball, disk, or cartridge, which operates against spring-loaded rubber seals or O-rings.

washer nail See *nails, common types.*

waste and overflow Of or relating to a T-shaped set of pipes and fittings used for a bathtub installation, in which one pipe fits the tub's overflow hole, another fits the tub's drain hole, and the two join a third pipe which connects to the trap. Waste and overflow kits also usually contain a pop-up stopper. See Fig. W-1.

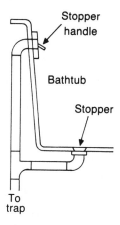

Fig. W-1. Waste and overflow setup for a bathtub

waste pipe Any pipe, such as that from a sink, that carries liquid waste only. See Fig. D-14.

waste tee A type of drainpipe that fits between a sink's tailpiece and the trap and has a smaller, side-outlet pipe to which the drain hose from a dishwasher is connected.

water-based paint See *paint resin: latex resin.*

water-based stain See *stain.*

water closet 1. A room or other enclosure that contains a toilet and often a sink. Abbreviated WC. 2. A toilet.

water hammer A harsh, banging noise occurring in plumbing pipes, caused by the abrupt moving and stopping of water in the pipes.

water heater Any of a variety of devices that use sunlight, electric elements or the combustion of fossil fuels in order to

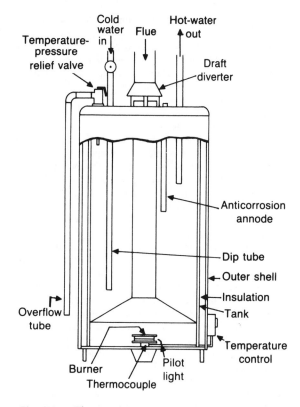

Fig. W-2. The components of a typical gas-fired water heater

heat water. See Fig. W-2.

water-heater damper A device used on a gas- or oil-fired water heater that automatically closes off the vent to prevent heat loss when the water heater's main burner is not on.

water-heater jacket See *insulation, thermal.*

water-heater timer A device that activates the burners or heating elements in a water heater at particular times of the day, rather than at particular temperatures.

waterproofing 1. A coating of tar, asphalt, or similar material applied to a foundation or other surface to render it impervious to water. 2. The act or process of making a surface waterproof.

waterproof plasterboard Plasterboard that has been chemically treated to resist moisture, for use in damp or high-humidity locations such as bathrooms. Usually covered on the face with green or blue paper for identification. Sometimes called greenboard or wetrock.

water repellent 1. Of or relating to a solution, usually paraffin wax and resin in mineral spirits, designed to penetrate wood and other porous materials to help repel water and provide a degree of preservative protection. 2. Of or relating to a material that will shed water for a reasonable period of time, but that is not totally waterproof.

water softener A device that substitutes sodium (salt) for the magnesium, iron, and other minerals found in water to help eliminate mineral deposits, bad taste and odor, and other potential health problems associated with hard water. However, the addition of sodium to the water supply creates another possible health problem for people on sodium-restricted diets.

water soluble Of or relating to a material that is capable of being dissolved in water.

water table 1. A redwood molding, basically square in section but with a slightly beveled top and often a drip groove underneath, commonly used over brick veneer wainscoting as a water-shedding cap. See Fig. M-4. 2. The upper level of the portion of the earth that is entirely saturated with water.

watertight Of or relating to an object or enclosure that is sealed to prevent the inflow or outflow of water or other liquids.

water wall A system used for passive solar heating in which rows of water-filled vertical tubes or barrels are stacked along a wall and exposed to the sun so that the water absorbs and retains the sun's heat during the day and radiates it into the house at night; sometimes called a drum wall.

watt The practical unit of electric and other power. One watt is dissipated by a resistance of 1 ohm through which a current of 1 ampere flows. Equal to the product of the total volts and amperes in a circuit (volts x amperes = watts). Abbreviated by the letter *W*.

watt hour See *kilowatt hour.*

wattle and daub 1. An ancient type of wall construction, in which thin sticks are woven or interlaced (wattle) and plastered over with mud, dung, or other materials (daub). 2. A relatively new type of construction using wooden lath strips covered with a plasterlike material in the same manner.

wax When used as a wood finish, most commonly a blend of beeswax, carnauba wax, paraffin, and turpentine; available as a liquid or paste.

wax ring A donut-shaped wax gasket used to form a seal between the base of the toilet and the closet flange.

WC See *water closet.*

weatherboard Exterior board siding.

weather head A curved cap that clamps to the top of the service entrance conduit; used to prevent rainwater from entering the conduit. See Fig. S-9.

weathering The mechanical or chemical disintegration or discoloring of the surface of a material, particularly wood, caused by exposure to sunlight, precipitation, dust, pollution, and temperature changes.

weatherproof box Any of a variety of one- or two-gang outlet boxes or round or octagonal fixture boxes made for exterior use; used in conjunction with various weatherproof covers that are sealed with rubber gaskets.

weatherstrip To apply weatherstripping materials.

weatherstripping Narrow strips of a variety of materials—among them metal, plastic, vinyl, felt, and foam rubber—used around doors, window sashes, and other areas to retard the passage of air, moisture, and dust.

web 1. The intermediate structural members in a truss that connect the top and bottom chords. Depending on the design of the truss, a web will act in compression or tension, and is named accordingly. See Fig. T-7. A flat canvas strap used in web-frame construction; also called webbing.

webbing See *web (2).*

web clamp A long, nylon or canvas belt with a racheting lock; used for clamping large or irregular surfaces.

web frame Of or relating to a type of furniture construction that uses flat canvas straps, called webs or webbing, stretched over a wood frame to support the padding and fabric.

wedge anchor See *anchor, masonry.*

wedged mortise-and-tenon Of or relating to a through mortise-and-tenon joint in which small wedges are driven into slots in the end of the tenon, expanding it for a tighter grip.

weep hole 1. One of the small openings cut or formed into the bottom track of a metal window frame to allow rainwater that collects in the tracks to drain off to the outside. 2. One of the small openings in a retaining wall that allows surface water to drain off. 3. One of the vertical joints in a masonry wall that are intentionally left unmortared to allow for the escape of moisture. See Fig. B-11.

weep screed A V-shaped sheet-metal strip with holes along one side; used at the bottom of a stuccoed wall to allow for the escape of accumulated moisture behind the stucco.

weight pocket A hollow space between the side jambs of a double-hung window and the adjacent wall framing in which the sash weight moves up and down.

weld 1. To join two pieces of metal using heat, often with the addition of a third metal in the form of a rod, causing the metal to fuse together into a solid mass. 2. To fuse or join two materials as a solid unit, as when gluing plastic pipe.

welded wire mesh A heavy wire mesh, having thick-gauge solid metal wires welded at right angles, usually in a grid 6 inches by 6 inches. Used primarily as a reinforcement in poured concrete to prevent cracking. Commonly abbreviated WW mesh. Also called highway mesh.

well A hole or shaft sunk into the earth in order to tap into underground water supplies.

DEEP WELL A shaft that is drilled to depths exceeding 25 feet.

SHALLOW WELL A well, often hand-dug, that does not exceed 25 feet in depth.

well hole The open shaft around which a circular staircase turns and ascends.

western framing See *platform framing.*

Western Wood Products Association A rules-writing and inspection agency, certified by the American Lumber Standards Committee, that oversees the production, inspection, and grading of about 40% of the lumber used in the United States, which is produced by mills in 13 western states and used throughout the country. Abbreviated WWPA.

wet-bulb temperature The temperature indicated on a thermometer with a wet cloth over the bulb and a measured amount of air flowing across it; used to measure the warmth of the moisture in the air.

wet lap   Of or relating to a method of applying paint, glue, or other liquid materials so that each new area of application overlaps and flows into the wet portion of the previously applied area.

wet location   See *location.*

wet or dry sandpaper   See *sandpaper.*

wetrock   See *waterproof plasterboard.*

wet vent   A plumbing vent arrangement in which a fixture's drainpipe is tied directly into a soil stack or branch drain, allowing the same pipe to act as both drain and vent. See Fig. D-14.

wet wall   Of or relating to interior wall finish that is applied wet, usually referring to plaster applied over gypsum lath.

wheel dresser   A hand tool containing a number of slotted, rotating metal disks attached to the end of a metal shaft and pressed against a rotating grinding wheel to clean and true the wheel's surface.

whet   To sharpen and finish the cutting edge of a tool or plane iron by rubbing it against an oilstone.

whetstone   See *oilstone.*

whip   A short, flexible piece of hose, normally used with paint sprayers and air tools. The whip is placed at the sprayer or compressor to help keep the hose from kinking at that point.

white glue   See *adhesive: polyvinyl resin adhesive.*

whiteprint   See *diazo print.*

whole-house fan   A large, switch-controlled electric fan covered by a spring-loaded, louvered faceplate and set into a ceiling in a central location. When on, suction from the fan pulls the louvers open and draws outside air from open windows through the house into the attic, where it exits through the attic vents, thus serving to cool and ventilate the house while forcing hot air out of the attic.

wicking   A capillary action that occurs between two dissimilar materials, causing one to pull heat and/or moisture out of the other one.

wiggle mold   A strip of wood having a regular series of convex and concave curves on its face; used under corrugated sheet materials such as plastic and fiberglass roof panels for equal support of the corrugations.

wiggle nail   See *corrugated fastener.*

wind   See *lumber defect, warp.*

wind break   A fence, wall, landscaping, or other object used to block the wind.

winder   One of the wedge-shaped treads used in making up a winding or spiral staircase.

winding stairs   See *stairs.*

window   An opening consisting of a window frame, sash, translucent material, weather stripping, and any mechanisms for moving and locking the sash. Windows differ primarily in the way the sash operates and in frame materials, which can be wood, steel, or aluminum. Some of the more common window types follow. See Fig. W-3.

AWNING WINDOW   A window having a sash hinged on pins at the two top or bottom corners of the window frame, and that opens outward.

BAY WINDOW   A window made up of three sash units that project out from the wall—a back sash parallel to the wall, and two side sashes set at 30-degree or 45-degree angles to the back sash. Usually includes the roof structure, which is commonly either copper or shingled to match the house roof; the head board, which is used like a ceiling within the bay to conceal the roof framing; and the seat board, used if the bay does not go all the way down to floor level.

BOW WINDOW   A type of bay window consisting of several window units set at slight angles to each other, so as to form a curved window.

BOX BAY WINDOW   A type of bay window in which the two side sashes are set at right angles to the back sash.

CASEMENT WINDOW   1. A window having a sash hinged on pins at the top and bottom corners of one side, and that opens outward by means of a crank. 2. Less commonly, an awning or hopper window.

CLAD WINDOW   For durability and ease of maintenance, a wood window frame covered on the outside with aluminum or vinyl in baked-on colors, while left in natural wood on the inside.

COMBINATION WINDOW   1. A unit combining one or more fixed sashes with one or more that open. Almost all types of sash operations are available in combination units. 2. A storm window and screen combined into one unit. The storm window can be slid down on a track in winter, while the screen is slid down in summer.

DOUBLE-HUNG WINDOW   A window having two movable sashes mounted in a channeled frame. The lower sash slides vertically, passing over and to the inside of the upper sash; the upper sash slides down on the outside. See Fig. W-3.

DOUBLE-PANED WINDOW   A window using two sheets of glass set in the frame with a sealed space between them to provide an area of dead air for thermal insulation. *Triple-paned windows* have three spaced panes. Also called a thermal or insulated window.

DROP WINDOW   A window that is opened by sliding the entire sash into a pocket located in the wall beneath it.

FIXED WINDOW   A sash that is permanently fixed in the window frame so as to be immovable. Available in a variety of shapes, including round, octagonal, triangular, and half-round.

FRENCH WINDOW   See *door.*

GLIDING WINDOW   See *window: sliding window.*

HOPPER WINDOW   A window in which the sash is hinged on pins at the two bottom corners, and which opens inward.

INSULATED WINDOW   See *window: double-paned window.*

JALOUSIE WINDOW   A window having a series of small, horizontal glass slats held at each end in a movable metal frame so that the slats overlap. A crank rotates all the slats simultaneously, like a Venetian blind.

MULTIPLE-USE WINDOW   A frame that contains a sash hinged by pins at two corners of one side, and that opens out. It can be

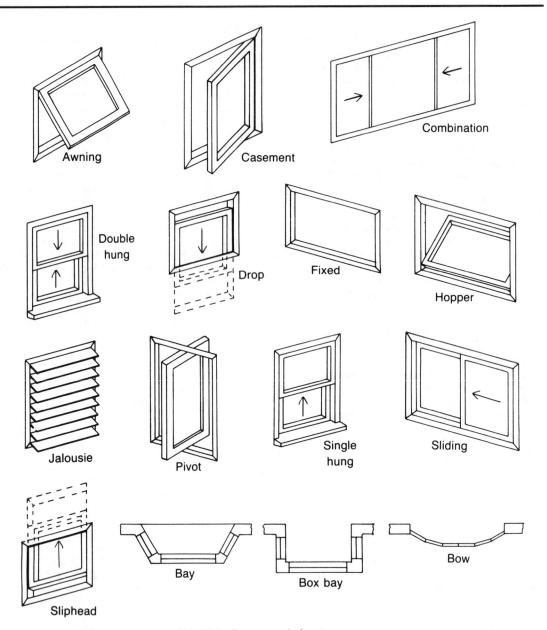

Fig. W-3. Common window types

installed vertically as a casement window or horizontally as an awning.

PICTURE WINDOW A relatively large, fixed window, sometimes divided into smaller lights.

PIVOT WINDOW A window in which the sash rotates from two center pivot points, located either on the top and bottom or on the sides, causing one-half of the window to open in, while the other half opens out.

QUAD-PANED WINDOW A newly developed type of insulated window, having two outside panes of glass and two inside panes of transparent plastic in a common sash.

SINGLE-HUNG WINDOW A window similar to a double-hung window, except that the upper sash is fixed.

SLIDING WINDOW A window having one fixed sash and one or more movable sashes that slide horizontally along a track across and to the inside of the fixed sash. A slider is specified as X,

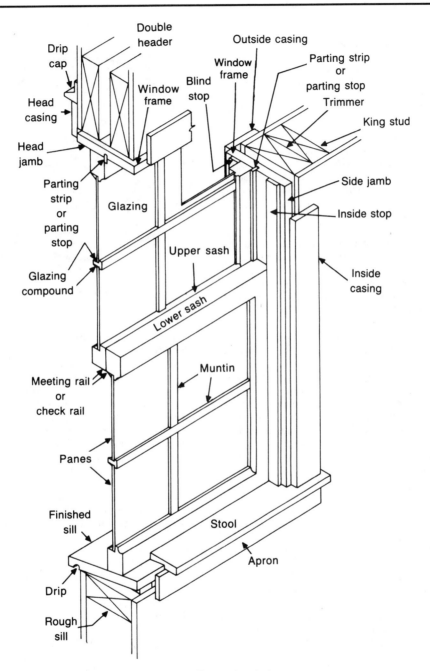

Fig. W-4. Parts of a window

and the fixed sash as *O*, listed left to right as viewed from outside. Also called a slider, a gliding window, or a traverse window.

SLIPHEAD WINDOW A type of window used primarily in outbuildings, in which the entire sash slides vertically into a recess in the wall above it.

STUDIO WINDOW A type of picture window, generally smaller and containing only one light.

TRAVERSE WINDOW See *window: sliding window.*

TRIPLE-PANED WINDOW See w*indow: double-paned window.*

window and door sealant See *sealant: silicone sealant.*

window frame The framework in which the sash is housed, and to which it is fixed or hinged. Window frames vary widely in makeup, depending on how the sash is to operate and the type of wall into which they are to be installed. See Fig. W-4.

window insulation Any of a variety of coverings having a relatively high R-value; used over a window to prevent heat loss or gain.

window seat A seating area built into a bay window recess.

window wall An exterior wall containing a large percentage of glass, often used in solar design.

wind shake See *lumber defect.*

windward On the side where the wind is blowing; exposed to the wind; the direction from which the wind is blowing. Opposite of leeward.

wing A secondary section of a building extending from a junction with the main portion of the building.

winter solstice The day, on or about December 21, on which the sun rises and sets farthest to the south and is at its lowest elevation of the year, appearing far down on the horizon; traditionally, the start of winter. See Fig. S-23.

wire glass Glass in which a layer of wire mesh has been embedded for security and protection against shattering.

wire hanger A device, used to support horizontal runs of pipe, consisting of a U-shaped wire in various sizes, the ends of which are bent over 90 degrees and pointed for nailing into wood. Also available with a padded coating to cushion soft pipes such as copper.

wire lath See *metal lath.*

wire nut A type of solderless connector, consisting of a color-coded plastic cap with a threaded insert, that is twisted onto the ends of two or more conductors to join and insulate them.

wire ties Steel wire that is formed in various shapes, used between or around two members for reinforcement.

wireway A metal channel, similar to a raceway, having an exposed cover that is hinged or screwed on; often used for exposed interior applications, such as utility or machinery rooms, where wiring changes are common.

wiring diagram A schematic drawing that uses standard symbols to show an electrical system, particularly that of an appliance or piece of machinery. Also called a schematic diagram.

withe See *tier.*

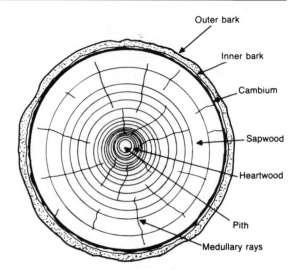

Fig. W-5. Wood terms

wood A strong, cellular material widely used in construction. The basic structure of wood includes the following. See Fig. W-5.

BARK Cells that grow into a tough outer layer to protect the new wood growth underneath.

CAMBIUM LAYER A layer just below the bark where new cells form.

CELLULOSE One of the two compounds that make up wood fibers.

EARLYWOOD See *springwood.*

FIBERS One of the long, narrow cells or tubes, about the size of a human hair and 1/25 to 1/8 inch in length, that comprise the tree's basic structure.

GROWTH RING A ring marking the yearly growth of the cambium layer.

HEARTWOOD The area at the center of the tree, formed as the sapwood nearest the center becomes inactive.

HEMICELLULOSE One of the two compounds that make up wood fibers.

LATEWOOD See *summerwood.*

LIGNIN A natural cement that binds wood fibers together.

MEDULLARY RAY One of the tiny cracks across the rings, radiating out from the center of the tree.

PITH The original, central core of the tree, around which wood formation takes place.

SAPWOOD The outer part of the tree, between the heartwood and the bark, containing the living cells.

SPRINGWOOD Large, thin-walled cells produced by rapid growth in the spring; also called earlywood.

SUMMERWOOD Small, thick-walled cells produced by slow growth during the summer; also called *latewood.*

wood-appearance shingle A composition strip shingle having an irregular placement of cutouts and an irregular bottom edge, designed to give the appearance of random wood shakes.

wood edging Strips of wood veneer used to cover the exposed edges of plywood or other building boards.

wooden brick A piece of dry, solid lumber cut to the size of a brick, used in place of a brick to provide a nailing surface in a masonry wall.

wood filler 1. A clear or pigmented liquid used for filling and leveling pores in open-grained woods such as oak. 2. A thick, heavily pigmented preparation used to fill nail holes and defects in finished wood.

wood fungus See *fungus, wood.*

wood hood A wall- or ceiling-mounted kitchen cabinet designed to receive and conceal a range hood.

wood joiner A small metal plate with eight spiked legs, driven into the faces of two adjacent pieces of wood that meet in a butt or miter joint. See Fig. W-6.

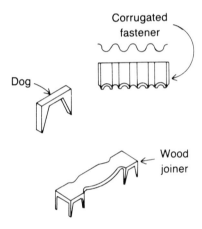

Fig. W-6. Common wood joiners

wood preservative A chemical used to preserve wood that will be buried or otherwise exposed to the elements, helping to prevent insect and dry-rot damage. Some of the more common wood preservatives follow.

CHROMATED COPPER ARSENATE A waterborne solution combining copper and arsenic to prevent fungus and insect decay, and chrome to stabilize the chemicals in the wood; commonly used in the making of pressure-treated wood.

COPPER NAPHTHENATE A greenish tinged liquid, applied by soaking. It is safe for plants and can be painted over.

CREOSOTE An oily liquid distilled from wood or coal tar, applied by soaking. In residential building, generally only used for below-ground applications, due to its heavy odor and the difficulty of painting over it.

PENTACHLOROPHENOL A solution containing an active chemical insecticide and fungicide dissolved in an organic solvent. Other ingredients, such as wax, can be added to help make the treated wood cleaner to handle. Abbreviated *penta* or *PCP.*

wood stove A generic term applied to a variety of airtight and non-airtight appliances that burn wood to produce heat.

wood thread A coarse, external screw thread that spirals to a pointed end, designed to be screwed into wood.

wood turning The process of rotating a piece of wood in a lathe and cutting or otherwise shaping it.

working drawings A set of scale drawings from which various tradesmen can take measurements during construction.

work triangle The total distance between the sink, refrigerator, and range in a kitchen, used as an indicator of the efficiency of a kitchen layout. The total distance between the three should be no less than 12 feet and no more than 22 feet. See Fig. K-2.

worm clamp See *hose clamp.*

woven valley Alternately overlapping rows of composition shingles where they meet at a valley, commonly used in reroofing applications.

wrecking bar Any of various types of tools used for demolition; typically, a steel bar from 18 to 30 inches long flattened on one end for prying, and curved and notched on the other end for pulling nails. Sizes and configurations vary widely. Also called a pry bar, ripping bar, or crow bar.

wrinkling A defect in a coat of paint, caused by applying too thick of a layer, in which the top of the paint film dries before the bottom does, causing the surface to wrinkle.

W-truss See *fink truss.*

WW mesh See *welded wire mesh.*

WWPA See *Western Wood Products Association.*

wye branch See *Y-branch.*

wythe See *tier.*

X 1. Symbol (ital) for reactance; used in electronics equations. 2. The letter designation of the operable sash in a sliding window.

x A symbol used in construction to indicate where to place a stud, joist, or rafter.

X-bracing Lumber that is placed diagonally between two posts or columns in the shape of an X, providing bracing against lateral stress.

X-bridge An ac bridge for measuring reactance.

Xc Symbol for capacitive reactance.

XL Symbol for inductive reactance.

X-O Standard designation for a sliding window. *X* is the designation for the operable sash, *O* is the designation for the fixed sash, listed from left to right as viewed from the outside. See *window: sliding window.*

X-O-X Standard designation for a combination sliding window, in which the two end sashes slide toward and over the middle sash, which is fixed. See *window: combination window.*

X-section Abbreviation for *cross-section.*

XT Symbol for total reactance.

yard lumber See *lumber.*

Y-branch A plumbing fitting having three outlets in a Y shape, with one outlet set at a 45-degree angle to the other two; sometimes spelled wye branch. See Fig. Y-1.

yard 1. A linear measurement equal to 3 feet or 36 inches, regardless of width or depth. 2. For bulk goods such as concrete or dirt, a cubic yard. See *cubic yard.*

yellowhead A nickname for a wire nut having a yellow outside jacket.

yoke vent A vent that connects a soil stack with a vent stack, designed to prevent pressure changes within the stacks.

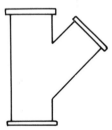

Fig. Y-1. Y-branch

Z

Z-bar   A narrow sheet-metal flashing having a cross-section similar to a letter Z; used where two boards or siding panels meet in a horizontal butt joint. The upper leg of the flashing slides behind the upper board, while the lower part of the flashing extends over and down the top edge of the lower board. See Fig. Z-1.

zero-clearance   Any combustion appliance so designed that it can be placed in direct contact with combustible materials.

zero-clearance fireplace   A prefabricated, insulated fireplace unit designed to be placed directly on or against combustible materials with little or no clearance. It is vented through the roof using triple-wall, insulated pipe.

zigzag rule   A type of wooden ruler, usually 6 feet long, that is constructed so as to fold up in a series of overlapping sections.

zone heating   Using individual heating units and controls to heat some portions of a building independently of others.

zoning   Of or relating to municipal regulations that govern the placement, style, type, and use of buildings constructed in various locations.

Zonolite concrete   The brand name for a lightweight, low-strength concrete combining portland cement and

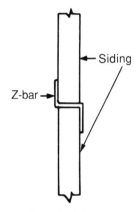

Fig. Z-1. Z-bar used as a flashing between two panels of siding

vermiculite, with or without sand; commonly used for floor slabs in multistory buildings and for various pre-cast shapes.

APPENDIXES

APPENDIXES

Appendix A
Common Abbreviations

A

a.	area
AAA	American Arbitration Association
AAMA	Architectural Aluminum Manufacturer's Association
abbr.	abbreviated
ABS	acrylonitrile-butadiene-styrene
abs.	absolute, absorption
A/C	air-conditioning
Ac	Actinium
ac	alternating current, asbestos cement, asphaltic concrete
acc.	accordian
acces.	accessory
ac/h	air changes per hour
act.	actual
AD	air-dried
ADA	Airtight Drywall Approach
adj.	adjusted
adpt.	adapt, adapter
aft.	after
Ag	silver
aggr.	aggregate
AHAM	Association of Home Appliance Manufacturers
AIA	American Institute of Architects
AIEE	American Institute of Electrical Engineers
air cond.	air-conditioning
Al	aluminum
allow.	allowance
alm.	alarm
ALS	American Lumber Standards
alt.	alter
altn.	alteration, alternate
alum.	aluminum
aly.	alloy
AM	Americium
amb.	amber
Amer. Std.	American Standard
amp	ampere
amp hr	ampere-hour
amt.	amount
anod.	anodized
ANSI	American National Standards Institute

ant.	antenna
AP	access panel
appd.	approved
approx.	approximate
apt	apartment
Ar	argon
arch.	architectural
arc/w	arc weld
As	arsenic
ASA	American Standards Association
asb.	asbestos
ASHRAE	American Society of Heating, Refrigerating and Air Conditioning Engineers
asph.	asphalt
assem.	assemble
assoc.	association
assy.	assembly
ASTM	American Society for Testing Materials
AT	ampere-turn
A.T.	air temperature, airtight
At	astatine
att.	attach
Au	gold
auto.	automatic
aux.	auxiliary
avg.	average
AW&L	all widths and lengths
AWG	American wire gauge

B

B	bathroom, baron
Ba	barium, bathroom
baf.	baffle
bal.	balance
batt.	battery
b-b	back to back
bbl.	barrel, barrels
BC	between centers
bd.	board
bd. ft.	board foot
bdl.	bundle
bdrm.	bedroom
Be	berylllium

bet.	between	cc	cement coated, cubic centimeter
bev.	beveled	c-c	center to center
b.f.	board foot	CCW	counterclockwise
B. H.	boxed heart	Cd	cadmium
Bi	bismuth	cdr.	cedar
Bk	berkelium	Ce	cerium
bk.	back	cem.	cement
bkd.	baked	cem. fl.	cement floor
bkr.	breaker	cem. mort.	cement mortar
BL	baseline, building line	cent.	centigrade, centrifugal
B/L	bill of lading	cer.	ceramic
bldg.	building	CF	cubic feet
blk.	black, blank, block, bulk	Cf	californium
blkg.	blocking	cfm	cubic feet per minute
blr.	boiler	cg	centigram
blu.	blue	cham.	chamfer
BM, B/M	bill of material	chem.	chemical
bm.	beam	chg.	change
b.m.	board measure	chk.	check
BOCA	Building Officials and Code Administrators International, Inc.	CI	cast iron
		CIP	cast-iron pipe
bot.	bottom	cir.	circle
BP	blueprint	circ.	circulate, circular, circumference
bp	base plate	CL	carload, centerline, critical level
Br	bromine	C/L	carload
brd.	board	Cl	chlorine, closet
brk.	brick	cl	centiliter
brn.	brown	clg.	ceiling
brs.	brass	clkg.	caulking
brz.	braze, bronze	clr.	clear, clearance
B&S	beams and stringers	C.M.	center matched
bsbd.	baseboard	Cm	curium
BSND	Bright sapwood, no defects	cm	centimeter
b to b	back to back	CMU	concrete, masonry unit
btr.	better	cnd.	conduit
Btu	British thermal unit	CO	change order
b/u	built-up	Co	cobalt
bvl.	bevel	col.	column
byp.	bypass	com.	common
		comb.	combination, combustion
		comm.	commercial

C

		comp.	component, composition
C	carbon, Celsius, centigrade, hundred	compd.	compound
c	coulomb	compl.	complete
Ca	calcium	conc.	concrete
CAB	concrete asbestos board	conc. b.	concrete block
cab.	cabinet	cond.	condition, conductor
CABO	Council of American Building Officials	conn.	connect, connection
CAD	computer-assisted drafting, computer-assisted design	const.	construction
		cont.	continue, continuous, control
CADD	computer-assisted drafting and design	contr.	contractor
cap.	capacity	cor.	correct, corrugated
CB	catch basin, center beaded	COS	color one side
CBS	color both sides	c.p.	candlepower
cbt.	cabinet	cplg.	coupling

cpm	cycles per minute
cps	cycles per second
CPSC	Consumer Product Safety Commission
CPVC	chlorinated polyvinyl chloride
Cr	chromium
cr.	center, circular
CS	cast steel, caulking seam
Cs	cesium
c/s	cycles per second
csg.	casing
csk.	countersink
ct.	coat
c to c	center to center
ctr.	center, counter
Cu	copper
cu.	cubic
cub.	cubicle
cu. ft.	cubic foot
cu. in.	cubic inch
cust.	custom
cu. yd.	cubic yard
cvr.	cover
CW	clockwise
cwt.	hundred weight
cx.	composite
CY	cubic yard
cy.	capacity, cycle
cyl.	cylinder

D

D	five hundred
d	penny
dB, db	decibel
dbl.	double
dc	direct current
dec.	decimal
deg.	degree
DET	double end trimmed
dev.	develop
DF	Douglas fir
DF-L	Douglas fir-larch
dia.	diameter
diag.	diagonal, diagram
diam.	diameter
dim.	dimension
disc.	discharged, disconnect
disch.	discharge, discharged
distr.	distribute
dk.	deck
dkg.	decking
DL	dead load, decking laminated
dlx.	deluxe
dn.	down
DOE	Department of Energy

D&M	dressed and matched
DR	dining room
dr.	drain, drill, drive
D/S, DS	double strength, drop siding
DSG	double-strength glass
dsp.	dispose
dup.	duplicate
dvtl.	dovetail
dwg.	drawing
dwl.	dowel
DWV	drain waste and vent
Dy	dysprosium

E

E	east, edge, modulus of elasticity
ea.	each
EB1S	edge bead one side
EB2S	edge bead two sides
e-c	edge to center, end to center
E&CB2S	edge and center bead two sides
econ.	economy
E&CV1S	edge and center vee one side
E&CV2S	edge and center vee two sides
edg.	edge
ee	eased edge
e-e	edge to edge, end to end
E.G.	edge glued, edge grain, end glued
E.J.	expansion joint
el.	elevation
elec.	electric
elev.	elevation, elevator
EM	end matched
EMC	equilibrium moisture content
en.	enamel
encl.	enclosed
ent.	entrance
EPA	Environmental Protection Agency
eq.	equal, equivalent
equip.	equipment
equiv.	equivalent
Er	erbium
ES	Engelmann spruce
Es	einsteinium
est.	estimate, estimated
e to c	edge to center, end to center
e to e	edge to edge, end to end
Eu	europium
EV1S	edge vee one side
EV2S	edge vee two sides
ew	each way
exc.	excavate
exh.	exhaust
exp.	expansion, exposed
ext.	exterior

F

F	Fahrenheit, fluorine
f, Fb	allowable fiber stress in bending
fab.	fabricate
FAO	finish all over
FAS	firsts and seconds
f.a.s.	free alongside
fbm	foot board measure
fdn.	foundation
Fe	iron
FG	field grade, fine grain
F.G.	flat grain
FHA	Federal Housing Administration
FHDA	Fir and Hemlock Door Association
fin.	finish
fix.	fixture
fj	finger jointed
fl.	flashing, fluid, flush
fL	foot-lambert
fld.	field
fldg.	folding
flex.	flexible
flg.	flange, flooring
flr.	floor
fluor.	fluorescent
Fm	fermium
fndn.	foundation
f.o.b.	free on board
FOHC	free of heart center
fp	freezing point
fpm	feet per minute
fprf.	fireproof
fps	feet per second
Fr	francium
fr.	front
freq.	frequency
frt.	freight
FSM	foot surface measure
ft.	foot, feet
FTC	Federal Trade Commission
ftg.	footing
f to f	face to face
fur.	furring
fwd.	forward
FX	fire blocking

G

g.	gram
Ga	gallium
ga.	gauge
gal.	gallon
galv.	galvanized
Gd	gadolinium
gd.	guard
Ge	germanium
gen.	general
GFCI, GFI	ground fault circuit interrupter
GI	galvanized iron
gl.	glass
govt.	government
gph	gallons per hour
gpm	gallons per minute
gps	gallons per second
gr.	grade
gran.	granite
grd.	ground
grn.	green
GS	galvanized steel
GSM	galvanized sheet metal
GWB	gypsum wallboard

H

H	hydrogen
HB	hardboard, hollow back, hose bibb
HC	hollow core
HD	heavy-duty
hdbd.	hardboard
hdwd.	hardwood
hdwe.,hdwre	hardware
He	helium
hem.	hemlock
hex.	hexagonal
Hf	hafnium
Hg	mercury
hgt.	height
HM	hollow metal
H&M	hit and miss
Ho	holmium
hol.	hollow
hor.	horizontal
hosp.	hospital
H.P.	high pressure
hp	horsepower
hr.	hour
HS	high speed
h.s.	high strength
h.t.	high tension
ht.	heat, height
htg.	heating
HUD	Department of Housing and Urban Development
HV	high-voltage
HVAC	heating, ventilation, and air conditioning
HVI	Home Ventilating Institute
hvy.	heavy
H.W.	hot water
hwy.	highway

I

I	Iodine
IAQ	indoor air quality
IC	incense cedar, integrated circuit, internal cooling
i.d.	inner diameter
i.e.	that is
IGCC	Insulated Glass Certification Council
In	Indium
in.	inch, inches
ind.	industrial
incr.	increase
in. lb.	inch-pound
ins.	inside, insulation, insurance
insp.	inspect, inspection
inst.	instant, instantaneous. institutional
instl.	installation
int.	interior, internal, intersect, intersection
intchg.	interchangeable
inter.	intermediate
ipm	inches per minute
ips	inches per second
Ir	iridium
ISDSI	Insulated Steel Door Systems Institute
IWP	Idaho white pine

J

j	jambs
jct., jctn.	junction
J&P	joists and planks
jt.	joint
jtd.	jointed

K

K	kelvin, kilogram, kitchen, potassium
kc	kilocycle
kc/s	kilocycles per second
KD	kiln-dried
K.D.	knocked down
kg	kilogram
kl	kiloliter
km	kilometer
Kr	krypton
kv	kilovolt
kw	kilowatt
kwh, kw:hr	kilowatt-hour

L

L	fifty, ladder stock, lambert, larch
l	liter, left, lumen
La	lanthanum

lam.	laminate, laminated
Lav.	lavatory
lb.	pound
lbr.	lumber
lbs.	pounds
LC	log cabin siding
LCL	less than carload
L Cl	linen closet
ldgr.	ledger
LF	light framing, linear foot
lg.	large
lgr.	longer
lgth.	length
LH	left hand
Li	lithium
lin.	linear
lino.	linoleum
liq.	liquid
LL	live load
LLC	log cabin logs
lng.	lining
LOA	length over all
long.	longitude
LP	lodgepole pine
l.p.	low pressure
L.P.G.	liquefied petroleum gas
lpw	lumens per watt
LR	living room, lower right
LS	low speed, lump sum
l.t.	low tension
lt.	light
ltbr.	lndscape timbers
ltwt.	lightweight
Lu	lutetium
LV	low voltage
Lw	lawrencium

M

M	thousand
m	milli-
ma	milliampere
mach.	machine
mat.	material
max.	maximum
MB	moisture barrier
MBF	thousand board feet
MBH	thousand Btus per hour
MBM	thousand-feet board measure
MBtu	thousand Btus
MC	medicine cabinet
mc	megacycle, moisture content
Md	mendelevium
mdo	medium-density overlay
med.	medium

met.	metal
M & F	male and female
mfd.	manufactured
mfr.	manufacturer
MG	mixed grain
Mg	magnesium
mg	milligram
mh	manhole
mi.	mile
mill.	million
min.	mineral, minimum, minute
misc.	miscellaneous
mixt.	mixture
ML	Material List
mL	millilambert
ml	milliliter
mld., mldg.	molding
MLF	thousand linear feet
mm	millimeter
Mn	manganese
mn	main
MO	masonry opening
Mo	molybdenum
mod.	modular
MOE	modulus of elasticity
mon.	month
mono.	monolithic
MOR	modulus of rupture
mos.	months
mp	melting point
mpg	miles per gallon
mph	miles per hour
ms	millisecond
MSF	thousand square feet
MSR	machine stress-rated
mtd.	mounted
mtg.	mortgage, mounting
mtl.	material, metal
MU	masonry unit
mull.	mullion
MW	megawatt, mixed widths
mw	milliwatt

N

N	nitrogen, north
Na	sodium
N1E	nose one edge
nat.	natural
natl.	national
Nb	niobium
NBM	net board measure
NBS	National Bureau of Standards
NC	national coarse, normally closed

NEC	National Electrical Code
Nd	neodymium
Ne	neon
neg.	negative
NEMA	National Electrical Manufacturers Association
neut.	neutral
NF	national fine
NFPA	National Fire Protection Association
Ni	nickel
nip.	nipple
NO	normally open
No	nobelium
no.	number
nom.	nominal
nor.	normal
nos.	nosing, numbers
Np	neptunium
NRC	noise reduction coefficient
NTS	not to scale
NWMA	National Woodwork Manufacturers Association

O

O	oxygen
oa.	overall
obs.	obscure
OC	on center
oc	off center
OCBW	on center both ways
oct.	octagonal
O.D.	outside diameter, outside dimension
OH&P	overhead and profit
o-o	outside to outside
opg., opng.	opening
opp.	opposite
opr.	operate
orig.	original
orn.	orange
Os	osmium
o to o	outside to outside
out.	outlet, outside
ovhd.	overhead
oz.	ounce

P

P	phosphorus
Pa	protactinium
PAD	partly air dried
pal.	pallet stock
par.	parallel
para.	paragraph
partn.	partition

patt.	pattern, special patterns
PB	particle board, polybutylene
Pb	lead
pbd.	particle board
pbo	prebored
pboard.	particle board
pc.	piece
pcf	pounds per cubic foot
Pd	palladium
PE	polyethylene
perf.	perforate, perforated
perp.	perpendicular
PET	precision end trimmed
pgbd.	pegboard
ph.	phase
PIP	poured in place
pl.	plate
plas.	plaster
plmg.	plumbing
pl. gl.	plate glass
plf	pounds per linear foot
pli	pounds per linear inch
plstc.	plastic
ply.	plywood
Pm	promethium
pneu.	pneumatic
pnl.	panel
Po	polonium
POC	point of connection
porc.	porcelain
pos.	position, positive
PP	preprimed, polypropylene, pondersoa pine
p-p	push-pull
Pr	praseodymium
pr.	pair
prcst, prec.	pre-cast
prefab.	prefabricated
prem.	premium
prep.	prepare
press.	pressure
pri.	primary
psf	pounds per square foot
psi	pounds per square inch
PSTC	Pressure-Sensitive Tape Council
PT	pipe tap, pipe thread
Pt	platinum
P&T	posts and timbers
ptd.	painted
PTL	pressure-treated lumber
P&TS	plugged and touch-sanded
Pu	plutonium
PVA	polyvinyl resin adhesive
PVC	polyvinyl chloride
pwr.	power

Q

qt.	quantity, quart
qtr.	quarter
qty.	quantity
qual.	quality

R

R	radius
r	right
Ra	radium
Rb	rubidium
RC	red cedar
RCS	Residential Conservation Service
rdm.	random
Re	rhenium
rec.	receptacle
recip.	reciprocating
recirc.	recirculate
recd.	received
red.	reduce
ref.	reference, refrigerator
refer.	refrigerator
reg.	regular, regulator
reinf.	reinforcing
rem.	removal, remove
rep.	repair
repl.	replace
repro.	reproduce, reproduction
req.	require, requisition
reqd.	required
res.	resawed, residential, resistor
ret.	retaining, return
rev.	reverse, revised, revision, revolution
rgh.	rough
rgh. opng.	rough opening
RH	right hand
Rh	rhodium
rig.	rigid
RL, R/L	random lengths
Rn	radon
RO	rough opening
rot.	rotate
rpm	revolutions per minute
rps	revolutions per second
RR	railroad
RS	rough-sawn
R/S	resawn
RTA	ready-to-assemble
Ru	ruthenium
rub.	red, rubber
RW	random widths, redwood
R/W R/L	random widths and random lengths

S

S	soft, south, sulfur, switch
s	solid
s.	side
S1E	surfaced one edge
S2E	surfaced two edges
S1S	surfaced one side
S2S	surfaced two sides
S4S	surfaced four sides
S1S&CM	surfaced one side and center-matched
S2S&CM	surfaced two sides and center-matched
S4S&CS	surfaced four sides and caulking seam
S1S1E	surfaced one side and one edge
S1S2E	surfaced one side and two edges
S2S1E	surfaced two sides and one edge
saf.	safety
sat.	saturate
Sb	antimony
SBCCI	Southern Building Code Congress International, Inc.
SB1S	single bead one side
Sc	scandium
sch.	schedule
schem.	schematic
SCR	Structural Clay Research
scr.	screen, screw
sd. bl.	sand blast
sdg.	siding
S&E	side and edge
Se	selenium
sec.	second, section, sector
sect.	section
sel.	select
sep.	separate
seq.	sequence
ser.	series
serr.	serrated
serv.	service
SF	square foot
SFCA	square-foot contact area
SFFA	square-foot face area
sft.	shaft
sftwd.	softwood
SG	slash grain
sgl.	single
shelv.	shelving
shld.	shield
shpt.	shipment
sht.	sheet
shthg.	sheathing
Si	silicon
sid.	siding
SIGMA	Sealed Insulated Glass Manufacturers Association
sim.	simulated

sk.	sack, sink
SL	snow load
sl.	slate, slide, slow
S/L, SL, Shlp.	shiplap
sld.	solder
slot.	slotted
slv.	sleeve
SM	sheet metal
Sm	samarium
sm.	small
S.m.	surface measure
smkls.	smokeless
smry.	summary
Sn	tin
snd.	sound
sol.	soluble, solution
SP	static pressure, steel pipe, sugar pine
sp.	space, spare, specific, speed
spec.	Specifications
specd.	specified
specl.	special
spl.	special
sp. ph.	split phase
spr.	spring
sq.	square, 100 square feet
sq. ft.	square feet
sq. in.	square inch
sq. yd.	square yard
SR	styrene rubber
Sr	strontium
SRB	stress-rated boards
SS	stainless steel
SSP	stainless steel pipe
st.	steel
sta.	station, stationary
STC	sound transmission coefficient
std.	standard
StdM	standard matched
stg.	storage
stk.	stock
stl.	steel
stn.	stainless
stp.	stepping
str.	straight, structural
stwy.	stairway
sub.	subcontractor, substitute
suc.	suction
sum.	summary
sup.	supply
supv.	supervise
sur.	surface
surv.	survey, surveyor
svy.	survey
sw.	switch
SWF	safe working pressure

SY	square yard
sym.	symbol, symmetrical
syn.	synthetic
sys.	system

T

T	tee, tension, time
t	ton
t.	table
Ta	tantalum
tab.	table
tach.	tachometer
TB	terminal board
Tb	terbium
tbl.	table
tbr.	timber
TC	thin coat, true course
Tc	technetium
TD	time delay
tdm.	tandem
TDC	time-delay closing
TDO	time-delay opening
Te	tellurium
tech.	technical
temp	temperature
tens.	tension
term.	terminal
text.	textured
T&G	tongue and groove
Th	thorium
thermo.	thermostat
thk.	thick
thrm.	thermal
thr.	through
Ti	titanium
TL	total load
Tl	thallium
TLbr.	treated lumber
Tm	thulium
tot.	total
tpi	teeth per inch
tr.	tread
trans.	transfer, transformer, transverse
transv.	transverse
trim.	trimmer
TS	tool steel
T-stat	thermostat
tub.	tubing
typ.	typical

U

U	uranium
u.	unit

UBC	Uniform Building Code
UCC	Uniform Commercial Code
UL	Underwriters Laboratories
ult.	ultimate
UMC	Uniform Mechanical Code
unexc.	unexcavated
univ.	universal
UPC	Uniform Plumbing Code
USDOE	United States Department of Energy
USG	United States gauge
USS	United States standard

V

V	vanadium, volt, volume
v	velocity
v.	valve, vertical
va	volt-ampere
vac.	vacuum
vap. prf.	vapor proof
var.	variable
v.c.	vinyl clad
ven.	veneer
vent.	ventilation
vert.	vertical
VF	very fine
V.F.	vertical feet
V.G.	vertical grain
vib.	vibrate
visc.	viscosity
vit.	vitreous
vol.	volume
vs.	versus
VWC	vinyl wallcovering

W

W	energy, tungsten, west
w	watt
w.	water, width, wire, with
w/	with
wash.	washer
WB	wallboard, wet bulb
w.c.	water closet
wd.	wind, wood
wdg.	winding
wdr.	wider
wdt.	width
W/E&SP	with equipment and spare parts
WF	white fir
WH	water heater
wh	watt-hour
whm	watt-hour meter
whr	watt-hour
wht.	white

W.I.	wrought iron
WL	wavelength, wind load
W.L.	water line
w/o	without
W/OE&SP	without equipment and spare parts
WP	weatherproof, wettable powder, working pressure
wp.	waterproof
wr.	water-resistant
WRC	western red cedar
WS	waterbed stock, water softener, water supply, weatherstripping
ws	water-soluble, wetted surface
W.T.	watertight
wt.	weight
wth.	width
Wtr	waters
WWPA	Western Wood Products Association

X

X	cross, exterior, extra

XConn.	cross connection
Xe	Xenon
X Hvy.	extra heavy
Xmtr.	transmitter
X-sect.	cross-section
X Str.	extra strong

Y

Y	yttrium
Yb	ytterbium
yd.	yard, yards
yel.	yellow
yr.	year

Z

Z	azimuth number, impedance
Zn	zinc
Zr	zirconium

Appendix B
Conversions, Tables, and Weights

CONVERSION TABLES

Multiply	By	To Obtain
Acres	43,560	Square feet
Acre-feet	43,560	Cubic feet
Acre-feet	325,851	Gallons
Atmospheres	76	Centimeters of mercury
Atmospheres	29.92	Inches of mercury
Atmospheres	33.90	Feet of water
Atmospheres	14.70	Pounds/square inch
Btu/hour	0.2931	Watts
Btu/minute	0.02356	Horsepower
Btu/minute	60	Btu/hour
Btu/minute	0.01757	Kilowatts
Btu/minute	17.57	Watts
Centimeters	0.3937	Inches
Centimeters of mercury	0.01316	Atmospheres
Centimeters of mercury	0.4461	Feet of water
Centimeters of mercury	27.85	Pounds/square foot
Centimeters of mercury	0.1934	Pounds/square inch
Cubic centimeters	0.06102	Cubic inches
Cubic feet	1728	Cubic inches
Cubic feet	0.02832	Cubic meters
Cubic feet	0.03704	Cubic yards
Cubic feet	7.48052	Gallons
Cubic feet	28.32	Liters
Cubic feet	29.92	Quarts (liquid)
Cubic feet/minute	472	Cubic centimeters/second
Cubic feet/minute	0.1274	Gallons/second
Cubic feet/minute	62.43	Pounds of water/minute
Cubic feet/second	0.646317	Million gallons/day
Cubic feet/second	448.831	Gallons/minute
Cubic inches	16.39	Cubic centimeters
Cubic inches	0.01639	Liters
Cubic meters	264.2	Gallons (U.S. liquid)
Cubic yards	27	Cubic feet
Cubic yards	46,656	Cubic inches
Degrees Centigrade + 273	1	Absolute temp (C)
Degrees Centigrade + 17.28	1.8	Degrees Fahrenheit
Degrees Fahrenheit + 460	1	Absolute temp (F)
Degrees Fahrenheit − 32	.5556	Degrees Centigrade
Feet	30.48	Centimeters
Feet	12	Inches
Feet	0.3048	Meters
Feet	304.8	Millimeters
Feet of water	0.02950	Atmospheres

Multiply	By	To Obtain
Feet of water	0.8826	Inches of mercury
Feet of water	62.43	Pounds/square foot
Feet of water	0.4335	Pounds/square inch
Feet/minute	0.01667	Feet/second
Feet/minute	0.01136	Miles/hour
Feet/second	0.6818	Miles/hour
Feet/second	0.01136	Miles/minute
Foot-pounds/minute	2.260×10^{-5}	Kilowatts
Foot-pounds/second	1.356×10^{-3}	Kilowatts
Gallons	3785	Cubic centimeters
Gallons	0.1337	Cubic feet
Gallons	231	Cubic inches
Gallons	3.785	Liters
Gallons	4	Quarts (dry)
Gallons of water	8.3453	Pounds of water
Gallons/minute	0.002228	Cubic feet/second
Gallons/minute	8.0208	Cubic feet/hour
Gallons of water/minute	6.0086	Tons of water/24 hours
Horsepower	745.7	Watts
Horsepower hours	0.7457	Kilowatt hours
Inches	2.540	Centimeters
Inches of mercury	0.03342	Atmospheres
Inches of mercury	1.133	Feet of water
Inches of mercury	0.4912	Pounds/square foot
Inches of water	0.002458	Atmospheres
Inches of water	0.07355	Inches of mercury
Inches of water	5.202	Pounds/square foot
Inches of water	0.03613	Pounds/square inch
Kilograms	2.205	Pounds
Kilograms	1.102×10^{-3}	Tons
Kilometers	0.6214	Miles
Kilometers/hour	0.6214	Miles/hour
Kilowatts	1.341	Horsepower
Kilowatt hours	3,413	Btus
Kilowatt hours	2.655×10^{6}	Foot pounds
Liters	1,000	Cubic centimeters
Liters	0.03531	Cubic feet
Liters	61.02	Cubic inches
Liters	0.2642	Gallons
Meters	3.281	Feet
Meters	39.37	Inches
Meters	1.094	Yards
Meters/second	3.281	Feet/second
Meters/second	2.237	Miles/hour
Miles	5,280	Feet
Miles	1.609	Kilometers
Miles/hour	1.467	Feet/second
Miles/hour	26.82	Meters/minute
Millimeters	0.1	Centimeters
Millimeters	0.03937	Inches
Million gallons/day	1.54723	Cubic feet/second
Ounces (fluid)	0.02957	Liters
Pints (liquid)	473.2	Cubic centimeters
Pounds	0.4536	Kilograms

Multiply	By	To Obtain
Pounds of water	0.01602	Cubic feet
Pounds of water	27.68	Cubic inches
Pounds of water	0.1198	Gallons
Pounds/cubic inch	1,728	Pounds/cubic foot
Pounds/square foot	0.01602	Feet of water
Pounds/square inch	0.06804	Atmospheres
Pounds/square inch	2.307	Feet of water
Pounds/square inch	2.036	Inches of mercury
Pounds/square inch	6,895	Pascals
Quarts (dry)	67.20	Cubic inches
Quarts (liquid)	57.75	Cubic inches
Quarts (liquid)	0.9463	Liters
Square feet	144	Square inches
Square inches	645.2	Square millimeters
Square meters	1,550	Square inches
Square miles	640	Acres
Square millimeters	1.550×10^{-3}	Square inches
Square yards	9	Square feet
Square yards	1,296	Square inches
Tons	2,000	Pounds
Watts	3.4129	Btu/hour
Watts	0.05688	Btu/minute
Watts	0.7378	Foot pounds/second
Watts	1.341×10^{-3}	Horsepower
Yards	3	Feet
Yards	36	Inches

WEIGHTS AND MEASURES

Metric System

Length

Unit			Metric Equivalent			U.S. Equivalent	
millimeter	(mm)	=	0.001	meter	=	0.03937	inch
centimeter	(cm)	=	0.01	meter	=	0.3937	inch
decimeter	(dm)	=	0.1	meter	=	3.937	inches
METER	(m)	=	1.0	meter	=	39.37	inches
dekameter	(dkm)	=	10.0	meters	=	10.93	yards
hectometer	(hm)	=	100.0	meters	=	328.08	feet
kilometer	(km)	=	1000.0	meters	=	0.6214	mile

Weight or Mass

Unit			Metric Equivalent			U.S. Equivalent	
milligram	(mg)	=	0.001	gram	=	0.0154	grain
centigram	(cg)	=	0.01	gram	=	0.1543	grain
decigram	(dg)	=	0.1	gram	=	1.543	grains
GRAM	(g)	=	1.0	gram	=	15.43	grains
dekagram	(dkg)	=	10.0	grams	=	0.3527	ounce avoirdupois
hectogram	(hg)	=	100.0	grams	=	3.527	ounces avoirdupois
kilogram	(kg)	=	1000.0	grams	=	2.2	pounds avoirdupois

Capacity

Unit			Metric Equivalent			U.S. Equivalent	
milliliter	(ml)	=	0.001	liter	=	0.034	fluid ounce
centiliter	(cl)	=	0.01	liter	=	0.338	fluid ounce
deciliter	(dl)	=	0.1	liter	=	3.38	fluid ounces
LITER	(l)	=	1.0	liter	=	1.05	liquid quarts
dekaliter	(dkl)	=	10.0	liters	=	0.284	bushel
hectoliter	(hl)	=	100.0	liters	=	2.837	bushels
kiloliter	(kl)	=	1000.0	liters	=	264.18	gallons

Area

Unit			Metric Equivalent			U.S. Equivalent	
square millimeter	(mm^2)	=	0.000001	centare	=	0.00155	square inch
square centimeter	(cm^2)	=	0.0001	centare	=	0.155	square inch
square decimeter	(dm^2)	=	0.01	centare	=	15.5	square inches
CENTARE also	(ca)	=	1.0	centare	=	10.76	square feet
square meter	(m^2)						
are also	(a)	=	100.0	centares	=	0.0247	acre
square dekameter	(dkm^2)						
hectare also	(ha)	=	10,000.0	centares	=	2.47	acres
square hectometer	(hm^2)						
square kilometer	(km^2)	=	1,000,000.0	centares	=	0.386	square mile

Volume

Unit			Metric Equivalent			U.S. Equivalent	
cubic millimeter	(mm^3)	=	0.001	cubic centimeter	=	0.016	minim
cubic centimeter	(cc, cm^3)	=	0.001	cubic decimeter	=	0.061	cubic inch
cubic decimeter	(dm^3)	=	0.001	cubic meter	=	61.023	cubic inches
STERE also	(s)	=	1.0	cubic meter	=	1.308	cubic yards
cubic meter	(m^3)						
cubic dekameter	(dkm^3)	=	1000.0	cubic meters	=	1307.943	cubic yards
cubic hectometer	(hm^3)	=	1,000,000.0	cubic meters	=	1,307,942.8	cubic yards
cubic kilometer	(km^3)	=	1,000,000,000.0	cubic meters	=	0.25	cubic mile

U.S. System

Liquid Measure

4	gills	=	1 pint (pt.)
2	pints	=	1 quart (qt.)
4	quarts	=	1 gallon (gal.)
31.5	gallons	=	1 barrel (bbl.)
2	barrels	=	1 hogshead
60	minims	=	1 fluid dram (fl. dr.)
8	fluid drams	=	1 fluid ounce (fl. oz.)
16	fluid ounces	=	1 pint

Linear Measure

1	mil	=	0.001 inch (in.)
12	inches	=	1 foot (ft.)
3	feet	=	1 yard (yd.)
6	feet	=	1 fathom
5.5	yards	=	1 rod (rd.)
40	rods	=	1 furlong
5280	feet	=	1 mile (mi.)
1760	yards	=	1 mile

Square Measure

144	square inches (sq. in.)	=	1 square foot (sq. ft.)
9	square feet	=	1 square yard (sq. yd.)
30.25	square yards	=	1 square rod (sq. rd.)
160	square rods	=	1 acre (A.)
640	acres	=	1 square mile (sq. mi.)

Apothecaries' Weight

20	grains (gr.)	=	1 scruple
3	scruples	=	1 dram (dr.)
8	drams	=	1 ounce (oz.)
12	ounces	=	1 pound (lb.)

Avoirdupois Weight

27.34	grains (gr.)	=	1 dram (dr. av.)
16	drams	=	1 ounce (oz. av.)
16	ounces	=	1 pound (lb. av.)
2000	pounds	=	1 short ton (sh. tn.)
2240	pounds	=	1 long ton (l. tn.)

Troy Weight

24	grains	=	1 pennyweight (dwt.)
20	pennyweight	=	1 ounce (oz. t.)
12	ounces	=	1 pound (lb. t.)

Cubic Measure

144	cubic inches (cu. in.)	=	1 board foot (bd. ft.)
1728	cubic inches	=	1 cubic foot (cu. ft.)
27	cubic feet	=	1 cubic yard (cu. yd.)
128	cubic feet	=	1 cord (cd.)

Dry Measure

2	pints	=	1 quart (qt.)
8	quarts	=	1 peck (pk.)
4	pecks	=	1 bushel (bu.)
3.28	bushels	=	1 barrel (bbl.)

WEIGHTS OF COMMON MATERIALS

Earth	Weight (lb/cu ft)	Masonry	Weight (lb/cu ft)
Clay, dry	63	Cement, portland	90
Clay, damp	110	Concrete, cinder	111
Clay and gravel, dry	100	Concrete, slag	138
Earth, dry	76 to 95	Concrete, sand and gravel	150
Earth, moist	78 to 96	Brick, common	112
Earth, packed	100	Brick, fire	144
Gravel	109	Limestone	163
Mud	108 to 115	Sandstone	144
Sand and gravel, dry	90 to 120		
Sand and gravel, wet	118 to 120		

Liquids	Weight (lb/cu ft)	Metals	Weight (lb/cu ft)
Gasoline	42	Aluminum alloy, cast	171
Water, fresh	62.4	Aluminum, cast	165
Water, ice	56	Brass, cast or rolled	526
Water, sea	64	Chromium	442
		Copper, cast	540
		Gold	1,204
		Iron, cast	450

Lumber (15% MC)	Weight (lb/sq ft) Nominal Size			Metals	Weight (lb/cu ft)
	2"	3"	4"	Iron wrought	480
Douglas fir	4.5	7.6	10.6	Lead	710
Hem-Fir	4.0	6.6	9.3	Magnesium	109
Englemann Spruce	3.1	5.1	7.2	Mercury	848
Pine: Lodgepole,				Silver	655
Ponderosa, Sugar	3.7	6.2	8.6	Steel, rolled	485
Western Cedar	3.3	5.5	7.7	Steel, tool	470
Western Hemlock	4.0	6.6	9.3	Tin	455
				Zinc	446

Roofing	Weight (lb/sq ft)	Wood	Weight (lb/cu ft)
Built-up, 3 ply	1.5	Ash	41
Built-up, 3 ply w/gravel	5.5	Balsa	10
Built-up, 5 ply	2.5	Cedar	29
Built-up 5 ply w/gravel	6.5	Cork	16
Roll Roofing	1.0	Hickory	51
Galvanized steel, 2 1/2" and 3"		Maple	43
corrugated, U.S. Standard Gauge,		Oak, white	48
including laps:		Pine, white	26
12 gauge	4.9	Pine, yellow	43
14 gauge	3.6	Poplar	31
16 gauge	2.9	Walnut, black	40
18 gauge	2.4		
20 gauge	1.8		
22 gauge	1.5		
24 gauge	1.3	Miscellaneous	Weight (lb/cu ft)
26 gauge	1.0		
Shingles:		Asbestos	175
Asphalt, 1/4"	2.0	Bakelite	79.5
Asbestos cement, 3/8"	4.0	Fiberglass, rigid	18
Clay tile w/mortar	19.0–24.0	Glass	156
Slate, 1/4"	10.0	Glass, plate	161
Spanish	19.0	Insulation, expanded polystyrene	2.4
Tile, 2"	12.0	Insulation, loose fiberglass	6
Tile, 3"	20.0	Pitch	69
Wood, 1"	3.0	Plexiglass	74.3
		Tar	74

Walls	Weight (lb/sq ft)
16" O.C.	1.0
24" O.C.	0.7
Glazed tile	18.0
Gypsum wallboard, 1/2"	2.5
Marble	15.0
Masonry, in 4" thickness:	
Brick	38.0
Concrete block	30.0
Concrete cinder block	20.0
Hollow clay tile, load-bearing	23.0
2 x 4 framing, bare:	
12" O.C.	1.3
Hollow clay tile, non-bearing	18.0
Limestone	53.0
Stone	55.0
Terra cotta tile	25.0
Plaster, 1"	8.0
Plaster, 1" on wood lath	10.0
Plaster, 1" on metal lath	8.5
Porcelain-enameled steel	3.0
Stucco, 7/8"	10.0
Windows, glass frame and sash	8.0
Wood paneling, 1"	2.5

DECIMAL EQUIVALENTS

Fraction	Decimal	Fraction	Decimal
1/32	.03125	17/32	.53125
1/16	.0625	9/16	.5625
3/32	.09375	19/32	.59375
1/8	.125	5/8	.625
5/32	.15625	21/32	.65625
3/16	.1875	11/16	.6875
7/32	.21875	23/32	.71875
1/4	.250	3/4	.750
9/32	.28125	25/32	.78125
5/16	.3125	13/16	.8125
11/32	.34375	27/32	.84375
3/8	.375	7/8	..875
13/32	.40625	29/32	.90625
7/16	.4375	15/16	.9375
15/32	.46875	31/32	.96875
1/2	.500	1	1.000

APPROXIMATE R VALUES FOR COMMON BUILDING MATERIALS

Air	R-Value
Heat Flow Up:	
Air film (still air, inside wall)	0.61
3/4" space	0.77
3 1/2" space	0.84
Heat Flow Down:	
Air film (still air, inside wall)	0.92
3/4" space	1.02
3 1/2" space	1.22
Heat Flow Horizontal:	
Air film (still air, inside wall)	0.68
3/4" space	0.94
3 1/2" space	0.91
Outside air film, 7.5 mph wind	0.25
Outside air film, 15 mph wind	0.17

Building Board	R-Value
Asbestos cement board, per inch	0.25
Asbestos cement board, 1/4"	0.07
Drywall, 3/8"	0.32
Drywall, 1/2"	0.45
Insulating board, 1/2" drop-in ceiling tiles	1.25
Insulating board, 1/2" sheathing	1.32
Hardboard, tempered, per inch	1.00
Particle board, medium density, per inch	1.06
Particle board, 5/8" underlayment	0.82
Plywood, per inch	1.25
Plywood, 1/4"	0.31
Plywood, 3/8"	0.47
Plywood, 1/2"	0.63
Wood sheathing, plywood or wood panels, 3/4"	0.93

Flooring	R-Value
Asphalt tile, 3/16"	0.04
Carpet and fiber pad	2.08
Carpet and rubber pad	1.23
Ceramic tile, 1/4"	0.04
Cork tile, 1/8"	0.28
Linoleum	0.08
Hardwood strips, 5/8"	0.68

Masonry	R-Value
Brick, per inch	0.20
Cement mortar, per inch	0.20
Concrete, per inch	0.08
Concrete blocks, 8"	1.00
Gypsum plaster, perlite aggregate, per inch	0.67
Gypsum plaster, sand aggregate, per inch	0.18
Limestone or sandstone, per inch	0.08
Pumice blocks, 8"	2.00
Stucco, per inch	0.20

Paper	R-Value
Felt building paper	0.06
Felt flooring paper	0.06

Roofing	R-Value
Asbestos-cement shingles	0.21
Asphalt roll roofing	0.15
Asphalt shingles	0.44
Built-up roofing, 3/8"	0.33
Slate shingles, 1/2"	0.05
Wood shingles	0.94

Siding	R-Value
Aluminum or steel, hollow back	0.61
Asbestos-cement shingles	0.21
Asbestos roll	0.15
Wood, bevel siding, 1/2" x 8", lapped	0.81
Wood, bevel siding, 3/4" x 10", lapped	1.05
Wood, drop siding, 1" x 8"	0.79
Wood, plywood, 3/8"	0.59
Wood siding shingles, 7 1/2" exposure	0.87

Thermal Insulation	R-Value
Batts:	
Fiberglass, per inch	3.20
Mineral wool, per inch	3.50
Loose Fill:	
Cellulose, per inch	3.70
Fiberglass, per inch	2.20
Mineral wool	3.00
Perlite, expanded, per inch	2.70
Vermiculite, expanded, per inch	2.20
Wood shavings, per inch	2.20

Thermal Insulation (Continued)	R-Value
Rigid:	
Polystyrene, expanded bead board, per inch	3.57
Polystyrene, extruded board, per inch	5.26
Polyurethane foam, per inch	6.25
Urea-formaldehyde foam, per inch	4.17

Wall Sections	R-Value
Standard frame wall, insulated	14.29
Standard frame wall, uninsulated	4.35

Windows	R-Value
Architectural glass	0.10
Double glass window, 1/4" air space	1.59
Double glass window, 1/2" air space	1.75
Single glass window	0.91
Single glass window, with storm, 1" to 4" air space	1.82

Wood	R-Value
Hardwood, per inch	0.91
Softwood, per inch	1.25
Softwood, 1 1/2"	1.89
Softwood, 3 1/2"	4.35

GEOMETRIC FORMULAS

Solid Geometry

Plane Geometry

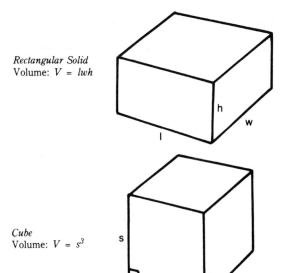

Rectangular Solid
Volume: $V = lwh$

Cube
Volume: $V = s^3$

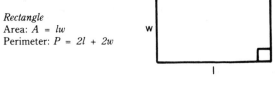

Rectangle
Area: $A = lw$
Perimeter: $P = 2l + 2w$

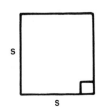

Square
Area: $A = s^2$
Perimeter: $P = 4s$

Solid Geometry

Plane Geometry

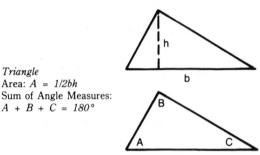

Right Circular Cylinder
Volume: $V = \pi r^2 h$
Lateral Surface Area: $L = 2\pi rh$
Total Surface Area:
$S = 2\pi rh + 2\pi r^2$

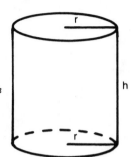

Triangle
Area: $A = 1/2 bh$
Sum of Angle Measures:
$A + B + C = 180°$

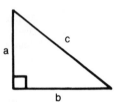

Right Triangle
Pythogorean Theorem:
$a^2 + b^2 = c^2$

Right Circular Cone
Volume: $V = 1/3\pi r^2 h$
Lateral Surface Area: $L = \pi rs$
Total Surface Area: $S = \pi r^2 + \pi rs$

Slant Height: $s = \sqrt{r^2 + h^2}$

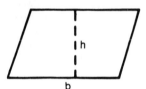

Parallelogram
Area: $A = bh$

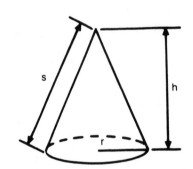

Trapezoid
Area: $A = 1/2h(a + b)$

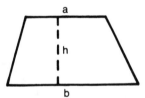

Sphere
Volume: $V = 4/3\pi r^3$
Surface Area: $S = 4\pi r^2$

Circle
Area: $A = \pi r^2$
Circumference:
$C = \pi D = 2\pi r$
(22/7 and 3.14 are different
approximations for π)

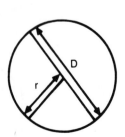

AREAS AND CIRCUMFERENCE OF CIRCLES

Diam.	Circum.	Area	Diam.	Circum.	Area
1/8	.39270	.01227	2 1/2	7.8540	4.9087
1/4	.78540	.04909	3	9.4248	7.0686
3/8	1.1781	.11045	4	12.566	12.566
1/2	1.5708	.19635	5	15.708	19.635
3/4	2.3562	.44179	6	18.850	28.274
1	3.1416	.7854	7	21.991	38.485
1 1/4	3.9270	1.2272	8	25.133	50.265
1 1/2	4.7124	1.7671	9	28.274	63.617
2	6.2832	3.1416	10	31.416	78.540

HEAT VALUE OF COMMON FIREPLACE FUELS

Fuel	Heat Value in Millions of Btus per cord	Btus per Pound
Ash, white	25.0	7,400
Beech	28.0	7,200
Birch, white	23.4	8,000
Cedar	16.5	7,900
Cherry	23.5	8,400
Chestnut	20.2	8,100
Coal, anthracite (hard)	—	15,000
Coal, bituminous (soft)	—	13,000
Elm, white	24.5	8,200
Fir, Douglas	21.4	8,600
Hickory	30.6	7,900
Maple, red	24.0	7,500
Maple, sugar	29.0	7,600
Oak, red	27.3	7,600
Oak, white	30.6	7,900
Pine, white	15.8	7,500
Sawdust logs	—	15,000
Spruce	17.5	8,300

Appendix C
Building and Framing Information

RECOMMENDED NAILING SCHEDULE
FOR FRAMING AND SHEATHING

Joining	Nailing Method	Nails		
		No.	Size	Placement
Header to joist	End-nail	3	16d	
Joist to sill or girder	Toenail	2	10d or	
		3	8d	
Header and stringer joist to sill	Toenail		10d	16 in. on center
Bridging to joist	Toenail each end	2	8d	
Ledger strip to beam, 2 in. thick		3	16d	At each joist
Subfloor, boards:				
1 by 6 in. and smaller		2	8d	To each joist
1 by 8 in.		3	8d	To each joist
Subfloor, plywood:				
At edges			8d	6 in. on center
At intermediate joists			8d	8 in. on center
Subfloor (2 by 6 in., T&G) to joist or girder	Blind-nail (casing) and face-nail	2	16d	
Soleplate to stud, horizontal assembly	End-nail	2	16d	At each stud
Top plate to stud	End-nail	2	16d	
Stud to soleplate	Toenail	4	8d	
Soleplate to joist or blocking	Face-nail		16d	16 in. on center
Doubled studs	Face-nail, stagger		10d	16 in. on center
End stud of intersecting wall to exterior wall stud	Face-nail		16d	16 in. on center
Upper top plate to lower top plate	Face-nail		16d	16 in. on center
Upper top plate, laps and intersections	Face-nail	2	16d	
Continuous header, two pieces, each edge			12d	12 in. on center
Ceiling joist to top wall plates	Toenail	3	8d	
Ceiling joist laps at partition	Face-nail	4	16d	
Rafter to top plate	Toenail	2	8d	
Rafter to ceiling joist	Face-nail	5	10d	
Rafter to valley or hip rafter	Toenail	3	10d	
Ridge board to rafter	End-nail	3	10d	
Rafter to rafter through ridge board	Toenail	4	8d	
	Edge-nail	1	10d	
Collar beam to rafter:				
2 in. member	Face-nail	2	12d	
1 in. member	Face-nail	3	8d	
1-in. diagonal let-in brace to each stud and plate (4 nails at top)		2	8d	
Built-up corner studs:				
Studs to blocking	Face-nail	2	10d	Each side
Intersecting stud to corner studs	Face-nail		16d	12 in. on center

		Nails		
Joining	Nailing Method	No.	Size	Placement
Built-up girders and beams, three or more members	Face-nail		20d	32 in. on center, each side
Wall sheathing:				
1 by 8 in. or less, horizontal	Face-nail	2	8d	At each stud
1 by 6 in. or greater, diagonal	Face-nail	3	8d	At each stud
Wall sheathing, vertically applied plywood:				
3/8 in. and less thick	Face-nail		6d	6 in. edge
1/2 in. and over thick	Face-nail		8d	12 in. intermediate
Wall sheathing, vertically applied fiberboard:				
1/2 in. thick	Face-nail			1 1/2 in. roofing nail 1 3/4 in. roofing nail
25/32 in. thick	Face-nail			3 in. edge and 6 in. intermediate
Roof sheathing, boards, 4-, 6-, 8-in. width	Face-nail	2	8d	At each rafter
Roof sheathing, plywood:				
3/8 in. and less thick	Face-nail		6d }	6 in. edge and 12 in. intermediate
1/2 in. and over thick	Face-nail		8d	

DEGREES OF SLOPE

Rise per 12" of Run	Degrees of Slope	Rise per 12" of Run	Degrees of Slope	Rise per 12" of Run	Degrees of Slope	Rise per 12" of Run	Degrees of Slope	Rise per 12" of Run	Degrees of Slope
1/2"	2 1/2	6 1/2"	28 1/4	12 1/2"	46 1/4	18 1/2"	57		
1"	4 1/2	7"	30 1/4	13"	47 1/4	19"	57 3/4		
1 1/2"	7	7 1/2"	32	13 1/2"	48 1/2	19 1/2"	58 1/2		
2"	9 1/2	8"	33 3/4	14"	49 1/2	20"	59		
2 1/2"	11 3/4	8 1/2"	35 1/4	14 1/2"	50 1/2	20 1/2"	59 3/4		
3"	14	9"	37	15"	51 1/2	21"	60 1/4		
3 1/2"	16 1/4	9 1/2"	38 1/2	15 1/2"	52 1/4	21 1/2"	61		
4"	18 1/2	10"	40	16"	53 1/4	22"	61 1/2		
4 1/2"	20 1/2	10 1/2"	41 1/4	16 1/2"	54	22 1/2"	62		
5"	22 1/2	11"	42 1/2	17"	54 3/4	23"	62 1/2		
5 1/2"	24 1/2	11 1/2"	43 3/4	17 1/2"	55 1/2	23 1/2"	63		
6"	26 1/2	12"	45	18"	56 1/4	24"	63 1/2		

SIZE, TYPE, AND USE OF NAILS[1]
(Courtesy of the United States Army)

Size	Lgth (in.)	Diam (in.)	Remarks	Where used
2d	1	.072	Small head	Finish work, shop work
2d	1	.072	Large flathead	Small timber, wood shingles, lathes
3d	1 1/4	.08	Small head	Finish work, shop work
3d	1 1/4	.08	Large flathead	Small timber, wood shingles, lathes
4d	1 1/2	.098	Small head	Finish work, shop work
4d	1 1/2	.098	Large flathead	Small timber, lathes, shop work
5d	1 3/4	.098	Small head	Finish work, shop work
5d	1 3/4	.098	Large flathead	Small timber, lathes, shop work
6d	2	.113	Small head	Finish work, casing, stops, etc., shop work
6d	2	.113	Large flathead	Small timber, siding, sheathing, etc., shop work

SIZE, TYPE, AND USE OF NAILS[1] (Continued)

Size	Lgth (in.)	Diam (in.)	Remarks	Where used
7d	2 1/4	.113	Small head	Casing, base, ceiling, stops, etc.
7d	2 1/4	.113	Large flathead	Sheathing, siding, subflooring, light framing
8d	2 1/2	.131	Small head	Casing, base, ceiling, wainscot, etc., shop work
8d	2 1/2	.131	Large flathead	Sheathing, siding, subflooring, light framing, shop work
8d	1 1/4	.131	Extra-large flathead	Roll roofing, composition shingles
9d	2 3/4	.131	Small head	Casing, base, ceiling, etc.
9d	2 3/4	.131	Large flathead	Sheathing, siding, subflooring, framing, shop work
10d	3	.148	Small head	Casing, base, ceiling, etc., shop work
10d	3	.148	Large flathead	Sheathing, siding, subflooring, framing, shop work
12d	3 1/4	.148	Large flathead	Sheathing, subflooring, framing
16d	3 1/2	.162	Large flathead	Framing, bridges, etc.
20d	4	.192	Large flathead	Framing, bridges, etc.
30d	4 1/2	.207	Large flathead	Heavy framing, bridges, etc.
40d	5	.225	Large flathead	Heavy framing, bridges, etc.
50d	5 1/2	.244	Large flathead	Extra-heavy framing, bridges, etc.
60d	6	.262	Large flathead	Extra-heavy framing, bridges, etc.

[1]This chart applies to wire nails, although it may be used to determine the length of cut nails.

CHARACTERISTICS OF WOODS FOR PAINTING AND FINISHING
(Courtesy of the U.S. Department of Agriculture)

Wood	Ease of keeping keeping well-painted 1—easiest 5—most exacting[1]	Weathering	
		Resistance to cupping 1—best 4—worst	Conspicuousness of checking 1—least 2—most
SOFTWOODS			
Cedar:			
Alaska	1	1	1
California incense	1		
Port-Orford	1		1
Western redcedar	1	1	1
White	1	1	
Cypress	1	1	1
Redwood	1	1	1
Pine:			
Eastern white	2	2	2
Sugar	2	2	2
Western white	2	2	2
Ponderosa	3	2	2
Fir, commercial white	3	2	2
Hemlock	3	2	2
Spruce	3	2	2
Douglas-fir (lumber and plywood)	4	2	2
Larch	4	2	2
Pine:			
Norway	4	2	2
Southern (lumber and plywood)	4	2	2
Tamarack	4	2	2

CHARACTERISTICS OF WOODS FOR PAINTING AND FINISHING (Continued)

| | Ease of keeping keeping well-painted 1—easiest 5—most exacting[1] | Weathering | |
		Resistance to cupping 1—best 4—worst	Conspicuousness of checking 1—least 2—most
Wood			
HARDWOODS			
Alder	3		
Aspen	3	2	1
Basswood	3	2	2
Cottonwood	3	4	2
Magnolia	3	2	
Poplar	3	2	1
Beech	4	4	2
Birch	4	4	2
Gum	4	4	2
Maple	4	4	2
Sycamore	4		
Ash	5 or 3	4	2
Butternut	5 or 3		
Cherry	5 or 3		
Chestnut	5 or 3	3	2
Walnut	5 or 3	3	2
Elm	5 or 4	4	2
Hickory	5 or 4	4	2
Oak, white	5 or 4	4	2
Oak, red	5 or 4	4	2

[1]Woods ranked in group 5 for *ease of keeping well-painted* are hardwoods with large pores that need filling with wood filler for durable painting. When so filled before painting, the second classification recorded in the table applies. NOTE: Omission in the table indicates inadequate data for classification.

ESTIMATING NEEDS FOR DRY PREMIXED CONCRETE

Approximate Number of Bags to Use

| Dimension in Feet: | | 3" Deep | | 4" Deep | |
Width	Length	60 lb bag	90 lb bag	60 lb bag	90 lb bag
1	1	1/2	1/2	2/3	1/2
1	2	1	3/4	1 1/3	1
1	3	1 1/2	1 1/4	2	1 1/2
2	2	2	1 1/2	2 2/3	2
2	3	3	2 1/4	4	3
2	4	4	3	5 1/3	4
2	5	5	3 3/4	6 2/3	5
2	6	6	4 1/2	8	6
2	7	7	5 1/4	9 1/3	7
2	8	8	6	10 2/3	8
2	9	9	6 3/4	12	9
2	10	10	7 1/2	13 1/3	10
3	3	4 1/2	3 1/3	6	4 1/2
3	5	7 1/2	5 2/3	10	7 1/2
3	7	10 1/2	8	14	10 1/2

DRILL AND AUGER BIT SIZES FOR WOOD SCREWS
(Courtesy of the United States Army)

Screw Size No.	1	2	3	4	5	6	7	8	9	10	12	14	16	18
Nominal screw	.073	.086	.099	.112	.125	.138	.151	.164	.177	.190	.216	.242	.268	.294
Body diameter	5/64	3/32	3/32	7/64	1/8	9/64	5/32	11/64	11/64	3/16	7/32	15/64	17/64	19/64
Pilot hole														
Drill size	5/64	3/32	7/64	7/64	1/8	9/64	5/32	11/64	3/16	3/16	7/32	1/4	17/64	19/64
Bit size	—	—	—	—	—	—	—	—	—	—	4	4	5	5
Starter hole														
Drill size	—	1/16	1/16	5/64	5/64	3/32	7/64	7/64	1/8	1/8	9/64	5/32	3/16	13/64
Bit size	—	—	—	—	—	—	—	—	—	—	—	—	—	4

LAG SCREW SIZES
(Courtesy of the United States Army)

Lengths (inches)	Diameters (inches)			
	1/4	3/8, 7/16, 1/2	5/8, 3/4	7/8, 1
1	x	x	—	—
1 1/2	x	x	x	—
2, 2 1/2, 3, 3/12, etc., 7 1/2, 8 to 10.	x	x	x	x
11 to 12	—	x	x	x
13 to 16	—	—	x	x

CONVERSION DIAGRAM FOR RAFTERS

(Courtesy of the National Forest Products Association, Washington, DC)

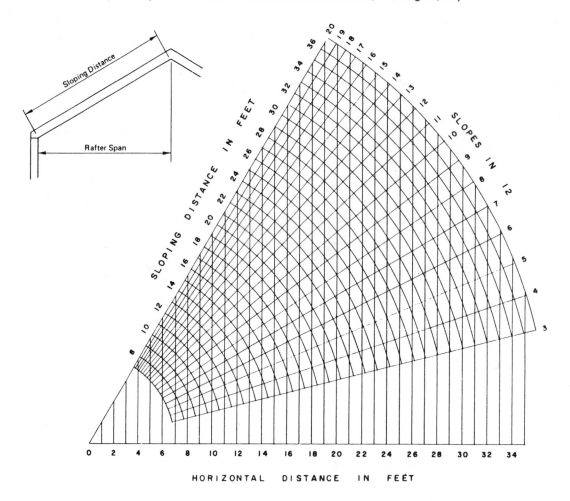

To use the diagram select the known horizontal distance and follow the vertical line to its intersection with the radial line of the specified slope, then proceed along the arc to read the sloping distance. In some cases it may be desirable to interpolate between the one foot separations. The diagram also may be used to find the horizontal distance corresponding to a given sloping distance or to find the slope when the horizontal distances are known.

Example: With a roof slope of 8 in 12 and a horizontal distance of 20 feet the sloping distance may be read as 24 feet.

Appendix D
Lumber and Plywood Information

COMMON SOFTWOOD GRADES

Wood	Top grade, best appearance, few if any defects	Slight defects okay for most exposed work	General use for framing, outdoor use	Many defects good for some rough usages
Redwood	Clear All Heart Clear	Select Heart Select	Constr. Heart Construction	Merchantable
Red Cedar	C and Better Finish	C Finish	Select Merchant. Construction	Standard
Douglas Fir	C and Better Finish	C Finish	Construction	Standard
Pine	C and Better 1 and 2 Clear Supreme	C Select Choice	Quality D Select	1–3 Dimension 1–5 Common Utility

TABLE OF BOARD MEASURE

Nominal Size of Piece	Board Feet Content When Length in Feet Equals											
	2	4	6	8	10	12	14	16	18	20	22	24
1 x 2	1/3	2/3	1	1 1/3	1 2/3	2	2 1/3	2 2/3	3	3 1/3	3 2/3	4
1 x 3	1/2	1	1 1/2	2	2 1/2	3	3 1/2	4	4 1/2	5	5 1/2	6
1 x 4	2/3	1 1/3	2	2 2/3	3 1/3	4	4 2/3	5 1/3	6	6 2/3	7 1/3	8
1 x 6	1	2	3	4	5	6	7	8	9	10	11	12
1 x 8	1 1/3	2 2/3	4	5 1/3	6 2/3	8	9 1/3	10 2/3	12	13 1/3	14 2/3	16
1 x 10	1 2/3	3 1/3	5	6 2/3	8 1/3	10	11 2/3	13 1/3	15	16 2/3	18 1/3	20
1 x 12	2	4	6	8	10	12	14	16	18	20	22	24
2 x 2	2/3	1 1/3	2	2 2/3	3 1/3	4	4 2/3	5 1/3	6	6 2/3	7 1/3	8
2 x 3	1	2	3	4	5	6	7	8	9	10	11	12
2 x 4	1 1/3	2 2/3	4	5 1/3	6 2/3	8	9 1/3	10 2/3	12	13 1/3	14 2/3	16
2 x 6	2	4	6	8	10	12	14	16	18	20	22	24
2 x 8	2 2/3	5 1/3	8	10 2/3	13 1/3	16	18 2/3	21 1/3	24	26 2/3	29 1/3	32
2 x 10	3 1/3	6 2/3	10	13 1/3	16 2/3	20	23 1/3	26 2/3	30	33 1/3	36 2/3	40
2 x 12	4	8	12	16	20	24	28	32	36	40	44	48
2 x 2	4 2/3	9 1/3	14	18 2/3	23 1/3	28	32 2/3	37 1/3	42	46 2/3	51 1/3	56
3 x 4	2	4	6	8	10	12	14	16	18	20	22	24
3 x 6	3	6	9	12	15	18	21	24	27	30	33	36
3 x 8	4	8	12	16	20	24	28	32	36	40	44	48
3 x 10	5	10	15	20	25	30	35	40	45	50	55	60
3 x 12	6	12	18	24	30	36	42	48	54	60	66	72
3 x 14	7	14	21	28	35	42	49	56	63	70	77	84
3 x 16	8	16	24	32	40	48	56	64	72	80	88	96
4 x 4	2 2/3	5 1/3	8	10 2/3	13 1/3	16	18 2/3	21 1/3	24	26 2/3	29 1/3	32
4 x 6	4	8	12	16	20	24	28	32	36	40	44	48
4 x 8	5 1/3	10 2/3	16	21 1/3	26 2/3	32	37 1/3	42 2/3	48	53 1/3	58 2/3	64
4 x 10	6 2/3	13 1/3	20	26 2/3	33 1/3	40	46 2/3	53 1/3	60	66 2/3	73 1/3	80
4 x 12	8	16	24	32	40	48	56	64	72	80	88	96
4 x 14	9 1/3	18 2/3	28	37 1/3	46 2/3	56	65 1/3	74 2/3	84	93 1/3	102 2/3	112
4 x 16	10 2/3	21 1/3	32	42 2/3	53 1/3	64	74 2/3	85 1/3	96	106 2/3	117 1/3	128
6 x 6	6	12	18	24	30	36	42	48	54	60	66	72
6 x 8	8	16	24	32	40	48	56	64	72	80	88	96
6 x 10	10	20	30	40	50	60	70	80	90	100	110	120
6 x 12	12	24	36	48	60	72	84	96	108	120	132	144
6 x 14	14	28	42	56	70	84	98	112	126	140	154	168
6 x 16	16	32	48	64	80	96	112	128	14	160	176	192
6 x 18	18	36	54	72	90	108	126	144	162	180	198	216
6 x 20	20	40	60	80	100	120	140	160	180	200	220	240
6 x 22	22	44	66	88	110	132	154	176	198	220	242	264
6 x 24	24	48	72	96	120	144	168	192	216	240	264	288
8 x 8	10 2/3	21 1/3	32	42 2/3	53 1/3	64	74 2/3	85 1/3	96	106 2/3	117 1/3	128
8 x 10	13 1/3	26 2/3	40	53 1/3	66 2/3	80	93 1/3	106 2/3	120	133 1/3	146 2/3	160
8 x 12	16	32	48	64	80	96	112	128	144	160	176	192
8 x 14	18 2/3	37 1/3	56	74 2/3	93 1/3	112	130 2/3	149 1/3	168	186 2/3	205 1/3	224
8 x 16	21 1/3	42 2/3	64	85 1/3	106 2/3	128	149 1/3	170 2/3	192	213 1/3	234 2/3	256
8 x 18	24	48	72	96	120	144	168	192	216	240	264	288
8 x 20	26 2/3	53 1/3	80	106 2/3	133 1/3	160	186 2/3	213 1/3	240	266 2/3	293 1/3	320
8 x 22	29 1/3	58 2/3	88	117 1/3	146 2/3	176	205 1/3	234 2/3	264	293 1/3	322 2/3	352
8 x 24	32	64	96	128	160	192	224	256	288	320	352	384
10 x 10	16 2/3	33 1/3	50	66 2/3	83 1/3	100	116 2/3	133 1/3	150	166 2/3	183 1/3	200
10 x 12	20	40	60	80	100	120	140	160	180	200	220	240
10 x 14	23 1/3	46 2/3	70	93 1/3	116 2/3	140	163 1/3	186 2/3	210	233 1/3	256 2/3	280
10 x 16	26 2/3	53 1/3	80	106 2/3	133 1/3	160	186 2/3	213 1/3	240	266 2/3	293 1/3	320
10 x 18	30	60	90	120	150	180	210	240	270	300	330	360
10 x 20	33 1/3	66 2/3	100	133 1/3	166 2/3	200	233 1/3	266 2/3	300	333 1/3	366 2/3	400
10 x 22	36 2/3	73 1/3	110	146 2/3	183 1/3	220	256 2/3	293 1/3	330	366 2/3	403 1/3	440
10 x 24	40	80	120	160	200	240	280	320	360	400	440	480

INTERPRETING GRADE MARKS

(Courtesy of the Western Wood Products Association)

Most grade stamps, except those for rough lumber or heavy timbers, contain 5 basic elements:

a. WWPA certification mark. Certifies Association quality supervision. Ⓦ® is a registered trademark.

b. Mill identification. Firm name, brand or assigned mill number. WWPA can be contacted to identify an individual mill whenever necessary.

c. Grade designation. Grade name, number or abbreviation.

d. Species identification. Indicates species by individual species or species combination. New* species identification marks

for groups to which design values are assigned are:

DOUG. FIR-L   D FIR S   HEM FIR

SPF S   WEST WOODS   WEST CDR

e. Condition of seasoning. Indicates condition of seasoning at time of surfacing:

MC-15 — 15% maximum moisture content
S-DRY — 19% maximum moisture content
S-GRN — over 19% moisture content (unseasoned)

*Effective upon publication of *Western Lumber Grading Rules '91*, September 15, 1991.

Inspection Certificate

When an inspection certificate issued by the Western Wood Products Association is required on a shipment of lumber and specific grade marks are not used, the stock is identified by an imprint of the Association mark and the number of the shipping mill or inspector.

Grade Stamp Facsimiles

WWPA uses a set of marks similar to the randomly selected examples shown on the reverse side, to identify lumber graded under its supervision.

Species Combinations

The species groupings for dimension lumber products are shown left and explained in the second box on the reverse side. When alternative species combinations, as shown in the third box on the reverse side, are used for structural applications, design values are controlled by the species with the lowest strength value within the combination.

PLYWOOD VENEER GRADES

(Courtesy of the American Plywood Association)

 A Smooth, paintable. Not more than 18 neatly made repairs, boat, sled, or router type, and parallel to grain, permitted. Wood or synthetic repairs permitted. May be used for natural finish in less demanding applications.

 B Solid surface. Shims, sled or router repairs, and tight knots to 1 inch across grain permitted. Wood or synthetic repairs permitted. Some minor splits permitted.

 C Plugged Improved C veneer with splits limited to 1/8 inch width and knotholes or other open defects limited to 1/4 x 1/2 inch. Admits some broken grain. Wood or synthetic repairs permitted.

C Tight knots to 1-1/2 inch. Knotholes to 1 inch across grain and some to 1-1/2 inch if total width of knots and knotholes is within specified limits. Synthetic or wood repairs. Discoloration and sanding defects that do not impair strength permitted. Limited splits allowed. Stitching permitted.

D Knots and knotholes to 2-1/2 inch width across grain and 1/2 inch larger within specified limits. Limited splits allowed. Stitching permitted. Limited to Interior, Exposure 1 and Exposure 2 panels.

Standard Lumber Sizes
Nominal & Dressed
Based on WWPA Rules

(Courtesy of the Western Wood Products Association)

Product	Description	Nominal Size		Dressed Dimensions		
				Thicknesses and Widths		
		Thicknesses	**Widths**	**Surfaced Dry**	**Surfaced Unseasoned**	**Lengths**
Dimension	S4S Other surface combinations are available. See "Abbreviations" next page	2" 3 4	2" 3 4 5 6 8 10 12 over 12	1 1/2" 2 1/2 3 1/2 4 1/2 5 1/2 7 1/2 9 1/2 11 1/2 3/4 off Nominal	1 9/16" 2 9/16 3 9/16 4 5/8 5 5/8 7 1/2 9 1/2 11 1/2 1/2 off Nominal	6' and longer, generally shipped in multiples of 2'
Scaffold Plank	Rough Full Sawn or S4S (Usually shipped unseasoned)	1 1/4" & Thicker	8" and Wider	If Dressed refer to "DIMENSION" sizes.		6' and longer, generally shipped in multiples of 2'
Timbers	Rough or S4S (Shipped unseasoned)	5" and Larger		1/2" off Nominal (S4S) See 3.20 of WWPA Grading Rules for Rough		6' and longer, generally shipped in multiples of 2'

Product	Description	Nominal Size		Dressed Dimensions (Dry)		
		Thicknesses	**Widths**	**Thicknesses**	**Face Widths**	**Lengths**
Decking	2" (Single T&G)	2"	5" 6 8 10 12	1 1/2"	4" 5 6 3/4 8 3/4 10 3/4	6' and longer, generally shipped in multiples of 2'
	3" and 4" (Double T&G)	3 4	6	2 1/2 3 1/2	5 1/4	
Flooring	(D&M), (S2S & CM)	3/8 1/2 5/8 1 1 1/4 1 1/2	2 3 4 5 6	5/16 7/16 9/16 3/4 1 1 1/4	1 1/8 2 1/8 3 1/8 4 1/8 5 1/8	4' and longer, generally shipped in multiples of 2'
Ceiling & Partition	(S2S & CM)	3/8 1/2 5/8 3/4	3 4 5 6	5/16 7/16 9/16 11/16	2 1/8 3 1/8 4 1/8 5 1/8	4' and longer, generally shipped in multiples of 2'
Factory & Shop Lumber	S2S	1 (4/4) 1 1/4 (5/4) 1 1/2 (6/4) 1 3/4 (7/4) 2 (8/4) 2 1/2 (10/4) 3 (12/4) 4 (16/4)	5" and wider (except 4" and wider in 4/4 No. 1 Shop and 4/4 No. 2 Shop, and 2" and wider in 5/4 & Thicker No. 3 Shop)	3/4 (4/4) 1 5/32 (5/4) 1 13/32 (6/4) 1 19/32 (7/4) 1 13/16 (8/4) 2 3/8 (10/4) 2 3/4 (12/4) 3 3/4 (16/4)	Usually sold random width	6' and longer, generally shipped in multiples of 2'

MINIMUM ROUGH SIZES. Thicknesses and Widths, Dry or Unseasoned, All Lumber. 80% of the footage in a shipment shall be at least 1/8" thicker than the standard surfaced size, the remaining 20% at least 3/32" thicker than the surfaced size. Widths shall be at least 1/8" wider than standard surfaced widths.

When specified to be full sawn, lumber may not be manufactured to a size less than the size specified.

ABBREVIATIONS

Abbreviated descriptions appearing in the table are defined at right.

S1S — Surfaced one side.
S2S — Surfaced two sides.
S4S — Surfaced four sides.
S1S1E — Surfaced one side, one edge.
S1S2E — Surfaced one side, two edges.
CM — Center matched.
D & M — Dressed and matched.
T&G — Tongued and grooved.
Rough Full Sawn — Unsurfaced lumber cut to full specified size.

Product	Description	Nominal Size		Dressed Dimensions		
		Thicknesses	Widths	Thicknesses	Widths	Lengths
Selects & Commons	S1S,S2S,S4S,S1S1E,S1S2E	4/4 5/4 6/4 7/4 8/4 9/4 10/4 11/4 12/4 16/4	2" 3 4 5 6 7 8 and wider	3/4" 1 5/32 1 13/32 1 19/32 2 13/16 2 3/32 2 3/8 2 9/16 2 3/4 3 3/4	1 1/2" 2 1/2 3 1/2 4 1/2 5 1/2 6 1/2 3/4 off Nominal	6' and longer in multiples of 1' except Douglas Fir and Larch Selects shall be 4' and longer with 3% of 4' and 5' permitted.
Finish & Boards	S1S,S2S,S4S,S1SE,S1S2E Only these thicknesses apply to Alternate Board Grades.	3/8" 1/2 5/8 3/4 1 1 1/4 1 1/2 1 3/4 2 2 1/2 3 3 1/2 4	2 3 4 5 6 7 8 and wider	5/16 7/16 9/16 5/8 3/4 1 1 1/4 1 3/8 1 1/2 2 2 1/2 3 3 1/2	1 1/2 2 1/2 3 1/2 4 1/2 5 1/2 6 1/2 3/4 off Nominal	3' and longer in multiples of 1' In Superior grade, 3% of 3' and 4' and 7% of 3' to 6' are permitted. In Prime Grade, 20% of 3' to 6' is permitted.
Rustic & Drop Siding	(D&M) If 3/8" or 1/2" T&G specified, same over-all widths apply. (Shiplapped, 3/8" or 1/2" lap.)	1	6 8 10 12	23/32	5 3/8 7 1/8 9 1/8 11 1/8	4' and longer in multiples of 1'
Paneling & Siding	T&G or Shiplap	1	6 8 10 12	23/32	5 7/16 7 1/8 9 1/8 11 1/8	4' and longer in multiples of 1'
Ceiling & Partition	T&G	5/8 1	4 6	9/16 23/32	3 3/8 5 3/8	4' and longer in multiples of 1'
Bevel Siding	Bevel or Bungalow Siding Western Red Cedar Bevel Siding available in 1/2", 5/8", 3/4" nominal thicknesses. Corresponding thick edge is 15/32", 9/16" and 3/4". Widths for 8" and wider, 1/2" off nominal.	1/2 3/4	4 5 6 8 10 12	15/32 butt, 3/16 tip 3/4 butt, 3/16 tip	3 1/2 4 1/2 5 1/2 7 1/4 9 1/4 11 1/4	3' and longer in multiples of 1' 3' and longer in multiples of 1'
				Surfaced Dry Green	**Surfaced** Dry Green	
Stress-Rated Boards	S1S,S2S,S4S,S1S1E,S1S2E	1 1 1/4 1 1/2	2 3 4 5 6 7 8 and wider	3/4 25/32 1 1 1/32 1 1/4 1 9/32	1 1/2 1 9/16 2 1/2 2 9/16 3 1/2 3 9/16 4 1/2 4 5/8 5 1/2 5 5/8 6 1/2 6 5/8 3/4 off 1/2 off Nominal Nominal	6' and longer in multiples of 1'

Facsimiles of Typical Grade Stamps

Perpetuating America's Forests for Products and the Environment

Dimension Grades

12 WWP® 2 S·DRY D/FIR

12 WWP® CONST S·DRY SPFᔆ

12 WWP® STAND &BTR S·DRY D/FIRˢ

Glued Products

12 WWP® STUD S·DRY STUD USE ONLY CERT GLUED JNTS WEST WOODS

12 WWP® 1 S·DRY CERT EXT JNTS HEM FIR

Finish Grade — Graded Under WCLIB Rules

12 WWP® C&BTR VG S·DRY WCLB RULES D/FIR

Cedar Grades

12 WWP® CLEAR VG HEART MC15 WR CDR

12 WWP® A MC15 WEST CDR

Commons

12 WWP® 2&BTR COM S·DRY ⫴

12 WWP® 4 COM S·DRY (ES)

12 WWP® STERLING S·DRY iWp

Machine Stress-Rated Products

MACHINE RATED WWP® 12 S·DRY D/FIR 1650 Fb 1.5E

MACHINE RATED WWP® 12 S·DRY D/FIR 1650Fb 1020Ft 1.5E

Finish & Select Grades

12 WWP® C&BTR SEL MC 15 ⫴

12 WWP® PRIME MC 15 D/FIR

12 WWP® D SEL MC 15 SP

Decking

12 WWP® SEL DECK MC 15 INC CDR

12 WWP® PATIO 1 S·DRY ⫴

Species Identification

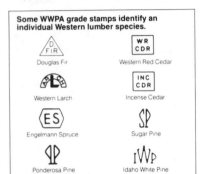

Some WWPA grade stamps identify an individual Western lumber species.

D/FIR Douglas Fir	WR CDR Western Red Cedar
Western Larch	INC CDR Incense Cedar
(ES) Engelmann Spruce	SP Sugar Pine
⫴ Ponderosa Pine	iWp Idaho White Pine

A number of Western lumber species have similar performance properties and are marketed with a common species designation. These species groupings are used for lumber to which design values are assigned.

DOUG. FIR-L Douglas Fir and Larch	D/FIRˢ Douglas Fir South*
HEM FIR California Red Fir, Grand Fir Noble Fir, Pacific Silver Fir, White Fir and Western Hemlock	SPFˢ Engelmann and Sitka Spruce Lodgepole Pine
WEST WOODS Alpine Fir, Ponderosa Pine, Sugar Pine, Idaho White Pine and Mountain Hemlock, plus any of the species in the other groupings except Western Cedars	WEST CDR Incense, Western Red, Port Orford and Alaska Cedar

*Lumber manufactured from Doulas Fir grown in Arizona, Colorado, Nevada, New Mexico and Utah.

Because of timber stand composition, some mills market additional species combinations.

ES LP Engelmann Spruce, Lodgepole Pine	⫴SP Ponderosa Pine, Sugar Pine
W W White Woods (used only for boards any true firs, spruces, hemlocks or pines)	A-F HEM FIR Alpine Fir, Hem-Fir
	ES-AF Engelmann Spruce, Alpine Fir
PP-LP Ponderosa Pine, Lodgepole Pine	ES LP AF Engelmann Spruce-Alpine Fir-Lodgepole Pine

(Courtesy of the Western Wood Products Association)

The *APA EWS* trademark appears only on products manufactured by *American Wood Systems* members. The mark signifies that the manufacturer is committed to a rigorous program of quality verification and testing and that products are manufactured in conformance with ANSI Standard A190.1, American National Standard for Structural Glued Laminated Timber.

The *APA EWS* mark includes the same information found in other engineered wood product trademarks. Product identification required in ANSI Standard A190.1 appears on all *APA EWS* trademarked glulams.

ACCEPTANCES

American Wood Systems is recognized by all major model building codes under CABO National Evaluation Service Committee Report NER— QA397.

(1) Indicates structural use:
B-Simple span bending member.
C-Compression member.
T-Tension member.
CB-Continuous or cantilevered span bending member.
(2) Mill number.
(3) Identification of ANSI Standard A190.1, Structural Glued Laminated Timber.

(4) Code recognition of *American Wood Systems* as a quality assurance agency for glued structural members.
(5) Applicable laminating specification.
(6) Applicable combination number.
(7) Species of lumber used.
(8) Designates appearance grade. INDUSTRIAL, ARCHITECTURAL, PREMIUM.

(Courtesy of the American Plywood Association)

GUIDE TO APA PERFORMANCE RATED PANELS
Trademarks Shown Are Typical Facsimiles

APA RATED SHEATHING

Specially designed for subflooring, wall sheathing and roof sheathing, but also used for broad range of other construction, industrial and do-it-yourself applications. Can be manufactured as conventional plywood, as a composite, or as oriented strand board. SPAN RATINGS: 12/0, 16/0, 20/0, 24/0, 24/16, 32/16, 40/20, 48/24, WALL-16 oc, WALL-24 oc. EXPOSURE DURABILITY CLASSIFICATIONS: Exterior, Exposure 1, Exposure 2. COMMON THICKNESSES: 5/16, 3/8, 7/16, 15/32, 1/2, 19/32, 5/8, 23/32, 3/4.

```
_____APA_____
RATED SHEATHING
24/16  7/16 INCH
  SIZED FOR SPACING
   EXPOSURE 1
_____000_____
NER-QA397  PRP-108
  HUD-UM-40C
```

APA RATED STURD-I-FLOOR

Specially designed as combination subfloor-underlayment. Provides smooth surface for application of carpet and pad, and possesses high concentrated and impact load resistance. Can be manufactured as conventional plywood, as a composite, or as oriented strand board. Available square edge or tongue-and-groove. SPAN RATINGS: 16, 20, 24, 32, 48. EXPOSURE DURABILITY CLASSIFICATIONS: Exterior, Exposure 1, Exposure 2. COMMON THICKNESSES: 19/32, 5/8, 23/32, 3/4, 1, 1-1/8.

```
_____APA_____
RATED STURD-I-FLOOR
20 oc  19/32 INCH
  SIZED FOR SPACING
   EXPOSURE 1
_____000_____
NER-QA397  PRP-108
  HUD-UM-40C
```

APA STRUCTURAL I RATED SHEATHING

Unsanded grade for use where cross-panel strength and stiffness or shear properties are of maximum importance, such as panelized roofs, diaphragms and shear walls. Can be manufactured as conventional plywood, a composite, or oriented strand board. All plies in Structural I plywood panels are special improved grades and panels marked PS 1 are limited to Group 1 species. (Structural II plywood panels are also provided for, but rarely manufactured. Application recommendations for Structural II plywood are identical to those for APA RATED SHEATHING plywood.) SPAN RATINGS: 20/0, 24/0, 24/16, 32/16, 40/20, 48/24. EXPOSURE DURABILITY CLASSIFICATIONS: Exterior, Exposure 1. COMMON THICKNESSES: 5/16, 3/8, 7/16, 15/32, 1/2, 19/32, 5/8, 23/32, 3/4.

```
_____APA_____
RATED SHEATHING
STRUCTURAL I
32/16  15/32 INCH
  SIZED FOR SPACING
   EXPOSURE 1
_____000_____
PS 1-83  C-D
NER-QA397  PRP-108
```

```
_____APA_____
RATED SHEATHING
32/16  15/32 INCH
  SIZED FOR SPACING
   EXPOSURE 1
_____000_____
STRUCTURAL I RATED
DIAPHRAGMS-SHEAR WALLS
PANELIZED ROOFS
NER-QA397  PRP-108
```

APA RATED SIDING

For exterior siding, fencing, etc. Can be manufactured as conventional veneered plywood, as a composite or as an overlaid oriented strand board siding. Both panel and lap siding available. Special surface treatment such as V-groove, shallow channel groove, deep groove (such as APA Texture 1-11), kerfed groove, brushed, rough sawn and texture-embossed (MDO). Span Rating (stud spacing for siding qualified for APA Sturd-I-Wall applications) and face grade classification (for veneer-faced siding) indicated in trademark. EXPOSURE DURABILITY CLASSIFICATION: Exterior. COMMON THICKNESSES: 11/32, 3/8, 15/32, 1/2, 19/32, 5/8.

```
_____APA_____
RATED SIDING
24 oc  15/32 INCH
  SIZED FOR SPACING
   EXTERIOR
_____000_____
NER-QA397  PRP-108
  HUD-UM-40C
```

```
_____APA_____
RATED SIDING
303-18-S/W
16 oc  11/32 INCH
      GROUP 1
  SIZED FOR SPACING
   EXTERIOR
_____000_____
PS 1-83  FHA-UM-64
NER-QA397  PRP-108
```

NOTE: Specify Performance Rated Panels by thickness and Span Rating. Span Ratings are based on panel strength and stiffness. Since these properties are a function of panel composition and configuration as well as thickness, the same Span Rating may appear on panels of different thicknesses. Similarly, panels of the same thickness may be marked with different Span Ratings.

(Courtesy of the American Plywood Association)

APA SANDED & TOUCH-SANDED PLYWOOD

Panels with B-grade or better veneer faces are sanded smooth in manufacture to fulfill the requirements of their intended end use – applications such as cabinets, shelving, furniture, built-ins, etc. These sanded panel grades – identified by face and back veneer – are widely used in a multitude of construction, industrial and do-it-yourself applications where strength and stiffness combined with premium appearance or surface smoothness is required.

When sanded plywood is manufactured with a special plugged inner-ply construction to resist dents and punctures from concentrated loads, it also may be used for nonstructural floor underlayment as a substrate for vinyl or other resilient flooring. The smooth, sanded surface and excellent dimensional stability makes these panels ideal for floor underlayment applications.

Touch-sanded panels – Underlayment, C-C Plugged, and C-D Plugged – are sanded only for "sizing" to assure uniform panel thickness.

Typical sanded and touch-sanded panel applications are described at right. For complete application recommendations for sanded plywood in industrial applications, write APA. Industrial use guides are available on **Materials Handling**, Form M200; **Slave Pallets**, Form S225; and **Transport Equipment**, Form G210.

NOTE: Exterior sanded panels, C-C Plugged, C-D Plugged and Underlayment grades can also be manufactured in Structural I (all plies limited to Group 1 species).

GUIDE TO APA-SANDED [TOUCH-SANDED PLYWOOD
Trademarks Shown Are Typical Facsimiles

APA A-A

Use where appearance of both sides is important for interior applications such as built-ins, cabinets, furniture, partitions; and exterior applications such as fences, signs, boats, shipping containers, tanks, ducts, etc. Smooth surfaces suitable for painting. EXPOSURE DURABILITY CLASSIFICATIONS: Interior, Exposure 1, Exterior. COMMON THICKNESSES: 1/4, 11/32, 3/8, 15/32, 1/2, 19/32, 5/8, 23/32, 3/4.

A·A·G·1·EXPOSURE1·APA·000·PS1-83

APA A-C

For use where appearance of one side is important in exterior applications such as soffits, fences, structural uses, boxcar and truck linings, farm buildings, tanks, trays, commercial refrigerators, etc.[1] EXPOSURE DURABILITY CLASSIFICATION: Exterior. COMMON THICKNESSES: 1/4, 11/32, 3/8, 15/32, 1/2, 19/32, 5/8, 23/32, 3/4.

APA

A-C   GROUP 1

EXTERIOR

000
PS 1-83

APA B-B

Utility panels with two solid sides. EXPOSURE DURABILITY CLASSIFICATIONS: Interior, Exposure 1, Exterior. COMMON THICKNESSES: 1/4, 11/32, 3/8, 15/32, 1/2, 19/32, 5/8, 23/32, 3/4.

B-B·G-2·EXPOSURE1-APA·000·PS1-83

APA A-B

For use where appearance of one side is less important but where two solid surfaces are necessary. EXPOSURE DURABILITY CLASSIFICATIONS: Interior, Exposure 1, Exterior. COMMON THICKNESSES: 1/4, 11/32, 3/8, 15/32, 1/2, 19/32, 5/8, 23/32, 3/4.

A-B·G-1·EXPOSURE1-APA·000·PS1-83

APA A-D

For use where appearance of only one side is important in interior applications, such as paneling, built-ins, shelving, partitions, flow racks, etc.[1] EXPOSURE DURABILITY CLASSIFICATIONS: Interior, Exposure 1. COMMON THICKNESSES: 1/4, 11/32, 3/8, 15/32, 1/2, 19/32, 5/8, 23/32, 3/4.

APA

A-D   GROUP 1

EXPOSURE 1

000
PS 1-83

(Courtesy of the American Plywood Association)

APA SANDED & TOUCH-SANDED PLYWOOD (Continued)

APA B-C

Utility panel for farm service and work buildings, boxcar and truck linings, containers, tanks, agricultural equipment, as a base for exterior coatings and other exterior uses.[1] EXPOSURE DURABILITY CLASSIFICATION: Exterior. COMMON THICKNESSES: 1/4, 11/32, 3/8, 15/32, 1/2, 19/32, 5/8, 23/32, 3/4.

APA C-C PLUGGED[3]

For use as an underlayment over structural subfloor, refrigerated or controlled atmosphere storage rooms, pallet fruit bins, tanks, boxcar and truck floors and linings, and other exterior applications. Provides smooth surface for application of carpet and pad, and possesses high concentrated and impact load resistance. Touch-sanded. For areas to be covered with resilient flooring, specify panels with "sanded face." EXPOSURE DURABILITY CLASSIFICATION: Exterior. COMMON THICKNESSES[4]: 11/32, 3/8, 1/2, 19/32, 5/8, 23/32, 3/4.

APA B-D

Utility panel for backing, sides of built-ins, industry shelving, slip sheets, separator boards, bins and other interior or protected applications.[1] EXPOSURE DURABILITY CLASSIFICATIONS: Interior, Exposure 1. COMMON THICKNESSES: 1/4, 11/32, 3/8, 15/32, 1/2, 19/32, 5/8, 23/32, 3/4.

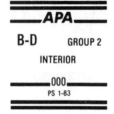

APA C-D PLUGGED

For open soffits, built-ins, cable reels, walkways, separator boards and other interior or protected applications. Not a substitute for Underlayment or APA Rated Sturd-I-Floor as it lacks their puncture resistance. Touch-sanded. EXPOSURE DURABILITY CLASSIFICATIONS: Interior, Exposure 1. COMMON THICKNESSES: 3/8, 1/2, 19/32, 5/8, 23/32, 3/4.

APA UNDERLAYMENT

For application over structural subfloor. Provides smooth surface for application of carpet and pad, and possesses high concentrated and impact load resistance. Touch sanded. For areas to be covered with resilient flooring, specify panels with "sanded face."[2] EXPOSURE DURABILITY CLASSIFICATIONS: Interior, Exposure 1. COMMON THICKNESSES[4]: 1/4, 11/32, 3/8, 1/2, 19/32, 5/8, 23/32, 3/4.

(1) For nonstructural floor underlayment, or other applications requiring improved inner-ply construction, specify panels marked either "plugged inner plies" (also may be designated plugged crossbands under face or plugged crossbands or core); or "meets underlayment requirements."

(2) Also available in Underlayment A-C or Underlayment B-C grades, marked either "touch sanded" or "sanded face."

(3) Also may be designated APA Underlayment C-C Plugged.

(4) Underlayment and C-C Plugged panels 1/2" and thicker are designated by Span Rating rather than species group number in trademark.

APA trademarked specialty grades include panels designed for specific applications (e.g., B-B Plyform for concrete forming, Marine), or with special surface treatments for applications with specific performance requirements (e.g., Medium and High Density Overlay, Plyron).

Complete concrete forming design data are contained in APA's *Concrete Forming*, Form V345. For additional information on High and Medium Density Overlay plywood, write for *HDO/MDO Plywood*, Form B360.

GUIDE TO APA SPECIALTY PANELS Trademarks Shown Are Typical Facsimiles

APA HIGH DENSITY OVERLAY (HDO)

Plywood panel manufactured with a hard, semi-opaque resin-fiber overlay on both sides. Extremely abrasion resistant and ideally suited to scores of punishing construction and industrial applications, such as concrete forms, industrial tanks, work surfaces, signs, agricultural bins, exhaust ducts, etc. Also available with skid-resistant screen-grid surface and in Structural I. EXPOSURE DURABILITY CLASSIFICATION: Exterior. COMMON THICKNESSES: 3/8, 1/2, 5/8, 3/4.

HDO · A-A · G-1 · EXT-APA · 000 · PS1-83

APA MARINE

Specially designed plywood panel made only with Douglas-fir or western larch, solid jointed cores, and highly restrictive limitations on core gaps and face repairs. Ideal for boat hulls and other marine applications. Also available with HDO or

APA SPECIALTY PANELS

MDO faces. EXPOSURE DURABILITY CLASSIFICATION: Exterior. COMMON THICKNESSES: 1/4, 3/8, 1/2, 5/8, 3/4.

MARINE · A-A · EXT-APA · 000 · PS1-83

APA B-B PLYFORM CLASS I

APA proprietary concrete form panels designed for high reuse. Sanded both sides and mill-oiled unless otherwise specified. Special restrictions on species. Also available in HDO for very smooth concrete finish, in Structural I, and with special overlays. EXPOSURE DURABILITY CLASSIFICATION: Exterior. COMMON THICKNESSES: 19/32, 5/8, 23/32, 3/4.

APA
PLYFORM
B-B CLASS I
EXTERIOR
000
PS 1-83

APA MEDIUM DENSITY OVERLAY (MDO)

Plywood panel manufactured with smooth, opaque, resin-treated fiber overlay providing ideal base for paint on one or both sides. Excellent material choice for shelving, factory work surfaces, paneling, built-ins, signs and numerous other construction and industrial applications. Also available as APA Rated Siding-303 with texture-embossed or smooth surface on one side only and in Structural I. EXPOSURE DURABILITY

CLASSIFICATION: Exterior. COMMON THICKNESSES: 11/32, 3/8, 15/32, 1/2, 19/32, 5/8, 23/32, 3/4.

APA
M. D. OVERLAY
GROUP 1
EXTERIOR
000
PS 1-83

APA DECORATIVE

Rough sawn, brushed, grooved, or other faces. For paneling, interior accent walls, built-ins, counter facing, exhibit displays, etc. Made by some manufacturers in Exterior for exterior siding, gable ends, fences and other exterior applications. Use recommendations for Exterior panels vary; check with the manufacturer. EXPOSURE DURABILITY CLASSIFICATIONS: Interior, Exposure 1, Exterior. COMMON THICKNESSES: 5/16, 3/8, 1/2, 5/8.

APA
DECORATIVE
GROUP 2
INTERIOR
000
PS 1-83

APA PLYRON

APA proprietary plywood panel with hardboard face on both sides. Faces tempered, untempered, smooth or screened. For countertops, shelving, cabinet doors, concentrated load flooring, etc. EXPOSURE DURABILITY CLASSIFICATIONS: Interior, Exposure 1, Exterior. COMMON THICKNESSES: 1/2, 5/8, 3/4.

PLYRON · EXPOSURE1-APA · 000

(Courtesy of the American Plywood Association)

Appendix E
Hardware Information

NAIL HEAD AND POINT TYPES

(Courtesy of Maze Nails)

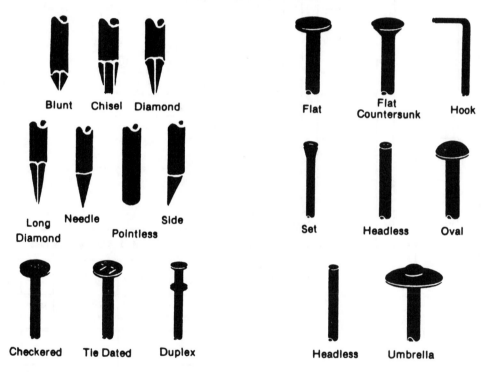

| Blunt | Chisel | Diamond | | Flat | Flat Countersunk | Hook |

| Long Diamond | Needle | Pointless | Side | | Set | Headless | Oval |

| Checkered | Tie Dated | Duplex | | Headless | Umbrella |

COMMON NAIL SIZES (Courtesy of Maze Nails)

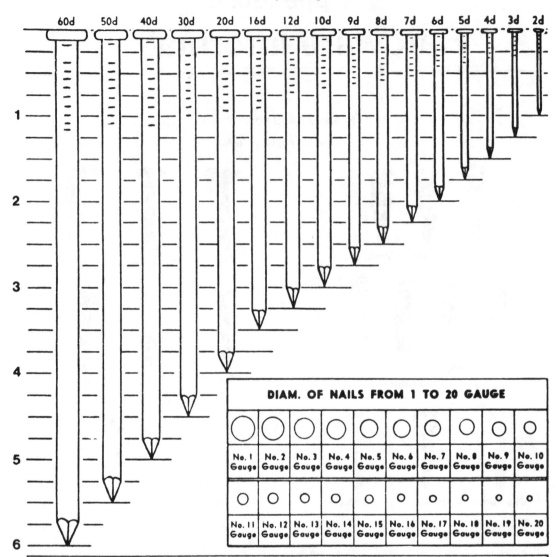

DIAM. OF NAILS FROM 1 TO 20 GAUGE

No. 1 Gauge	No. 2 Gauge	No. 3 Gauge	No. 4 Gauge	No. 5 Gauge	No. 6 Gauge	No. 7 Gauge	No. 8 Gauge	No. 9 Gauge	No. 10 Gauge
No. 11 Gauge	No. 12 Gauge	No. 13 Gauge	No. 14 Gauge	No. 15 Gauge	No. 16 Gauge	No. 17 Gauge	No. 18 Gauge	No. 19 Gauge	No. 20 Gauge

COMMON NAILS REFERENCE TABLE

SIZE	LENGTH AND GAUGE	DIAMETER HEAD	APPROX. NO. TO POUND	SIZE	LENGTH AND GAUGE	DIAMETER HEAD	APPROX. NO TO POUND
2d	1 inch..........No. 15	$^{11}/_{64}$	845	10d	3 inch..........No. 9	$^{9}/_{16}$	65
3d	$1^{1}/_{4}$ inch..........No. 14	$^{13}/_{64}$	540	12d	$3^{1}/_{4}$ inch..........No. 9	$^{9}/_{16}$	60
4d	$1^{1}/_{2}$ inch.........No. $12^{1}/_{2}$	$^{1}/_{4}$	290	16d	$3^{1}/_{2}$ inch..........No. 8	$^{11}/_{32}$	45
5d	$1^{3}/_{4}$ inch.........No. $12^{1}/_{2}$	$^{1}/_{4}$	250	20d	4 inch..........No. 6	$^{13}/_{32}$	30
6d	2 inch.........No. $11^{1}/_{2}$	$^{17}/_{64}$	165	30d	$4^{1}/_{2}$ inch..........No. 5	$^{7}/_{16}$	20
7d	$2^{1}/_{4}$ inch ....No. $11^{1}/_{2}$	$^{17}/_{64}$	150	40d	5 inch..........No. 4	$^{15}/_{32}$	17
8d	$2^{1}/_{2}$ inch.........No. $10^{1}/_{4}$	$^{9}/_{32}$	100	50d	$5^{1}/_{2}$ inch..........No. 3	$^{1}/_{2}$	13
9d	$2^{3}/_{4}$ inch..........No. $10^{1}/_{4}$	$^{9}/_{32}$	90	60d	6 inch..........No. 2	$^{17}/_{32}$	10

NAIL TYPES (Courtesy of Maze Nails)

STORMGUARD® NAILS FOR EXTERIOR APPLICATIONS
(Hot-dipped zinc-coated twice in molten zinc)

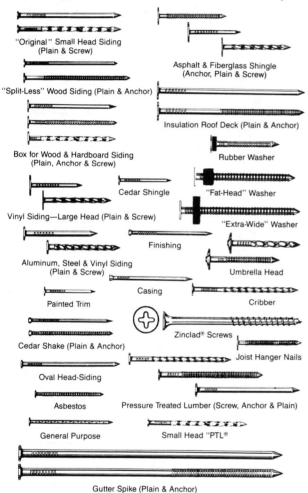

"Original" Small Head Siding (Plain & Screw)

Asphalt & Fiberglass Shingle (Anchor, Plain & Screw)

"Split-Less" Wood Siding (Plain & Anchor)

Insulation Roof Deck (Plain & Anchor)

Box for Wood & Hardboard Siding (Plain, Anchor & Screw)

Rubber Washer

Cedar Shingle

"Fat-Head" Washer

Vinyl Siding—Large Head (Plain & Screw)

"Extra-Wide" Washer

Aluminum, Steel & Vinyl Siding (Plain & Screw)

Finishing

Umbrella Head

Casing

Painted Trim

Cribber

Cedar Shake (Plain & Anchor)

Zinclad® Screws

Joist Hanger Nails

Oval Head-Siding

Pressure Treated Lumber (Screw, Anchor & Plain)

Asbestos

General Purpose

Small Head "PTL®"

Gutter Spike (Plain & Anchor)

INTERIOR & OTHER NAILS

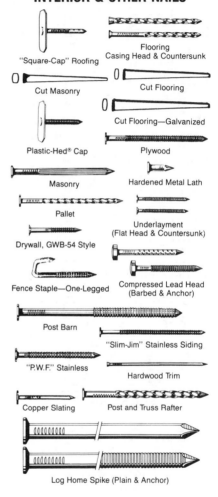

"Square-Cap" Roofing

Flooring Casing Head & Countersunk

Cut Masonry

Cut Flooring

Plastic-Hed® Cap

Cut Flooring—Galvanized

Masonry

Plywood

Hardened Metal Lath

Pallet

Underlayment (Flat Head & Countersunk)

Drywall, GWB-54 Style

Fence Staple—One-Legged

Compressed Lead Head (Barbed & Anchor)

Post Barn

"Slim-Jim" Stainless Siding

"P.W.F." Stainless

Hardwood Trim

Copper Slating

Post and Truss Rafter

Log Home Spike (Plain & Anchor)

NAIL TYPES (Continued)

"SPLIT-LESS"® WOOD SIDING NAILS

Stock Numbers		Length			Nails
Plain	Ring	Inch	Gauge	Head	Per Lb.
S-225	S-225-A	2	14	3/16"	283
S-226	S-226-A	2 1/4	14	3/16"	248
S-227	S-227-A	2 1/2	13	7/32"	189
S-228	S-228-A	2 3/4	13	7/32"	171
S-229	S-229-A	3	13	7/32"	153
S-2290	S-2290-A	3 1/4	12	1/4"	113
S-2291	S-2291-A	3 1/2	12	1/4"	104
S-2292	S-2292-A	4	12	1/4"	89
S-2293	— —	4 1/2	12	1/4"	62

Specially designed for split-prone woods. Slim, sturdy shank and special blunt point virtually eliminate splits. Double-dipped in molten zinc. Checkered head enhances paint adhesion on color-matched nails. High count greatly reduces per nail cost. Especially popular for sidings scheduled for finishing with paints or pigmented stains. For jobs that will be left natural or treated with a clear or semi-transparent finish, we recommend our stainless steel "Slim-Jim" nails.

BOX NAILS FOR WOOD & HARDBOARD SIDING

Stock Numbers			Length			Nails
Plain	Ring	Spiral	Inch	Head		Per Lb.
S-202	S-202-A	— —	1 1/4	3/16"		442
S-203	S-203-A	— —	1 1/2	7/32"		392
S-205	S-205-A	S-205-S	2	17/64"		194
S-206	S-206-A	— —	2 1/4	19/64"		172
S-207	S-207-A	S-207-S	2 1/2	19/64"		123
S-209	S-209-A	S-209-S	3	19/64"		103
S-2090	S-2090-A	— —	3 1/4	9/16"		93
S-2091	S-2091-A	S-2091-S	3 1/2	9/16"		80
S-2092	S-2092-A	— —	4	11/32"		53

Sturdy shank drives easily. Double-dipped in molten zinc. Checkered head enhances paint adhesion on color-matched nails. Type recommended by FHA for plywood sheathing. Large head preferred by many hardboard siding applicators and manufacturers. Meets the specifications of the American Hardboard Association.

"ORIGINAL" SMALL HEAD SIDING NAILS

Stock Numbers		Length			Nails
Plain	Spiral	Inch	Gauge	Head	Per Lb.
S-255	S-255-S	2	12	3/16"	181
S-257	S-257-S	2 1/2	12	3/16"	146
S-258	S-258-S	2 3/4	12	3/16"	136
S-259	S-259-S	3	11	7/32"	98
S-2590	S-2590-S	3 1/4	11	7/32"	90
S-2591	S-2591-S	3 1/2	11	7/32"	82
S-2592	S-2592-S	4	11	7/32"	73

This nail, made originally to the specifications of the early producers of hardboard siding, now has a multitude of other uses with other types of siding, trim, fencing, decking, etc. It's a great all-around nail. Extra stiff shank for easy driving. Self-seating head. Double-dipped in molten zinc. Refer to Wood Siding Box Nails for specifications requiring a larger diameter head.

OVAL HEADED SIDING NAILS

Stock Numbers	Length			Nails
Plain Shank	Inch	Gauge	Head	Per Lb.
S70	2 1/4	12 1/2	7/32"	159

This plain shank nail has many uses with siding and trim. Many prefer the "finished" look of an oval head. Double dipped in molten zinc.

CEDAR SHAKE SIDING NAILS

Stock Numbers		Length			Nails
Plain	Ring	Inch	Gauge	Head	Per Lb.
S-233	S-233-A	1 1/2	14	1/8"	402
S-234	S-234-A	1 3/4	14	1/8"	346
S-235	S-235-A	2	14	1/8"	310
— —	*S-285-S	2	12 1/2	3/16"	209
S-237	S-237-A	2 1/2	13	9/64"	196

Ideally suited for shake siding applications. Slim shank, blunt point, small checkered brad head. Double-dipped in molten zinc. Available painted. *HDF Nail - Type recommended over high density fiberboard sheathing.

NAILS FOR ALUMINUM, STEEL AND VINYL SIDING

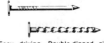

Stock Numbers		Length			Nails
Plain	Spiral	Inch	Gauge	Head	Per Lb.
S-262	— —	1 1/4	12	5/16"	270
S-263	S-263-S	1 1/2	12	5/16"	228
S-265	S-265-S	2	12	5/16"	178
S-267	— —	2 1/2	11 1/2	19/64"	123
S-269	— —	3	11 1/2	19/64"	103

Easy driving. Double-dipped zinc coating completely compatible with aluminum. Also popular for application of steel and vinyl siding. Economical too — — cost less per nail than aluminum and drive far better.

LARGE HEAD FOR VINYL

Stock Numbers		Length			Nails
Plain	Spiral	Inch	Gauge	Head	Per Lb.
S-103	— —	1 1/2	11	7/16"	180
S-104	— —	1 3/4	11	7/16"	156
S-105	S-105-S	2	11	7/16"	136
S-107	— —	2 1/2	11	7/16"	112
S-109	— —	3	10	7/16"	72

Larger 7/16" heads are preferred by many applicators. Double hot-dipped zinc-coated.

MINERAL FIBER (ASBESTOS) SIDING NAILS

Stock Numbers	Length			Nails
Ring Shank	Inch	Gauge	Head	Per Lb.
S-211-A	1	12 1/2	3/16"	387
S-212-A	1 1/4	12 1/2	3/16"	304
S-213-A	1 7/16	12 1/2	3/16"	262
S-214-A	1 3/4	12 1/2	3/16"	209
S-215-A	2	12 1/2	3/16"	209
S-217-A	2 1/2	12 1/2	3/16"	160

Originally designed for asbestos siding, but also useful in many other applications where extra holding power is required. Available in wide range of colors. Heads are striated or checkered (depending on length) to enhance paint adhesion. Double-dipped in molten zinc.

"GENERAL PURPOSE" NAILS

Stock Numbers	Length			Nails
Inch		Gauge	Head	Per Lb.
S-240	7/8	14	3/32"	665
S-245	2	12 1/2	3/16"	210
S-247	2 1/2	12 1/2	3/16"	172

These general purpose nails are popular for a wide range of uses, where good driving and rust resistance are important. The 7/8" length is ideal for screen trim, lattice work, etc., while the longer lengths are widely used for wood sidings, facia, soffits, trim, etc. Checkered heads enhance paint adhesion. Double-dipped in molten zinc.

CRIBBER NAILS

Stock Numbers	Length			Nails
Spiral Shank	Inch	Gauge	Head	Per Lb.
S-277-S	2 1/2	11	3/8"	106
S-279-S	3	11	3/8"	86

Specially designed for corn crib erection and repair. Double-dipped in molten zinc.

— BE SAFE —
SAFETY GLASSES ARE RECOMMENDED WHEN DRIVING ANY NAILS.

ASPHALT & FIBERGLASS SHINGLE NAILS

Stock Numbers Plain	Stock Numbers Ring	Stock Numbers Spiral	Length Inch	Head	Nails Per Lb.
R-100	R-100-A	R-100-S	7/8	7/16"	272
R-101	R-101-A	——	1	7/16"	250
R-102	R-102-A	——	1 1/4	7/16"	202
R-103	R-103-A	R-103-S	1 1/2	7/16"	180
R-104	R-104-A	R-104-S	1 3/4	7/16"	156
R-105	R-105-A	R-105-S	2	7/16"	136
R-107	R-107-A	——	2 1/2	7/16"	112

Designed for the secure application of asphalt and fiberglass shingles to solid wood sheathing, old roofing, decking, and plywood. High holding ring shanks recommended for decking and plywood applications. Large 7/16" head provides ample hold down surface. Double-dipped in molten zinc. 11 gauge shanks.

INSULATION ROOF DECK NAILS (LONG ASPHALT SHINGLE NAILS)

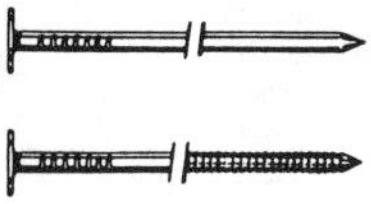

Stock Numbers Plain	Stock Numbers Ring	Length Inch	Gauge	Head	Nails Per Lb.
R-159	R-159-A	3	10	7/16"	72
R-1591	R-1591-A	3 1/2	10	7/16"	62
R-1592	R-1592-A	4	9	7/16"	46
R-1593	R-1593-A	4 1/2	9	7/16"	40

Well suited to decking applications. Large head prevents pull through. Extra long nails accommodate rigid foam insulation in roofing applications. Double-dipped in molten zinc.

CEDAR SHINGLE & SHAKE NAILS

Stock Numbers Plain Shank	Length Inch	Gauge	Head	Nails Per Lb.
R-112	1 1/4	14	7/32"	515
R-113	1 1/2	14	1/4"	392
R-114	1 3/4	14	1/4"	344
R-115	2	13	1/4"	232
R-116	2 1/4	13	1/4"	185

Specially designed for shingle & shake applications. Slim shank minimizes splitting. Long, sharp points. Double-dipped in molten zinc. Meets Cedar Shake & Shingle Bureau specifications.

FINISHING NAILS

Stock Numbers Plain Shank	Length Inch	Gauge	Head	Nails Per Lb.
T-315	2	14	1/8"	310
T-317	2 1/2	13	9/64"	196
T-319	3	11 1/2	5/32"	109
T-321	3 1/2	11 1/2	5/32"	95

Designed for exterior trim applications. Brad head sets easily and insures neat finished appearance. Double-dipped in molten zinc.

CASING NAILS

Stock Numbers Plain Shank	Length Inch	Gauge	Head	Nails Per Lb.
T-305	2	12 1/2	9/64"	212
T-307	2 1/2	11 1/2	5/32"	131
T-309	3	10	11/64"	85
T-3091	3 1/2	10	11/64"	67
T-3092	4	9	3/16"	48

Designed for the secure application of window and door frames, cornices and exterior molding. Eliminate need for countersinking and puttying. Double-dipped in molten zinc.

PAINTED TRIM NAILS FOR SOFFIT AND FASCIA

Stock Numbers Plain Shank	Length Inch	Gauge	Head	Nails Per Lb.
R112-M61 White	1 1/4	14	7/32"	450
R112-M22 Brown	1 1/4	14	7/32"	450

Maze Trim Nails are ideal for applying painted aluminum soffit and fascia. The thick, uniform zinc coating is compatible with aluminum, and the strong steel shank drives much better than aluminum nails. White and Brown trim nails are stock items - in 1 lb. boxes, 50 per master carton. Double-dipped in molten zinc.

GUTTER SPIKES

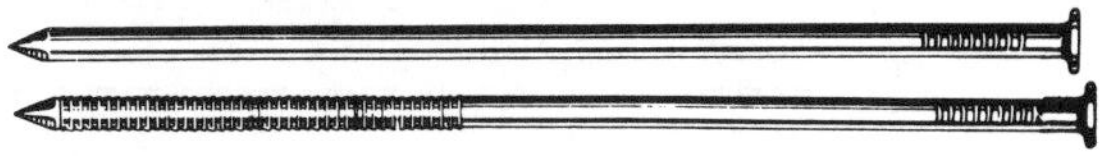

Stock Numbers Plain	Stock Numbers Ring	Length Inch	Gauge	Head	Nails Per Lb.
T-364	T-364-A	7	3/16"	7/16"	16
T-365	T-365-A	8	3/16"	7/16"	14
T-366	T-366-A	8	1/4"	9/16"	8
T-368	T-368-A	9	1/4"	9/16"	7
T-369	T-369-A	10	1/4"	9/16"	6.5

Ideal for both aluminum and galvanized gutters. Drive easily. Hot-dipped in molten zinc. Checkered head. Meets FHA specifications. Available painted.

PRESSURE TREATED LUMBER "P.T.L.® " NAILS

Stock Numbers Plain	Stock Numbers Ring	Stock Numbers Spiral	Length Inch	Gauge	Nails Per Lb.
T445	T445-A	T445-S	2	11	137
T447	T447-A	T447-S	2 1/2	11	111
T449	T449-A	T449-S	3	10	75
T4490	T4490-A	T4490-S	3 1/4	10	69
T4491	T4491-A	T4491-S	3 1/2	9	54
T4492	T4492-A	T4492-S	4	7	33
T4493	T4493-A	T4493-S	4 1/2	7	29
T4494	T4494-A	T4494-S	5	5 1/2	22
T4496	T4496-A	T4496-S	6	5 1/2	18

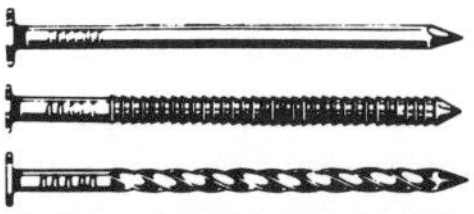

Producers of wood treating chemicals recommend that hot-dipped zinc-coated nails be used with P.T.L. Other nails may rust prematurely and spoil treated wood projects. High carbon steel, **double** hot-dipped MAZE PTL nails drive with minimal nail bending and hold tight.

SMALL HEAD "PTL® " NAILS

Stock Numbers Spiral Shank	Length Inch	Gauge	Head	Nails Per Lb.
T57-S	2 1/2	12	3/16"	146
T59-S	3	11	7/32"	98
T591-S	3 1/2	11	7/32"	82

These small-headed "PTL" Nails are less noticeable resulting in a more finished look. A slimmer shank made from extra stiff wire drives easier in treated lumber and causes fewer splits especially on red-wood and cedar decks. Spiral shanks hold tighter than smooth shank nails. Double-dipped in molten zinc.

JOIST HANGER NAILS

Stock Numbers Ring Shank	Length Inch	Gauge	Head	Nails Per Lb.
XR-162-A	1 1/4	10	3/8"	155
XR-163-A	1 1/2	10	3/8"	128

These well-threaded 10 gauge nails are *STORMGUARD* double hot-dipped zinc-coated to insure superior holding power and long term rust-protection. The hot-dipped coating is the type of galvanized coating which is recommended by the makers of CCA lumber for use with their products.

Zinclad

ZINCLAD® SCREWS FOR DECKING, ETC.

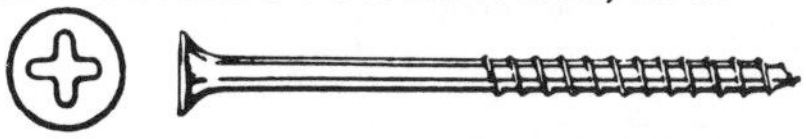

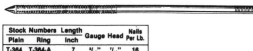

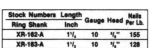

Stock Numbers	Length	Gauge	Head	Screws Per Lb.
DZ163	1 5/8"	6	11/32"	180
DZ200	2"	8	11/32"	117
DZ250	2 1/2"	8	11/32"	100
DZ300	3"	8	11/32"	90

These case-hardened screws are hot-dipped in molten zinc to meet the specifications of producers of wood treatment chemicals. Excellent for a wide variety of outdoor uses such as decks, boat docks, steps, patios, furniture, fences, flower boxes, or any project where long-term corrosion-resistance and exceptional holding power are required.

MAZEMADE®

LOG HOME SPIKES — PLAIN SHANK

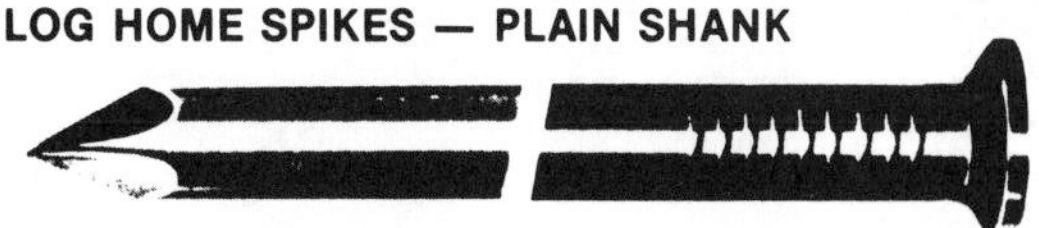

Stock Numbers Bright	Stock Numbers Zinclad®	Length Inch	Gauge	Head	Nails Per Lb.
L8	ZL8	8	3/8"	11/16"	3.9
L10	ZL10	10	3/8"	11/16"	3.1
L12	ZL12	12	3/8"	11/16"	2.7

Heavy gauge plain shank spikes are ideal for log home construction, securing landscape timbers, or wherever a sturdy, long spike is required. Available bright or Zinclad (hot-dipped zinc-coated).

LOG HOME SPIKES — RING SHANK

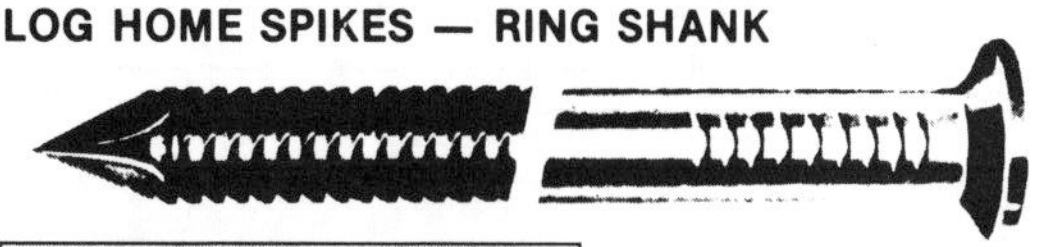

Stock Numbers Bright	Stock Numbers Zinclad®	Length Inch	Gauge	Head	Nails Per Lb.
L8-A	ZL8-A	8	3/8"	11/16"	3.9
L10-A	ZL10-A	10	3/8"	11/16"	3.1
L12-A	ZL-12-A	12	3/8"	11/16	2.7

These spikes are the same style as those above, but with ring shanks which increase holding power. Available bright or Zinclad (hot-dipped zinc-coated).

NAIL TYPES (Continued)

STORMGUARD FLAT RUBBER WASHER NAILS

Stock Numbers	Length	Gauge	Head	Nails Per Lb.
Ring Shank	Inch			
R-163-AF	1 1/2	10	3/8"	126
R-164-AF	1 3/4	10	3/8"	111
R-165-AF	2	10	3/8"	100
R-166-AF	2 1/4	10	3/8"	91
R-167-AF	2 1/2	10	3/8"	82
R-169-AF	3	10	3/8"	70
R-1691-AF	3 1/2	10	3/8"	61
R-1692-AF	4	9	7/16"	45
R-1693-AF	4 1/2	9	7/16"	39

Soft, flat washers provide fine seal and avoid sheet dimpling or denting. Checkered head enhances paint adhesion on color-matched nails. Sturdy ring shanks hold tight. Double-dipped in molten zinc. These washers are made of an EPDM compound.

STORMGUARD UMBRELLA HEAD NAILS

Stock Numbers		Length	Gauge	Head	Nails Per Lb.
Ring	Spiral	Inch			
R-133-A	— —	1 1/2	10	7/16"	125
R-134-A	R-134-S	1 3/4	10	7/16"	112
R-135-A	R-135-S	2	10	7/16"	99

Concave umbrella head caps nail hole. Threaded shanks provide greatly increased holding power. Double-dipped in molten zinc.

STORMGUARD "FAT HEAD" GRAY SILICONE WASHER NAILS

Stock Numbers	Length	Gauge	Head	Nails Per Lb.
Ring Shank	Inch			
R163-FH	1 1/2	10	7/16"	119
R164-FH	1 3/4	10	7/16"	107
R165-FH	2	10	7/16"	96
R167-FH	2 1/2	10	7/16"	79
R169-FH	3	10	7/16"	69
R1691-FH	3 1/2	10	7/16"	60
R1692-FH	4	9	7/16"	45

These gray silicone washers are twice as thick as our black EPDM washers. The silicone washers themselves are many times more expensive than the black EPDM washers, but many roofers like this style because their resiliency allows them to seal the nail hole even if not driven at right angles to the sheet. Double-dipped in molten zinc.

STORMGUARD "EXTRA-WIDE" RUBBER WASHER NAILS
(For Corrugated Asphalt Roofing)

Stock Numbers	Length	Gauge	Head	Nails Per Lb.
Ring Shank	Inch			
R167-EW	2 1/2	10	5/8"	75
R169-EW	3	10	5/8"	66
R1691-EW	3 1/2	10	5/8"	57

These "Extra-Wide" 9/16" rubber (EPDM) washers are being widely used for application of Corrugated Asphalt Roofing and Siding. Double-dipped in molten zinc.

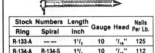

MAZEMADE®

COMPRESSED LEAD HEAD NAILS — BRIGHT

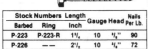

Stock Numbers		Length	Gauge	Head	Nails Per Lb.
Barbed	Ring	Inch			
P-223	P-223-R	1 3/4	10	3/8"	90
P-226	— —	2 1/2	10	3/8"	72

Lead-over-the-head style. Bright finish. Threaded shanks provide greatly increased holding power.

"SKINNY SPIKES" — RING SHANK

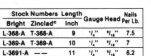

Stock Numbers		Length	Gauge	Head	Nails Per Lb.
Bright	Zinclad®	Inch			
L-368-A	T-368-A	9	1/4"	9/16"	7.5
L-369-A	T-369-A	10	1/4"	9/16"	7
L-3691-A	— —	11	1/4"	9/16"	6.2
L-3692-A	— —	12	1/4"	9/16"	5.8

These thin gauge ring shank spikes give a higher nail count per pound than 3/8" spikes - and the thinner diameter helps to reduce wood splits. Available bright or Zinclad® (hot-dipped zinc-coated).

ONE-LEGGED FENCE STAPLES

Stock Numbers		Length	Gauge	Head	Staples Per Lb.
Bright	Stormguard	Inch			
F3	SF3	1 1/2	9	1/2"	72
F4	SF4	1 3/4	9	1/2"	61

These ring shank staples have excellent holding power, and are offered in bright steel or Stormguard (double hot-dipped zinc-coated) finish.

— AMERICAN MADE —

MAZEMADE®

SQUARE CAP® NAILS — STEEL HEAD

Stock Numbers		Length	Gauge	Head	Nails Per Lb.
Bright	Zinclad®	Inch			
853	— —	3/4	12	1"	100
854	— —	7/8	12	1"	98
855	855-Z	1	12	1"	94
856	— —	1 1/4	12	1"	89
857	857-Z	1 1/2	12	1"	84
858	— —	1 3/4	12	1"	81
859	859-Z	2	12	1"	77
861	861-Z	2 1/2	12	1"	74
863	863-Z	3	12	1"	64
865	865-Z	3 1/2	12	1"	57

Specially designed for built-up roofing, furniture packing, freight car lining, sheathing and veneering. Popular for applying rigid insulation. Meet federal and industry specifications. Full 1" steel head provides large bearing surface. Heavily barbed shank adds extra holding power. Available hot-dipped zinc-coated (Zinclad) in popular sizes.

PLASTIC-HED® CAP NAILS
Stocked in Handy 2000-Count Cartons

Stock Numbers	Length	Gauge	Head	Weight Per Box
Bright Ring Shank	Inch			
PH-01	3/4	12	15/16"	9 lbs.
PH-02	7/8	12	15/16"	9 lbs.
PH-03	1	12	15/16"	9 lbs.
PH-05	1 1/4	12	15/16"	10 lbs.
PH-06	1 1/2	11	15/16"	15 lbs.
PH-07	1 3/4	11	15/16"	17 lbs.
PH-08	2	11	15/16"	18 lbs.
PH-09	2 1/2	11	15/16"	23 lbs.

Maze-Yellow Plastic-Hed Cap Nails allow for more nails per pound than metal cap nails. Siding nails drive right through the Plastic-Hed nail caps to eliminate bending problems often encountered when a skinny siding nail strikes a metal cap. The annular ring shank holds tight, and the clipped corners eliminate damage to roofing felt, housewraps and foam insulation.

Stainless Steel Nails — Type 302/304

STAINLESS STEEL "SLIM-JIM"™ WOOD SIDING NAILS

Stock Numbers Ring Shank	Length Inch	Gauge	Head	Nails Per Lb.
SS-6-WS	2	13	5/32"	245
SS-8-WS	2 1/2	13	5/32"	196
SS-10-WS	3	12	7/32"	120
SS-16-WS	3 1/2	11	15/64"	91

These slender, blunt-pointed wood siding nails are made from high tensile stainless steel to minimize bending and reduce splitting. Excellent for redwood and cedar siding, especially on wood to be left natural or to receive treatment with semi-transparent coatings or clear finishes (to give the "natural" look). Small head permits blind nailing. Also available from stock-painted in cedar/redwood color.

STAINLESS STEEL "P.W.F." NAILS

Stock Numbers Ring Shank	Length Inch	Gauge	Head	Nails Per Lb.
SS-6	2	11	9/32"	145
SS-8	2 1/2	10	5/16"	94
SS-10	3	9	5/16"	64
SS-16	3 1/2	8	3/8"	44
SS-20	4	6	7/16"	25

Ring shank Stainless Steel nails are the type recommended for below grade use in "PWF" permanent wood foundation systems. Excellent for use in corrosive environments such as salt and fertilizer bins, etc. Fine for cedar and redwood decks too.

MAZEMADE®

MASONRY NAILS — HARDENED — ROUND WIRE TYPE

Stock Numbers Fluted	Length Inch	Gauge	Head	Nails Per Lb.
H-52-S	3/4	9	5/16"	230
H-53-S	1	9	5/16"	176
H-54-S	1 1/4	9	5/16"	150
H-55-S	1 1/2	9	5/16"	129
H-56-S	1 3/4	9	5/16"	108
H-57-S	2	9	5/16"	95
H-58-S	2 1/4	9	5/16"	85
H-59-S	2 1/2	9	5/16"	77
H-60-S	2 3/4	9	5/16"	70
H-61-S	3	9	5/16"	62
H-62-S	3 1/4	9	5/16"	57
H-63-S	3 1/2	7	3/8"	41
H-65-S	4	7	3/8"	35

Designed for fastening furring strips, floor plates, partition walls. Hardened steel. For best results drive nails 3" or more from edge of masonry or block allowing 3/4" penetration. Safety glasses recommended when driving hardened nails. For 1/2" size, see Hardened Metal Lath Nails (below).

HARDENED METAL LATH NAILS

Stock Numbers Plain Shank Hardened	Length Inch	Gauge	Head	Nails Per Lb.
HML-50	1/2	9	3/8"	324

These 1/2" Hardened Metal Lath nails are ideal for attaching metal lath to masonry surfaces. The head is larger than standard 1/2" masonry nails to insure that the lath doesn't pull through.

These nails may also be used as 1/2" Masonry Nails.

FLOORING NAILS — CASING HEAD

Stock Numbers Stiff	Stock Numbers Hardened	Length Inch	Gauge	Head	Nails Per Lb.
——	HF-164	1 1/2	13	1/8"	369
——	HF-166	2	11 1/2	11/64"	164
F-167	HF-167	2 1/4	11 1/2	11/64"	152
F-168	HF-168	2 1/2	11 1/2	11/64"	138

Same as flooring nails below except features casing type head. Recommended for hand driving. Safety glasses should be used when driving hardened nails.

FLOORING NAILS — COUNTERSUNK HEAD

Stock Numbers Stiff	Stock Numbers Hardened	Length Inch	Gauge	Head	Nails Per Lb.
F-67	HF-67	2 1/4	11 1/2	13/64"	152
F-68	HF-68	2 1/2	11 1/2	13/64"	138

Sharp spiral threads insure maximum holding power. Available stiff stock or hardened. Flat countersunk head suitable for either hand or machine driving. Safety glasses recommended when using hardened nails.

UNDERLAYMENT NAILS — FLAT HEAD

Stock Numbers Ring Shank	Length Inch	Gauge	Head	Nails Per Lb.
F-141	1	14	3/16"	640
F-142	1 1/4	14	3/16"	528

Sharp ring shanks insure extra holding power. Regular steel shank drives easily.

UNDERLAYMENT NAILS — COUNTERSUNK HEAD

Stock Numbers Hardened	Length Inch	Gauge	Head	Nails Per Lb.
HF-32	1 1/4	14	3/16"	528

Special countersunk head sets easily in underlayment and insures clean finished appearance. Hardened nails popular for application over old oak and maple flooring.

TREMONT

HARDENED STEEL CUT NAILS

CUT MASONRY — HARDENED

Stock Numbers Hardened	Length Inch	Nails per Pound
CMH-4	1 1/2	125
CMH-6	2	85
CMH-7	2 1/4	78
CMH-8	2 1/2	64
CMH-10	3	48
CMH-12	3 1/4	43
CMH-16	3 1/2	34

These Hardened Steel Cut Masonry nails, manufactured by the 170 year old Tremont Nail Co. of Wareham, Mass., are ideal for attaching wood members to cinder block walls as well as to fresh concrete. The blunt point and tapered shank cause minimal spalling during penetration and offer even greater pull-out resistance than wire masonry nails. 3/4" penetration is recommended for satisfactory holding power. Use safety glasses.

CUT FLOORING — HARDENED

Stock Numbers Hardened	Length Inch	Nails per Pound
CFH-4	1 1/2	226
CFH-6	2	104
CFH-7	2 1/4	88
CFH-8	2 1/2	78
CFH-10	3	53
CFH-16	3 1/2	35

Tremont Hardened Steel Cut Flooring nails are ideal for laying hardwood tongue and groove floors. The slim shanks avoid splitting the groove and the holding power is excellent. Nails should be spaced 8 inches apart on 3/8" thick material and 12" apart on 3/4" flooring. Note: the wide side of the nail should always be driven parallel to the grain of the wood. Use safety glasses.

CUT FLOORING — GALVANIZED

Stock Numbers Hot-Dip Galvanized	Length Inch	Approximate Nails per Pound
CF6-Z	2	93
CF8-Z	2 1/2	79
CF10-Z	3	48

Tremont Hot-Dip Galvanized Cut Flooring Nails are great for applications where rust-resistance is important. The square shank holds tight, and the head works well in both face and blind nailing applications. Also known as hot-dip galvanized cut finish nails.

— BE SAFE —
SAFETY GLASSES ARE RECOMMENDED WHEN DRIVING ANY NAILS.

COPPER SLATING NAILS

Stock Numbers Plain Shank	Length Inch	Gauge	Head	Nails Per Lb.
CU-125	1 1/4	11	3/8"	187
CU-150	1 1/2	11	3/8"	155
CU-175	1 3/4	11	3/8"	139
CU-200	2	11	3/8"	124

These solid copper nails are the type recommended for use with many major brands of tile and slate roofing — and are the most compatible nails for use with copper flashing.

The full 3/8" head gives good bearing pressure on roofing and flashing material. Diamond point and smooth shank make driving easy.

NAIL TYPES (Continued)

PLYWOOD NAILS

Stock Numbers	Length			Nails
Ring Shank	Inch	Gauge	Head	Per Lb.
F-143	1¹/₂	12¹/₂	¹/₄"	285
F-144	1³/₄	12¹/₂	¹/₄"	250
F-145	2	12	¹/₄"	190
F-147	2¹/₂	11	⁹/₃₂"	118
F-149	3	10	⁹/₃₂"	78

Type recommended by plywood producers for underlayment, sheathing and sub-floor applications. Sharp ring-shanks insure extra holding power.

DRYWALL NAILS

Stock Numbers		Length			Nails
Bright	Zinclad®	Inch	Gauge	Head	Per Lb.
D-191	D-191-Z	1¹/₄	12¹/₂	¹⁹/₆₄"	352
D-192	D-192-Z	1³/₈	12¹/₂	¹⁹/₆₄"	321
D-1925	D-1925-Z	1¹/₂	12¹/₂	¹⁹/₆₄"	302
D-193	D-193-Z	1⁵/₈	12¹/₂	¹⁹/₆₄"	274
D-194	D-194-Z	1³/₄	12¹/₂	¹⁹/₆₄"	259
D-196	D-196-Z	2	12¹/₂	¹⁹/₆₄"	228

These GWB-54 style nails have long sharp diamond points for quick and easy driving. Heads are slightly countersunk with feathered edge to seat properly. Manufactured to meet government and Gypsum Assoc. guidelines. Available with Zinclad (hot-dipped zinc-coated) finish for use in high humidity areas.

HARDWOOD TRIM NAILS

Stock Numbers	Length			Nails
Plain Shank	Inch	Gauge	Head	Per Lb.
HT100	1	.068	.089	925
HT125	1¹/₄	.068	.089	795
HT150	1¹/₂	.068	.089	666
HT200	2	.072	.106	472
HT250	2¹/₂	.083	.120	256

These slim diameter Hardwood Trim Nails are made from a high carbon steel wire to insure good driving into Oak, Birch, Walnut and other Hardwoods. There are many more nails per pound than with ordinary finish nails - so they cost less to use. And, there are fewer wood splits!

POST BARN NAILS

Stock Numbers		Length			Nails
Stiff	Hardened	Inch	Gauge	Head	Per Lb.
519-A	—	2	11	⁹/₃₂"	144
521-A	H-521-A	2¹/₂	11	⁹/₃₂"	115
523-A	H-523-A	3	10	⁹/₃₂"	78
524-A	H-524-A	3¹/₄	10	⁹/₃₂"	73
525-A	H-525-A	3¹/₂	9	⁵/₁₆"	57
526-A	H-526-A	4	7	³/₈"	35
527-A	H-527-A	4¹/₂	7	³/₈"	30
528-A	H-528-A	5	5¹/₂	²⁷/₆₄"	23
—	H-676-A	5	7	³/₈"	26
530-A	H-530-A	6	5¹/₂	²⁷/₆₄"	19
—	H-680-A	6	7	³/₈"	22
532-A	H-532-A	7	5¹/₂	²⁷/₆₄"	16
534-A	H-534-A	8	5¹/₂	²⁷/₆₄"	14

Sharp ring shanks provide excellent holding power in treated posts and timbers. Safety glasses recommended when driving hardened nails. Also excellent for common nail applications. Most sizes available Stormguard double hot-dipped zinc-coated. See Pressure Treated Lumber Nails on page 2.

HOT-DIPPED GALVANIZED / HARDENED POST BARN NAILS

Stock Numbers	Length			Nails
Hot-Dipped/Hard	Inch	Gauge	Head	Per Lb.
TH4491-A	3¹/₂	9	⁵/₁₆"	54
TH4492-A	4	7	¹/₄"	33
TH4493-A	4¹/₂	7	³/₈"	29
TH4495-A	5	7	³/₈"	24
TH4497-A	6	7	¹/₄"	21

These ring shank nails are Stormguard hot-dipped galvanized after hardening - and drive much better than regular commons with fewer wood splits. The superior hot-dip zinc-coating, recommended for pressure treated lumber, is especially popular in farm building construction or for tough, hard lumber. Safety glasses recommended when driving hardened nails.

POST AND TRUSS NAILS

Stock Numbers		Length			Nails
Stiff	Hardened	Inch	Gauge	Head	Per Lb.
521-S	H-521-S	2¹/₂	11	⁹/₃₂"	115
523-S	H-523-S	3	10	⁹/₃₂"	78
524-S	H-524-S	3¹/₄	10	⁹/₃₂"	73
525-S	H-525-S	3¹/₂	9	⁵/₁₆"	57
526-S	H-526-S	4	7	³/₈"	35
527-S	H-527-S	4¹/₂	7	³/₈"	30
528-S	H-528-S	5	5¹/₂	²⁷/₆₄"	23
530-S	H-530-S	6	5¹/₂	²⁷/₆₄"	19

Spiral shanks hold well and greatly assist in preventing racking of nailed truss and post members. Stiff stock or hardened. Safety glasses recommended when driving hardened nails. Also excellent for common nail applications. Most sizes available Stormguard double hot-dipped zinc-coated. See Pressure Treated Lumber Nails on page 2.

PALLET NAILS — SPIRAL SHANK

Stock Numbers		Length			Nails
Stiff	Hardened	Inch	Gauge	Head	Per Lb.
S-14	—	1¹/₂	11	⁹/₃₂"	188
S-145	—	1³/₄	11	⁹/₃₂"	175
S-16	HS-16	2	11	⁹/₃₂"	144
S-17	HS-17	2¹/₄	11	⁹/₃₂"	130
S-18	HS-18	2¹/₂	11	⁹/₃₂"	115
S-20	—	3	10	⁹/₃₂"	78
S-21	—	3¹/₄	10	⁹/₃₂"	73

Spiral shank type designed for dependable construction of wood pallets. Excellent for pallet repair, renailing wagon beds, trailers. Safety glasses recommended when driving hardened nails. Stiff stock available cement coated.

— BE SAFE —
SAFETY GLASSES ARE RECOMMENDED WHEN DRIVING ANY NAILS.

COMMONLY AVAILABLE WOOD SCREW SIZES

Length   **Gauge Number**

Length	0	1	2	3	4	5	6	7	8	9	10	11	12	14	16	18	20	24
1/4"	0	1	2	3														
3/8"			2	3	4	5	6	7										
1/2"			2	3	4	5	6	7	8									
5/8"				3	4	5	6	7	8	9	10							
3/4"					4	5	6	7	8	9	10	11						
7/8"							6	7	8	9	10	11	12					
1"							6	7	8	9	10	11	12	14				
1 1/4"							6	7	8	9	10	11	12	14	16			
1 1/2"							6	7	8	9	10	11	12	14	16	18		
1 3/4"									8	9	10	11	12	14	16	18	20	
2"									8	9	10	11	12	14	16	18	20	
2 1/4"										9	10	11	12	14	16	18	20	
2 1/2"													12	14	16	18	20	
2 3/4"														14	16	18	20	
3"															16	18	20	
3 1/2"																18	20	24
4"																18	20	24

SANDPAPER GRADES

Grit Number	0 Series Number	Abrasive Properties
600	—	Super Fine
500	—	Super Fine
400	10/0	Super Fine
360	—	Super Fine
320	9/0	Super Fine
280	8/0	Very Fine
240	7/0	Very Fine
220*	6/0	Very Fine

*(Grit usually supplied when paper is simply called "very fine")

180	5/0	Fine
150*	4/0	Fine

*(Grit usually supplied when paper is simply called "fine")

120	3/0	Medium
100*	2/0	Medium

*(Grit usually supplied when paper is simply called "medium")

80	1/0	Medium
60*	1/2	Coarse

*(Grit usually supplied when paper is simply called "coarse")

50	1	Coarse
40	1 1/2	Coarse
36	2	Very Coarse
30	2 1/2	Very Coarse
24	3	Very Coarse

WIRE AND SHEET-METAL GAUGES
Dimensions in decimal parts of an inch

No. of gauge	American or Brown and Sharpe[a]	Washburn & Moon or American Steel[b]	Birmingham or Stubs Iron wire[c]	Music wire[d]	Imperial wire gage[e]	U.S. Std. for plate[f]
0000000	—	0.4900	—	—	0.5000	0.5000
000000	0.5800	0.4615	—	0.004	0.4640	0.4688
00000	0.5165	0.4305	0.500	0.005	0.4320	0.4375
0000	0.4600	0.3938	0.454	0.006	0.4000	0.4063
000	0.4096	0.3625	0.425	0.007	0.3720	0.3750
00	0.3648	0.3310	0.380	0.008	0.3480	0.3438
0	0.3249	0.3065	0.340	0.009	0.3240	0.3125
1	0.2893	0.2830	0.300	0.010	0.3000	0.2813
2	0.2576	0.2625	0.284	0.011	0.2760	0.2656
3	0.2294	0.2437	0.259	0.012	0.2520	0.2500
4	0.2043	0.2253	0.238	0.013	0.2320	0.2344
5	0.1819	0.2070	0.220	0.014	0.2120	0.2188
6	0.1620	0.1920	0.203	0.016	0.1920	0.2031
7	0.1443	0.1770	0.180	0.018	0.1760	0.1875
8	0.1285	0.1620	0.165	0.020	0.1600	0.1719
9	0.1144	0.1483	0.148	0.022	0.1440	0.1563
10	0.1019	0.1350	0.134	0.024	0.1280	0.1406
11	0.0907	0.1205	0.120	0.026	0.1160	0.1250
12	0.0808	0.1055	0.109	0.029	0.1040	0.1094
13	0.0720	0.0915	0.095	0.031	0.0920	0.0938
14	0.0641	0.0800	0.083	0.033	0.0800	0.0781
15	0.0571	0.0720	0.072	0.035	0.0720	0.0703
16	0.0508	0.0625	0.065	0.037	0 0640	0 0625
17	0.0453	0.0540	0.058	0.039	0.0560	0.0563
18	0.0403	0.0475	0.049	0.041	0.0480	0.0500
19	0.0359	0.0410	0.042	0.043	0.0400	0.0438
20	0.0320	0.0348	0.035	0.045	0.0360	0.0375
21	0.0285	0.0317	0.032	0.047	0.0320	0.0344
22	0.0253	0.0286	0.028	0.049	0.0280	0.0313
23	0.0226	0.0258	0.025	0.051	0 0240	0.0281
24	0.0201	0.0230	0.022	0.055	0.0220	0.0250
25	0.0179	0.0204	0.020	0.059	0.0200	0.0219
26	0.0159	0.0181	0.018	0.063	0.0180	0.0188
27	0.0142	0.0173	0 016	0.067	0 0164	0.0172
28	0.0126	0.0162	0.014	0.071	0.0148	0.0156
29	0.0113	0.0150	0.013	0.075	0.0136	0.0141
30	0.0100	0.0140	0.012	0.080	0.0124	0.0125
31	0.0089	0.0132	0.010	0 085	0.0116	0.0109
32	0.0080	0.0128	0.009	0.090	0.0108	0.0102
33	0.0071	0 0118	0.008	0.095	0.0100	0.0094
34	0.0063	0.0104	0.007	0.100	0.0092	0.0086
35	0.0056	0.0095	0.005	0.106	0.0084	0.0078
36	0.0050	0.0090	0.004	0.112	0.0076	0.0070
37	0.0045	0.0085	—	0.118	0.0068	0.0066
38	0.0040	0.0080	—	0.124	0.0060	0.0063
39	0.0035	0.0075	—	0.130	0.0052	
40	0.0031	0.0070	—	0.138	0.0048	

[a] Recognized standard in the United States for wire and sheet metal of copper and other metals except steel and iron.
[b] Recognized standard for steel and iron wire. Called the "U.S. steel wire gage."
[c] Formerly much used, now nearly obsolete.
[d] American Steel & Wire Co.'s music wire gage. Recommended by U.S. Bureau of Standards.
[e] Official British Standard.
[f] Legalized U.S. Standard for iron and steel plate, although plate is now always specified by its thickness in decimals of an inch. Preferred thicknesses for uncoated thin flat metals (under 0.250 in.): ASA B32.1—1959 gives recommended sizes for sheets.

Appendix F
Electrical Information

(The following tables are reprinted with permission from NFPA 70-1993, National Electrical Code®, Copyright © 1992. National Fire Protection Association, Quincy, MA 02269)

METAL BOXES

Box Dimension, Inches Trade Size or Type	Min. Cu In. Cap.	Maximum Number of Conductors						
		No. 18	No. 16	No. 14	No. 12	No. 10	No. 8	No. 6
4 x 1 1/4 Round or Octagonal	12.5	8	7	6	5	5	4	2
4 x 1 1/2 Round or Octagonal	15.5	10	8	7	6	6	5	3
4 x 2 1/8 Round or Octagonal	21.5	14	12	10	9	8	7	4
4 x 1 1/4 Square	18.0	12	10	9	8	7	6	3
4 x 1 1/2 Square	21.0	14	12	10	9	8	7	4
4 x 2 1/8 Square	30.3	20	17	15	13	12	10	6
4 11/16 x 1 1/4 Square	25.5	17	14	12	11	10	8	5
4 11/16 x 1 1/2 Square	29.5	19	16	14	13	11	9	5
4 11/16 x 2 1/8 Square	42.0	28	24	21	18	16	14	8
3 x 2 x 1 1/2 Device	7.5	5	4	3	3	3	2	1
3 x 2 x 2 Device	10.0	6	5	5	4	4	3	2
3 x 2 x 2 1/4 Device	10.5	7	6	5	4	4	3	2
3 x 2 x 2 1/2 Device	17.5	8	7	6	5	5	4	2
3 x 2 x 2 3/4 Device	14.0	9	8	7	6	5	4	2
3 x 2 x 3 1/2 Device	18.0	12	10	9	8	7	6	3
4 x 2 1/8 x 1 1/2 Device	10.3	6	5	5	4	4	3	2
4 x 2 1/8 x 1 7/8 Device	13.0	8	7	6	5	5	4	2
4 x 2 1/8 x 2 1/8 Device	14.5	9	8	7	6	5	4	2
3 3/4 x 2 x 2 1/2 Masonry Box/Gang	14.0	9	8	7	6	5	4	2
3 3/4 x 2 x 3 1/2 Masonry Box/Gang	21.0	14	12	10	9	8	7	4
FS—Minimum Internal Depth 1 3/4 Single Cover/Gang	13.5	9	7	6	6	5	4	2
FD—Minimum Internal Depth 2 3/8 Single Cover/Gang	18.0	12	10	9	8	7	6	3
FS—Minimum Internal Depth 1 3/4 Multiple Cover/Gang	18.0	12	10	9	8	7	6	3
FD—Minimum Internal Depth 2 3/8 Multiple Cover/Gang	24.0	16	13	12	10	9	8	4

For SI units: one cubic inch = 16.4 cm.

VOLUME REQUIRED PER CONDUCTOR

Size of Conductor	Free Space Within Box for Each Conductor
No. 18	1.5 cubic inches
No. 16	1.75 cubic inches
No. 14	2 cubic inches
No. 12	2.25 cubic inches
No. 10	2.5 cubic inches
No. 8	3 cubic inches
No. 6	5 cubic inches

For SI units: one cubic inch = 16.4 cm.

MAXIMUM NUMBER OF FIXTURE WIRES IN TRADE SIZES OF CONDUIT OR TUBING
(40 PERCENT FILL BASED ON INDIVIDUAL DIAMETERS)

Conduit Trade Size (Inches) Wire Types	1/2					3/4					1					1 1/4					1 1/2					2				
	18	16	14	12	10	18	16	14	12	10	18	16	14	12	10	18	16	14	12	10	18	16	14	12	10	18	16	14	12	10
PTF, PTFF, PGFF, PGF, PFF, PF, PAF, PAFF, ZF, ZFF	23	18	14			40	31	24			65	50	39			115	90	70			157	122	95			257	200	156		
TFFN, TFN	19	15				34	26				55	43				97	76				132	104				216	169			
SF-1	16					29					47					83					114					186				
SFF-1	15					26					43					76					104					169				
TF	11	10				20	18				32	30				57	53				79	72				129	118			
RFH-1	11					20					32					57					79					129				
TFF	11	10				20	17				32	27				56	49				77	66				126	109			
AF	11	9	7	4	3	19	16	12	7	5	31	26	20	11	8	55	46	36	19	15	75	63	49	27	20	123	104	81	44	34
SFF-2	9	7	6			16	12	10			27	20	17			47	36	30			65	49	42			106	81	68		
SF-2	9	8	6			16	14	11			27	23	18			47	40	32			65	55	43			106	90	71		
FFH-2	9	7				15	12				25	19				44	34				60	46				99	75			
RFH-2	7	5				12	10				20	16				36	28				49	38				80	62			
KF-1, KFF-1, KF-2, KFF-2	36	32	22	14	9	64	55	39	25	17	103	89	63	41	28	182	158	111	73	49	248	216	152	100	67	406	353	248	163	110

233

MAXIMUM NUMBER OF CONDUCTORS IN TRADE SIZES OF CONDUIT OR TUBING

Type Letters	Conductor Size AWG/kcmil	1/2	3/4	1	1 1/4	1 1/2	2	2 1/2	3	3 1/2	4	5	6
TW, XHHW (14 through 18), RH (14 + 12)	14	9	15	25	44	60	99	142	171				
	12	7	12	19	35	47	78	111	131	176			
	10	5	9	15	26	36	60	85					
	8	2	4	7	12	17	28	40	62	84	108		
RHW and RHH (without outer covering), RH (10 + 8) TWH, THHW	14	6	10	16	29	40	65	98	143	192			
	12	4	8	13	24	32	53	76	117	157			
	10	4	6	11	19	26	43	61	95	127	163		
	8	1	3	5	10	13	22	32	49	66	85	133	
TW,	6	1	2	4	7	10	16	23	36	48	62	97	141
	4	1	1	3	5	7	12	17	27	36	47	73	106
THW,	3	1	1	2	4	6	10	15	23	31	40	63	91
	2	1	1	2	4	5	9	13	20	27	34	54	78
	1		1	1	3	4	6	9	14	19	25	39	57
FEPB (6 through 2), RHW and RHH (without outer covering)	1/0		1	1	2	3	5	8	12	16	21	33	49
	2/0		1	1	1	3	5	7	10	14	18	29	41
	3/0		1	1	1	2	4	6	9	12	15	24	35
	4/0			1	1	1	3	5	7	10	13	20	29
RH, THHW	250			1	1	1	2	4	6	8	10	16	23
	300			1	1	1	2	3	5	7	9	14	20
	350				1	1	1	3	4	6	8	12	18
	400				1	1	1	2	4	5	7	11	16
	500				1	1	1	1	3	4	6	9	14
	600					1	1	1	3	4	5	7	11
	700					1	1	1	2	3	4	7	10
	750					1	1	1	2	3	4	6	9

Note 1. This table is for concentric stranded conductors only.

Note 2. Conduit fill for conductors with a -2 suffix is the same as for those types without the suffix.

MAXIMUM NUMBER OF CONDUCTORS IN TRADE SIZES OF CONDUIT OR TUBING

Type Letters	Conductor Size AWG/kcmil	1/2	3/4	1	1 1/4	1 1/2	2	2 1/2	3	3 1/2	4	5	6
THWN,	14	13	24	39	69	94	154						
	12	10	18	29	51	70	114	164					
	10	6	11	18	32	44	73	104	160				
	8	3	5	9	16	22	36	51	79	106	136		
THHN, FEP (14 through 2), FEPB (14 through 8), PFA (14 through 4/0), PFAH (14 through 4/0), Z (14 through 4/0), XHHW (4 through 500 kcmil).	6	1	4	6	11	15	26	37	57	76	98	154	
	4	1	2	4	7	9	16	22	35	47	60	94	137
	3	1	1	3	6	8	13	19	29	39	51	80	116
	2	1	1	3	5	7	11	16	25	33	43	67	97
	1	1	1	1	3	5	8	12	18	25	32	51	72
	1/0		1	1	3	4	7	10	15	21	27	42	61
	2/0		1	1	2	3	6	8	13	17	22	35	51
	3/0		1	1	1	3	5	7	11	14	18	29	42
	4/0		1	1	1	2	4	6	9	12	15	24	35
	250			1	1	1	3	4	7	10	12	20	28
	300			1	1	1	3	4	6	8	10	17	24
	350			1	1	1	2	3	5	7	9	15	21
	400				1	1	1	3	5	6	8	13	19
	500				1	1	1	2	4	5	7	11	16
	600				1	1	1	1	3	4	5	9	13
	700					1	1	1	3	4	4	8	11
	750					1	1	1	2	3	3	7	11
XHHW	6	1	3	5	9	13	21	30	47	63	81	128	185
	600				1	1	1	1	1	3	4	5	13
	700					1	1	1	1	3	4	5	11
	750					1	1	1	1	2	3	4	10

Note 1. This table is for concentric stranded conductors only.

Note 2. Conduit fill for conductors with a -2 suffix is the same as for those types without the suffix.

(Continued)

MAXIMUM NUMBER OF CONDUCTORS IN TRADE SIZES OF CONDUIT OR TUBING (Continued)

Type Letters	Conductor Size AWG/kcmil	1/2	3/4	1	1 1/4	1 1/2	2	2 1/2	3	3 1/2	4	5	6
RHW,	14	3	6	10	18	25	41	58	90	121	155		
	12	3	5	9	15	21	35	50	77	103	132		
	10	2	4	7	13	18	29	41	64	86	110		
	8	1	2	4	7	9	16	22	35	47	60	94	137
RHH	6	1	1	2	5	6	11	15	24	32	41	64	93
	4	1	1	1	3	5	8	12	18	24	31	50	72
(with	3	1	1	1	3	4	7	10	16	22	28	44	63
outer	2		1	1	3	4	6	9	14	19	24	38	56
covering)	1		1	1	1	3	5	7	11	14	18	29	42
	1/0		1	1	1	2	4	6	9	12	16	25	37
	2/0			1	1	1	3	5	8	11	14	22	32
	3/0			1	1	1	3	4	7	9	12	19	28
	4/0			1	1	1	2	4	6	8	10	16	24
	250				1	1	1	3	5	6	8	13	19
	300				1	1	1	3	4	5	7	11	17
	350				1	1	1	2	4	5	6	10	15
	400				1	1	1	1	3	4	6	9	14
	500					1	1	1	3	4	5	8	11
	600					1	1	1	2	3	4	6	9
	700					1	1	1	1	3	3	6	8
	750					1	1	1	1	3	3	5	8

Note 1. This table is for concentric stranded conductors only.
Note 2. Conduit fill for conductors with a -2 suffix is the same as for those types without the suffix.

SOURCES

Adams, James E. *Electrical Principals and Practices.* New York: McGraw-Hill Book Company.

Adams, Jeannette T., and Stieri, Emanuele. *The Complete Woodworking Handbook.* New York: Arco Publishing Co., Inc.

Alth, Max. *Masonry, A Homeowner's Bible.* Garden City, New York: Doubleday & Company, Inc.

_____. *Masonry and Concrete Work.* New York: Popular Science Books.

_____. *The Handbook of Do-It-Yourself Materials.* New York: Crown Publishers, Inc.

American Institute of Timber Construction. *Glulam Systems.* Englewood, Colorado: American Institute of Timber Construction.

American Plywood Association. *APA Design/Construction Guide, Residential & Commercial.* Tacoma, Washington: American Plywood Association.

_____. *APA Product Guide, Grades & Specifications.* Tacoma, Washington: American Plywood Association.

Anderson, Bruce. *Solar Energy-Fundamentals in Building Design.* New York: McGraw-Hill Book Co.

Anderson, Edwin P. *Home Workshop & Tool Handy Book.* Indianapolis: Howard W. Sams & Company, Inc.

Arco How-To Library. *Handy Man's Plumbing and Heating Guide.* New York: Arco Publishing Co., Inc.

Asimov, Isaac. *Understanding Physics.* New York: Dorset Press.

Barnow, Benjamin. *Basic Roof Framing.* Blue Ridge Summit, Pennsylvania: Tab Books, Inc.

Beckman, William A., Klein, Sanford A., and Duffie, John A. *Solar Heating by the F-Chart Method.* New York: John Wiley & Sons.

Bianchina, Paul. *How to Hire the Right Contractor.* Yonkers, New York: Consumer Reports Books.

Bibliographisches Institut and Simon + Schuster Inc., American ed. *How Things Work.* Vols. 1-4. Geneva: Edito-Service S.A.

Bonneville Power Administration. Department of Energy. *Environment and Power, Home Weatherization & Indoor Air Pollutants.* Washington: Government Printing Office.

Brann, Donald R. *Bricklaying Simplified.* New York: Directions Simplified.

Burbank, Nelson L. *House Carpentry Simplified.* Simmons-Boardman Publishing Corp. USA.

Cheever, Ellen. *Beyond the Basics, Advanced Kitchen Design.* Hackettstown, New Jersey: The National Kitchen and Bath Association.

Council of American Building Officials, The. *CABO One and Two Family Dwelling Code.* Falls Church, Virginia: The Council of American Building Officials.

Dagostino, Frank R. *Mechanical & Electrical Systems in Building.* Reston, Virginia: Reston Publishing Co., Inc.

Dalzell, J. Ralph. *Simplified Concrete Masonry Planning and Building.* 2d ed. New York: McGraw-Hill Book Co.

Dalzell, J. Ralph, and Townsend, Gilbert. *Masonry Simplified.* Vol. 1. Chicago: American Technical Society.

Daniels, George. *Home Guide to Plumbing, Heating, Air Conditioning.* New York: Popular Science.

Delair Publishing Co., *New Webster's Dictionary of the English Language.* Delair Publishing Co. USA.

Department of the Army. *Carpenter.* Washington: Government Printing Office.

Diamond, R.C., and Grimsrud, D.T. Lawrence Berkeley Laboratory. University of California. Applied Science Division. *Manual on Indoor Air Quality.* U.S. Department of Energy. Washington: Government Printing Office.

Durbahn, Walter E., and Putnam, Robert E. *Fundamentals of Carpentry, Tools, Materials, Practices.* Alsip, Illinois: American Technical Publishers, Inc.

Feirer, John L., and Hutchings, Gilbert R. *Carpenters and Building Construction.* 3d ed. New York: Popular Science Books.

French, Thomas E., and Vierck, Charles J. *Engineering Drawing.* New York: McGraw-Hill Book Co.

Funk & Wagnalls, *New Encyclopedia.* Vols. 1-29. Funk & Wagnalls, Inc. USA.

_____. *Standard College Dictionary.* Sacramento: California State Department of Education.

Groneman, Chris H. *General Woodworking.* New York: McGrawHill Book Co.

Guralnik, David B., editor-in-chief. *Webster's New World Dictionary.* 2d ed. New York: Prentice Hall Press.

Harris, Charles O. *Structural Design.* Chicago: American Technical Society.

Hayward, Charles H. *Woodwork Joints.* New York: Sterling Publishing Co., Inc.

Hazen, Jon Brenner. *The Woodheat Handbook.* Salem, Oregon: Oregon Department of Energy.

Hornung, William J. *Builder's Vest Pocket Reference Book.* Englewood Cliffs, New Jersey: Prentice-Hall, Inc.

_____. *Plumbing and Heating.* Englewood Cliffs, New Jersey: Prentice-Hall, Inc.

International Association of Plumbing & Mechanical Engineers. *Uniform Plumbing Code.* Los Angeles: International Association of Plumbing & Mechanical Engineers.

International Conference of Building Officials. *Uniform Building Code.* Whittier, California: International Conference of Building Officials.

Jarvis, Wm. Don, ed. *Goodheart-Wilcox's Painting & Decorating Encyclopedia*. Chicago: The Goodheart-Wilcox Co., Inc.

KC Metal Products. *Handbook of Structural Designs and Load Values*. San Jose: KC Metal Products.

Kitchen, Judith L. *Caring for Your Old House*. Washington: The Preservation Press.

Lawrence Berkeley Laboratory. University of California. Energy & Environment Division. *Energy Efficient Buildings, Program Chapter*. U.S. Department of Energy. Washington: Government Printing Office.

Louisiana-Pacific. *Pocket Guide to Windows and Doors*. Barberton, Ohio: Louisiana-Pacific.

Love, T.W. *Construction Manual: Finish Carpentry*. Solana Beach, California: Craftsman Book Company of America.

Lstiburek, P., and Lischkoff, James K. *A New Approach to Affordable Low Energy House Construction*. Alberta: Alberta Department of Housing, Edmonton.

McConnell, Charles. *Plumbers and Pipefitters Library*. Vol. 3. New York: The Bobbs-Merrill Company, Inc.

R. S. Means Company, Inc. *Home Improvement Cost Guide*. Mount Vernon, New York: Consumers Union.

Miller, Martin, and Miller, Judith. *Period Details: A Sourcebook for House Preservation*. New York: Crown Publishers, Inc.

Mix, Floyd M. *House Wiring Simplified*. S. Holland, Illinois: The Goodheart-Wilcox Co., Inc.

Morgan, Alfred P. *Tools & How To Use Them for Woodworking and Metal Working*. New York: Crown Publishers, Inc.

National Fire Protection Association. *National Electrical Code*. Quincy, Massachusetts: National Fire Protection Agency.

National Forest Products Association. *Manual for House Framing; Wood Construction Data #1*. Washington: National Forest Products Association.

_____. *Span Tables for Joists and Rafters*. Washington: National Forest Products Association.

Nunn, Richard V. *Popular Mechanics Guide to Do-It-Yourself Materials*. New York: Hearst Books.

Oregon State University Extension Energy Program. *Super Good Cents Technical Reference Manual*. Bonneville Power Administration. U.S. Department of Energy, Washington: Government Printing Office.

Palmquist, Roland E. *House Wiring*. Indianapolis: Howard W. Sams and Company, Inc.

Patton, W.J. *Construction Materials*. Englewood Cliffs, New Jersey: Prentice-Hall, Inc.

Philbin, Tom, and Ettlinger, Steve. *The Complete Illustrated Guide to Everything Sold in Hardware Stores*. New York: Macmillan Publishing Company.

Phillips Drill Division. *Red Head Anchoring Systems Specification Guide*. Michigan City, Indiana: ITT Phillips Drill Division.

Reader's Digest. *Complete Do-It-Yourself Manual*. Pleasantville, New York: The Reader's Digest Association, Inc.

Richter, H.P. *Wiring Simplified*. St. Paul, Minnesota: Park Publishing, Inc.

Rooney, William F. *Practical Guide to Home Lighting*. New York: Bantam/Hudson Idea Books.

Sabo, Werner. *A Legal Guide to AIA Documents*. Chicago, Illinois: AE LawNews.

Salaman, R. A. *Dictionary of Woodworking Tools*. Newtown, Connecticut: The Taunton Press.

Sandreuter, Gregg E. *The Complete Painters Handbook*. Emmaus, Pennsylvania: Rodale Press.

Scharff, Robert. *Math for Construction, Workshop and the Home*. New York: Popular Science Books.

Shelton, Jay. *Solid Fuels Encyclopedia*. Charlotte, Virginia: Garden Way Publishing.

Sunset Books. *Basic Home Wiring Illustrated*. Menlo Park, California: Lane Publishing Co.

_____. *Basic Plumbing Illustrated*. Menlo Park, California: Lane Publishing Co.

_____. *Homeowner's Guide to Solar Heating*. Menlo Park, California: Lane Publishing Co.

Time-Life Books. *Complete Home Repair Manual*. New York: Prentice Hall Press.

_____. *Heating and Cooling*. Alexandria, Virginia: Time-Life Books.

_____. *Working with Plastics*. Alexandria, Virginia: Time-Life Books, Inc.

Universal Forest Products, Inc. *Lumber: Basic Engineering Principles & Wood Preservation*. Grand Rapids: Universal Forest Products, Inc.

U.S. Department of Agriculture. Forest Service. *Wood-Frame House Construction*. Washington: Government Printing Office.

Wagner, Willis H. *Modern Carpentry*. S. Holland, Illinois: The Goodheart-Wilcox Co., Inc.

Walker, John R. *Modern Metalworking*. Homewood, Illinois: The Goodheart-Wilcox Co., Inc.

Watson Donald. *Designing and Building a Solar House—Your Place in the Sun*. Charlotte, Virginia: Garden Way Publishing.

Western Wood Products Association. *1985 Buyers Manual*. Portland, Oregon: Western Wood Products Association.

_____. *Western Woods Use Book*. Portland, Oregon: Western Wood Products Association.

Weygers, Alexander G. *The Modern Blacksmith*. New York: Van Norstrand Reinhold Co.

World Information Systems. *The Almanac of Science and Technology: What's New and What's Known*. Edited by Richard Golob and Eric Brus. . New York: Harcourt Brace Jovanonich.